Ecological Genetics

David J. Merrell

D0209758

University of Minnesota Press • Minneapolis

Library of Congress Cataloging in Publication Data

Merrell, David J.
    Ecological genetics.

    Bibliography: p.
    Includes index.
    1. Ecological genetics. I. Title. [DNLM:
1. Genetics, Population. 3. Ecology. QH 456 M568e]
QH456.M47   575.1'5   81-14789
ISBN 0-8166-1019-3   AACR2

The University of Minnesota
is an equal-opportunity
educator and employer.

Preface

Recently a new approach to the study of populations has appeared under a variety of labels, among them ecological genetics, evolutionary biology, Darwinian ecology, evolutionary genetics, and population biology. This approach is an effort to merge the previously separate fields of population genetics and population ecology into a common discipline. Thus far, the union between population genetics and population ecology has been an uneasy one, but it seems inevitable that studies of populations must move in this direction, and this book is an attempt to further the process.

My preference for "ecological genetics" as the term to characterize the emergent discipline undoubtedly reflects my own background in genetics. Each of the labels cited above has somewhat different connotations, and the diversity of names used reflects to some extent the differences in background and interests of the authors, most of whom were trained either as ecologists or as geneticists. Another dichotomy among students of populations is that between the mathematical theorists on the one hand and the experimental naturalists on the other. Although some of the underlying theory is indicated, the book is focused primarily on the results of studies of actual populations.

In one sense ecological genetics is a methodology. The combination of laboratory and field research provides insights into the way populations adapt to their environments that can be gained in no other way. In another sense ecological genetics is a state of mind,

v

for it provides a different perspective on the biological world and makes possible the study of problems of both theoretical and practical interest, ranging from the origin of pesticide resistance to the origin of species.

In writing this book on ecological genetics, I have not attempted an exhaustive review of the literature, but have instead tried to cite pertinent references to illustrate particular points. One problem is that the literature in ecological genetics is widely scattered and often does not travel under that label. In the process I have undoubtedly omitted some significant work, and have probably cited my own work more than necessary simply because it was familiar and came quickly to mind.

The background assumed for readers of the book is some knowledge of plants and animals, a familiarity with the principles of genetics, and some understanding of elementary mathematics and statistics. Perhaps it should be added that we do not deal with what is usually referred to as "the evidence for evolution," but this is hardly necessary, for ecological genetics is the study of evolution in progress.

D. J. M.

Acknowledgments

For more than a decade Dr. James C. Underhill and I have held a seminar on ecological genetics, and my thinking on the subject has undoubtedly been influenced by the many graduate students and faculty who have participated. Dr. Underhill, and Charles R. Rodell and Phillip T. Barnes, were kind enough to read and make helpful comments on the entire manuscript, but the ultimate responsibility for its contents, of course, is mine.

My wife, Jessie Clark Merrell, has been my amanuensis, making invaluable contributions throughout the course of the work, which was begun during a sabbatical leave at the University of Florida. Our thanks go to Dr. and Mrs. Frank G. Nordlie and the members of the Zoology Department there for their kindness during our stay in Gainesville. Working with the staff of the University of Minnesota Press was a pleasure.

The illustrations were done by Kris Kohn, whose cheerful competence was much appreciated. Where material in the tables or figures has come from other sources, permission for use has been obtained from the sources cited in the captions and is gratefully acknowledged. In addition, Figure 7-1 is derived from L. B. Slobodkin, *Growth and Regulation of Animal Populations,* now available in a Second Enlarged Edition (1980) from Dover Press, N. Y. and is being used by permission of the author; Figure 12-1 is based on a figure in Kenneth Mather and John L. Jinks: *Biometrical Genetics.* Copyright © 1971 by Kenneth Mather and John L. Jinks. Used by permission of the publisher, Cornell University Press; Table 16-1 is adapted from *Fundamentals of Ecology*, Third Edition, by Eugene P. Odum. Copyright © 1971 by W. B. Saunders Co. Copyright 1953 and 1959 by W. B. Saunders Co. Reprinted by permission of Holt, Reinhart and Winston.

Contents

Ecological Genetics

CHAPTER 1

The Nature
of Ecological Genetics

Ecological genetics represents a union between population genetics and population ecology, combining certain aspects of each discipline, but also differing in certain respects from both. Even though both population ecology and population genetics are concerned with populations, the two fields developed independently until quite recently. Population genetic theory, developed initially by R. A. Fisher, J. B. S. Haldane, and S. Wright, is based on the principles of heredity established by Mendel, Morgan, and their successors. Population ecology developed in the absence of a comparable set of general ecological principles. The union of population genetics and population ecology has been an uneasy one and, according to Levin (1978) and Lewontin (1979), has yet to be consummated. As Lewontin wrote:

> Despite the pious hopes and intellectual convictions of evolutionary geneticists and ecologists, evolutionary genetics and ecology remain essentially separate disciplines, traveling separate paths while politely nodding to each other as they pass. The functional separation of population genetics and ecology is immediately obvious in books on "population biology," as, for example, the superb introductory text by Wilson and Bossert (1971), in which the sections on population genetics, ecology and biogeography are totally independent entities, each standing on its own feet, each quite self-contained in its analysis.

The difficulty was identified by Lerner (1965) with a delightful analogy drawn from Dickens, which he called Pott's synthesis. Mr. Pickwick, when informed by Mr. Pott of the *Eatanswill Gazette* of an extensive review of a work on Chinese metaphysics in the *Gazette,* was interested to learn the source of the author's information about such an abstruse subject. He was quite amazed to learn that it had come from the *Encyclopedia Britannica,* for he was not aware that the encyclopedia contained any information whatever on Chinese metaphysics. However, Mr. Pott explained that the author "read for metaphysics under the letter M, and for China under the letter C, and combined his information." Thus it has often seemed with ecological genetics.

In 1968 Waddington wrote, "The two major, long-standing problems of evolution are speciation and adaptation." The origin of species and the origin of adaptations have occupied this central position ever since 1859 when Darwin proposed, in *The Origin of Species,* that natural selection is the mechanism of evolution. Darwin's thesis was that the adaptation of populations to their environments resulted from natural selection and that if this process continued long enough, it could ultimately lead to the origin of new species. In short, those individuals with traits most favorable under the existing environmental conditions would survive to reproduce and, to the extent that these favorable characteristics were hereditary, would pass on their favorable genes to their offspring. These offspring would then, on the average, be somewhat better adapted to the environmental conditions than the previous generation. If the environmental conditions changed, the favored traits might change, and the adaptations in the population would tend to track the changes in the environment. This process, continued without limit in time or space, eventually could lead to the origin of distinct new species.

Darwin's theory of evolution is obviously a genetic theory, but only after 1900, when understanding of the principles of genetics began to emerge, did it become possible to frame evolutionary theory in quantitative terms according to known principles of heredity. The influence of Fisher, Haldane, and Wright, the three scientists primarily responsible for developing the mathematical theory of population genetics and evolution, was so pervasive that Lewontin (1965) wrote:

> In many ways the lot of the theoretical population geneticist of 1963 is a most unhappy one. For he is employed, and has been employed for the last thirty years, in polishing with finer and finer grades of jeweller's rouge those three colossal monuments of mathematical biology *The Causes of Evolution, The Genetical Theory of Natural Selection* and *Evolution in Mendelian Populations.* By the end of 1932 Haldane, Fisher and Wright had said everything of truly fundamental importance about the theory of genetic change in populations and it is due mainly to man's infinite capacity

to make more and more out of less and less, that the rest of us are not currently among the unemployed.

Somewhat earlier, however, Waddington (1953a) had written of the mathematical theory:

> Examined after this lapse of time it has the peculiar character of not having achieved either of the two results which one normally expects from a mathematical theory. It has not, in the first place, led to any noteworthy quantitative statements about evolution. The formulae involve parameters of selective advantage, effective population size, migration and mutation rates, etc., most of which are still too inaccurately known to enable quantitative predictions to be made or verified. But even when this is not possible, a mathematical treatment may reveal new types of relation and of process, and thus provide a more flexible theory, capable of explaining phenomena which were previously obscure. It is doubtful how far the mathematical theory of evolution can be said to have done this. Very few qualitatively new ideas have emerged from it. Wright's theory of drift has, perhaps, the most convincing claim to be something quite fresh and novel, but several other authorities express grave doubts whether it plays any important role in nature.

Perhaps the sentiment expressed by Waddington can be better understood in the light of a comment by Lewontin (1968):

> There is . . . a wide misunderstanding of the function of theoretical studies in population biology. It is *not* the function of theory to describe what has happened in a particular instance. Only observation can do that. The purpose of theoretical studies in population biology is *to set limits.* . . . Theoretical population biology is the science of the *possible;* only direct observation can yield a knowledge of the actual. But theoretical studies can then put limits on the experimental and observational procedures of observers and can also "explain" the results of experiments and observations.

However, the words of Crow (1955) are worth recalling:

> A full quantitative theory of evolution would be impossibly complex. For example, it would have to consider adaptability as well as adaptedness, for in the long view the former must also be important. Such a complete description is far beyond the capacity of workable mathematical models, and in so far as evolution depends on essentially unique events it is even in principle incapable of mathematical analysis.

In other words, the mathematical theory may aid in the interpretation and understanding of the evolutionary process, but it should not be mistaken for the evolutionary process itself.

A difficulty with the mathematical theory is that simplifying assumptions are usually made in order to keep the mathematics more tractable. For example, population geneticists often assume a constant environment while population ecologists usually assume that all members of a population have identical genotypes. Seldom is either assumption true. Thus, the validity of the theory and of the limits set by the theory depend on the validity of the assumptions made. As it is easier to make assumptions than it is to collect the data needed to verify them, the theory has often seemed to develop in its own merry way, unconstrained by the limitations imposed by the real world. A theory is most useful when it approximates reality; otherwise it may be irrelevant. The great need at present is for more and better estimates of the parameters involved in population biology. This call for more empirical evidence is hardly new; it was sounded by Timoféeff-Ressovsky (1940a, p. 104) more than a generation ago, but the need still exists.

In contrast to the approach to the study of populations taken by the mathematical theorists is that of the "experimental naturalists" (Waddington, 1953a). In recent years this area of research has come to be identified as ecological genetics. The first symposium on ecological genetics at an International Genetics Congress was held at the Hague in 1963, and the first edition of E. B. Ford's book entitled *Ecological Genetics* appeared in 1964. Ford stated that he had employed the term "ecological genetics" for many years in lectures and scientific discussions and indeed it appeared in his introductory remarks (1960) to his paper at the Darwin Centennial. His colleague, P. M. Sheppard, had a brief chapter entitled Ecological Genetics in his book *Natural Selection and Heredity* (1958), but used the term in a matter-of-course way without definition. Lerner (1965), in his synthesis at the conclusion of the symposium on Ecological Genetics at the Hague, attempted to trace the origins and meaning of the term, and found that it had been used by various workers with somewhat different shades of meaning and that priority in the use of the term was somewhat difficult to establish. One reason is that research in ecological genetics had been carried on for some time before the 1960s when "ecological genetics" began to come into general use.

Certainly one of the first, if not the first, to carry out research in ecological genetics and to set forth in some detail the nature and objectives of such research was Turesson (1922a, 1922b, 1923, 1925, 1930). He coined a number of terms related to this work, most of which have fallen into disuse, among them the word *genecology*. In doing so, he wrote (1922a), "The species problem is thus seen to be in large measure an ecological problem," and in 1923, "It seems appropriate for several reasons to denote this study of species-ecology by the term *genecology* (from the Greek 'genos,' race, and 'ecology') as distinct from the ecology of the individual organism, for which study the old term autecology seems to be the adequate expression." Thus he reserved autecology for the ecology of individuals, genecology for the ecology of species, and synecology for the ecology of communities. He also wrote (1923), "The Linnaean species represents as such a much (*sic*) important ecological unit, to which unit the name *ecospecies* has been given by the present writer." Furthermore, he wrote (1922a), "The term *ecotype* is proposed here as (an) ecological unit to cover the product arising as the result of the genotypical response of an ecospecies to a particular habitat. The *ecotypes* are then the ecological sub-units of the ecospecies." Of these terms, and others, proposed by Turesson, only ecotype remains in widespread use. However, not only his writing but the nature of his research with plants make clear that his conceptual approach to his research incorporated both genetics and ecology in a manner we now recognize as typical of studies in ecological genetics.

Population ecology is concerned with the kinds of organisms in an area and with their distribution and numbers, and may deal in statics, the description of a population at a single point in time, or in dynamics, the assessment of the physical and biological factors that produce changes in species composition, distribution, or numbers. In population genetics, the unit of study is the breeding population, which may be as small as a local breeding population (or deme) or as large as an entire species. Statics in population genetics usually involves the description of some form of gene-frequency equilibrium; dynamics involves the study of gene-frequency change due to mutation, selection, migration, and random genetic drift. The major requirement for research in population genetics is the presence of detectable genetic variation.

Ecological genetics is the study of the adaptation of natural populations to their physical and biological environments, and the mechanisms by which they respond to environmental change. It requires an awareness that populations are dynamic units very precisely adapted physiologically and genetically to their environments and sensitive to, and within limits responsive to, any change in their environmental conditions. The interplay between a genetically variable population and its ever-changing environment is the focus of attention in ecological genetics. Thus, the ecological geneticist must be concerned not just with the kinds of organisms present and their distribution and numbers, but also with the gene pools of the populations under study. Ecological genetics is, in fact, the direct study of evolution at the level at which it actually occurs.

In one sense ecological genetics is a methodology. To gain insight into the dynamics of a population, a combination of field and laboratory research is ordinarily required. Such research provides the following information.

1. **Distribution.** Knowledge of the geographical range of a species may provide information about the climatic or other limits to the distribution of the species. Similarly, knowledge of the microgeographical or topographical distribution of a species in a given area may help in identifying the "preferred" habitat of the species. Together, the information on the geographical and topographical species distribution may help to delimit the biological and physical dimensions of the space within which the species lives, that is, its *ecological niche.* A comparison of the physical and biological conditions at the center of the species range, in marginal populations, and just beyond the limits of the species distribution may aid in the identification of those factors crucial to the existence of the species.

Exact information on distribution, and hence on ecological requirements, is much easier to obtain in plants (and sedentary species of animals) than it is for most species of animals, which are active and may move about for a variety of reasons. Perhaps this difference is responsible for the fact that many of the classical studies in ecological genetics have been done with plants.

2. **Population number.** *Population density* is the number of individuals of a species per unit area or per unit volume and may be estimated from samples taken at various points within the range of the species. Areas of high density are presumably favorable for the species, and thus density estimates may also be helpful in the identification of the ecological requirements of the species.

Estimates of *population size* are more difficult to make than estimates of population density. Ideally, one would like to count all of the individuals in a population, but this is seldom possible. Therefore, capture-mark-release-recapture experiments are often used to estimate population size. A further problem is that unless the boundaries of a population are well defined, as they are, for example, in a population on an island, it may be difficult to determine the exact limits of the particular demes under study.

Of even greater interest than population size, which, in effect, is a census of the population, is *"effective" population size,* which is usually an abstract number of crucial importance to the estimation of random genetic drift. As might be imagined, "effective" population size is even more difficult to ascertain in natural populations than is population size. A living population, of course, does not remain constant in number. In order to study *population dynamics,* the way population size changes through time, it is necessary to estimate a population's *"birth" rate* and *death rate,* and the rates of *immigration* and *emigration* of individuals to and from the population. Fluctuations in population number will result from the net effect of these four factors.

Migration also has genetic implications. In order to estimate *gene flow* into a population, it is necessary to estimate not just the number of individuals entering the population but also the extent of their genetic contribution to the gene pool of the next generation. The *dispersal* of individuals from a population may be a significant factor in the dynamics of the population, especially if these individuals differ in age or in genotype from those remaining in the population. Dispersal may be difficult to measure because disappearance of individuals from the population under study may be due to predation rather than to dispersal.

The word *migration* is used in a variety of ways by population biologists. In population genetics it is synonymous with gene flow. It may also be used in relation to irregular or sporadic movements of individuals or groups that may occur in times of environmental stress such as drought or famine. It may refer to regular movements such as the seasonal migrations of birds between the northern latitudes and warmer regions or to shorter-range regular movements in other groups of animals. All such movements may have genetic as well as numerical implications for the populations involved. The particular meaning of "migration" can usually be inferred from the context.

3. **Genetics.** To study the genetics of a population, estimates of *gene frequencies* are first required. For these estimates to be of interest, genetically variable loci are necessary. Then it will be possible to determine whether gene-frequency equilibria exist, or whether the gene frequencies change through time or space.

To assess the causes of observed gene-frequency changes, estimates of *mutation rates, fitness* (or *selection coefficients), migration coefficients,* and *effective population size* are desirable. Estimates of migration coefficients and of effective population size must normally be made in natural populations, but estimates of mutation rates and of fitness may be possible under experimental conditions.

Further information useful in assessing the evolutionary capability of a population is knowledge of its *mating system,* the *sex ratio,* and its *genetic system,* that is, the way its genetic material is organized and transmitted.

These estimates provide the working material in ecological genetics. Information is required not only on the distribution and abundance of a species but on its genetic characteristics as well. Some of these parameters may be devilishly difficult to estimate. For example, following the 1970 census, the U.S. Census Bureau estimated that there had been an error of five percent in their census of the city of St. Paul, Minnesota. If an error of this magnitude is possible in a simple enumeration of a human population in a relatively stable situation where individuals can be identified and counted within a well-defined geographic area, how much greater is the potential for error in estimates for these parameters in the much more fluid situation that exists in wild populations, where neither the individuals nor the geographic limits of the population can ordinarily be identified.

Studies in ecological genetics may be made with populations in the field or with experimental populations in the laboratory. They may involve one, two, or even more species, but we shall focus attention primarily on studies with a single species. Although species interactions have often been studied, seldom have the genotypes of these species been monitored along with the changes in their numbers or distribution. Some work has been done on genotypic co-evolution in host and parasite in certain plant and animal diseases (flax and flax rust, the rabbit and the myxoma virus in Australia), but most research in ecological genetics has dealt with a single species, a trend likely to continue for some time.

Although everyone would probably prefer to study wild populations living in their natural environments, much work has been done with laboratory populations. The value of laboratory studies has sometimes been disparaged because the situation is considered too artificial to have much relevance to natural populations. However, laboratory populations are natural populations living in an unnatural environment, and many questions in ecological genetics can be tackled better in the laboratory than in the field. Given the number of genetic and ecological variables that must be dealt with, ecological geneticists will often learn more by working in the laboratory where they have a greater degree of control, for example, over the environmental conditions or over the amount of migration, than they will by working in the field. The best approach to research in ecological genetics is a combination of field and laboratory work: to study in the field those questions that can best be answered under natural conditions, and in the laboratory those questions most easily answered there.

Thus, one will need to study distribution, abundance, and migration or dispersal in the field. Also gene frequencies may be estimated in populations living in similar and in different habitats or in relation to particular environmental factors. Within a given population, gene-frequency change may be monitored seasonally or annually, or in different age classes. These analyses may provide clues to the selection pressures operating in the population.

Experimental field studies may involve the release of marked individuals or mutants into existing populations to estimate, for example, selection pressures or rates of dispersal. It is also possible to establish entirely new colonies with known genotypes and then to follow the fate of these colonies. Reciprocal translocation of organisms from different populations—that is, rearing each in the other's environment—is a standard method for estimating the relative influence of genetic and environmental factors on their phenotypes. Similar information may be obtained by rearing them together in a common environment. This approach has been used primarily with plants, but might prove equally fruitful if used more often with animals.

A variety of information is best provided by laboratory studies. The study of genetic variation in natural populations, whether morphological, lethal, cytogenetic, electrophoretic, or other variants are involved, usually requires some form of laboratory analysis to detect the variants, and crosses to determine their mode of inheritance. Useful information on viability, fecundity, longevity and other traits related to fitness can often be obtained by studies of these genetic variants under laboratory conditions. Moreover, the possible range of phenotypic expression for a given genotype and its adaptive limits can best be ascertained by rearing the individuals under a variety of controlled environmental conditions in the laboratory.

Laboratory populations can be established in which some of the biological or physical variables can be controlled or eliminated. The course of events in these experimental populations, which simulate the natural populations but under more controlled conditions, may lead to fruitful inferences about events in the field. Laboratory populations, for example, have been useful in the study of certain aspects of population dynamics such as population growth, competition, selection, mutation, and drift.

However, ecological genetics is more than just a methodology, for it also provides a different perspective on the biological world. The combination of field and laboratory research in ecological genetics soon leads to the realization that populations are not fixed, static entities, but instead are dynamic, and capable of changing, not just in distribution and numbers but in genetic composition as well. Studies in ecological genetics have application to a variety of fascinating biological problems of both theoretical and practical interest, among which are the following:

1. The origin of adaptations.
2. The role of the high levels of genetic variability (i.e., genetic polymorphism) in natural populations.
3. The causes of divergence between populations of the same species, which may lead to the formation of geographic races and ecotypes.
4. The mechanisms responsible for the origin of species.
5. The evolution of species under domestication.
6. The evolution of resistance to pesticides and antibiotics.
7. The response of populations to herbicides, chemical and physical mutagens, and other types of environmental pollutants. Equally important are studies of populations in undisturbed habitats to provide the base-line data against which to measure the effects of pollution.
8. The effects of species introductions, both accidental and intentional, some of which, such as the gypsy moth, Japanese beetle, chestnut blight, and Dutch elm disease have had dire consequences in the United States.
9. Evolutionary trends in *Homo sapiens.* Ignorance about present trends in human evolution is profound. For example, the invention of agriculture some 400 generations ago and the conquest of infectious diseases in the past century undoubtedly changed the selection pressures at work on the human gene pool, but little is known about the evolutionary forces now affecting human populations.

Thus, in a world where rapid enviromental change, often due to human activities, seems to be the order of the day, ecological genetics can provide us with insight into the ways species can respond and adapt to their changing environments and also into the limitations on their ability to adapt. Even though many studies in ecological genetics have been made on flowers and flies, snails and butterflies, the implications of the research reach far beyond the few species on which most of the work has been done.

CHAPTER 2

Adaptation

With the approach of winter in the northern United States, some species migrate south. The nighthawks, chimney swifts, and swallows start their migrations in late August and are followed by other species in September and October, with some of the hawks and eagles and waterfowl not leaving until November when the lakes begin to freeze. Other species, especially small mammals, go into hibernation, a period of dormancy that lasts through the winter. The snowshoe hare remains active, but the prospect of a northern winter turns his coat white from its drab summer brown. Although they do not breed in winter, the chickadees and nuthatches remain active and are seemingly unaffected by the change from tropical to arctic temperatures. All of these species have adapted to the onset of the northern winter.

Adaptation is a central fact of biology. All organisms are adapted to both their physical and biological environments. If they were not, they could not survive. Thus biological adaptation can be considered to be a biological axiom. The origin of these adaptations, manifested in many ways, is a central question in biology. The adaptive response of the migratory birds to the advent of winter is essentially behavioral, that of the hibernating mammals is physiological, while that of the snowshoe hare is biochemical, a change in the pigment produced.

The word *adaptation* is so commonly used by biologists that it is seldom defined because its meaning is considered to be generally understood. However, adaptation is actually used in several different but interrelated ways with various nuances. Hence, the more carefully one tries to define it, the more elusive it becomes. Adaptation is often used, for example, with respect to a trait. A cat's eyes are an adaptation for seeing, a swallow's wings are an adaptation for flight, and the fins of a smallmouth bass are an adaptation for swimming. These are all distinctive morphological traits with recognizable functional, and hence adaptive, significance. A blind cat or a wingless bird will not long survive.

Adaptation is also used to describe the physiological adjustments made by individuals in response to changes in their environment. In bright light, for instance, the pupils of a cat's eyes narrow to fine slits, but in the dark they dilate greatly to admit more light. In this case, the adaptation involves a marked change in appearance. In other cases, the adaptations involve adjustments that permit the organism to remain unchanged in the face of environmental changes. This type of individual adaptation is known as physiological homeostasis, which involves self-regulating physiological processes that help to maintain a constant internal environment. The maintenance of blood sugar and blood pH levels within very narrow limits despite dietary and other factors tending to change the glucose and CO_2 levels in the blood of mammals provides examples of physiological homeostasis. Similarly, a change from life in the lowlands to the mountains leads to an increase in red cell number, which ensures a constant supply of oxygen to the cells in the rarified atmosphere at the higher elevations. When you are cold, you shiver; when you are hot, you sweat. The end result of these responses is the maintenance of a constant body temperature despite different air temperatures.

Light-skinned human races exposed to sunlight tend to tan, that is, increased amounts of melanin are deposited in their skin, which serves as an adaptive protection against sunburn. Dark-skinned human races do not require the stimulus of exposure to sunlight to develop their pigmented skins; the entire population is already adapted in this way. This case introduces the distinction between individual adaptation or *adaptability*, resulting from an individual's response to the environment, and population adaptation or *adaptedness*, where the population as a whole manifests the adaptation even in the absence of any environmental stimulus. It should be realized that both adaptability and adaptedness are under genetic control. Albino individuals lack melanin in their skin and cannot tan no matter how long they are exposed to sunlight. One of the more fascinating questions in evolutionary biology is the relationship, if any, between individual and population adaptation. Is individual adaptability a necessary preliminary step toward the development of population adaptation or are the two phenomena quite independent?

Individual adaptations may occur in a matter of seconds or minutes or hours, as in the essentially regulatory pupillary response to light or the body's response to varying temperatures. The more gradual individual adaptations to changing seasons or to life at higher elevations are usually categorized as acclimatization. Both of these processes are reversible.

Another form of individual adaptation is developmental adaptation, an essentially irreversible process, which occurs during the growth and differentiation of the individual. In many species, especially among animals, development is canalized or channeled so that the adult form produced is much the same under a variety of environmental conditions. In other species, especially among plants, there is considerable developmental flexibility, and the same plant genotype gives rise to quite different phenotypes as a consequence of different environmental conditions during development. The phenotypes not only differ, but each phenotype is better adapted to the particular set of environmental conditions it was subject to during development than the other phenotypes possible with that genotype. Hence, this too is an adaptive response, slow compared to regulatory responses but more permanent. A dandelion grown in the lowlands is a large plant with large leaves and flowers and an erect growth habit. A clone of the same plant grown in an alpine environment takes on the aspect of a typical alpine species, a small compact dwarf plant growing close to the ground.

Differences such as these make clear that there is not a one-to-one correspondence between genotype and phenotype. Instead, a genotype sets limits within which development can proceed. The particular phenotype to emerge is the result of the interaction between the genotype and the particular set of environmental conditions in which that genotype develops.

In contrast to individual adaptations, the adaptation of populations extends over many generations and hence is clearly an evolutionary phenomenon. In the process of becoming better adapted to existing environmental conditions, the average characteristics of the population shift, which, by definition, is evolution. Nevertheless, the adaptability of individuals is also under genetic control, as noted above, so that it too is subject to evolutionary change.

By now it must have become apparent that adaptation is always used with reference to some set of environmental conditions. It involves not just an organism or a group of organisms but an environment as well. Thus to study adaptation requires a knowledge and understanding of both the organism and its environment. The environment includes not only the physical environment (temperature, humidity, available O_2, etc.), but also the biological environment (conspecifics, predators, prey, parasites, etc.). When we say that a species is well adapted, we mean that it is well adapted to a particular environment. The whale, for instance, is adapted for life in the oceans; the elephant, another large mammal, is not.

Exactly what is meant when we say an organism is well adapted to its environment? Ordinarily we mean that it is able to survive in that environment. However, when we say that a population is well adapted to its environment, the usual implication is that not only are the individual members of that population able to survive but they are able to reproduce successfully. This added requirement introduces a new dimension to the concept. A well-adapted individual may live a long and happy life in its environment, but as a member of a population it is not well adapted if it fails to reproduce and contribute progeny to subsequent generations of that population. Suddenly we are confronted by a dilemma. Which is more important to our concept of adaptation: survival or reproduction? Strange though it may seem at first, in an evolutionary sense, reproductive success is the primary measure of well-adapted organisms; survival is secondary.

At this point, it may be well to introduce another widely used biological term, fitness. The expression "the survival of the fittest" has come down to us from Darwin's time. The linkage in this expression between survival and fitness is unfortunate because in an evolutionary sense, and this was the sense of most interest to Darwin, it has become increasingly clear that the most fit, which to most biologists is the same as saying the best adapted, are those individuals who contribute relatively the greatest proportion of progeny to the next generation. From a practical or operational standpoint, this conclusion is unfortunate because survival is relatively easy to measure in terms of either viability or longevity, but reproductive success is not. Efforts to measure fitness, or reproductive success, are fraught with pitfalls. For example, in species with separate sexes, the reproductive success of an individual will in part be dependent on the characteristics of its mate or mates, an immediate complication to any simple measure of the fitness of an individual.

Another truism in biology is that natural selection is the mechanism that increases fitness and enhances the adaptation of individuals or of populations to the existing environmental conditions. This concept leads us into another logical tangle, namely, which individuals in a population are best adapted or most fit to survive and reproduce? The tautological answer suggested above is those that do survive and reproduce. The best escape from this tautology is to regard the relationship between fitness and reproductive success simply as a definition rather than as an assumption or as an hypothesis to be proved. This definition then means that natural selection will not act to favor the strongest, the healthiest, the most attractive, or those who live the longest, but simply those who leave the most offspring.

From the above discussion it should be clear that there is an intimate relationship among the concepts of adaptation, fitness, and natural selection. If we accept the idea that fitness should be measured in terms of reproductive fitness, then the question becomes how to measure reproductive success. One way would be to measure the number of fertilized eggs or zygotes produced. The obvious drawback to such a measure is that if few or none of these zygotes reaches sexual maturity, the contribution to the next generation of the individual whose fitness we are measuring is minimal. Therefore, it has been suggested that a better measure of fitness is the number or relative proportion of sexually mature progeny that an individual contributes to the next generation. From a practical standpoint, especially in species with long generation times, the use of such a criterion may introduce considerable complexity into the measurement of fitness. Moreover, a mutant in *Drosophila* known as *grandchildless* permits the production of normal F_1 progeny, but, as its name suggests, inhibits the production of F_2 offspring. A true measure of its effect on fitness, then, would require study not only of the F_1, but of the F_2. Considerations such as these led Thoday (1953, 1958a) to postulate that a truer measure of fitness would require knowledge of the persistence of evolutionary lineages over much longer periods of time, say 10^8 years. However, the practical problems become so intractable that measures of fitness involving long periods of time are unrealistic and useless. Therefore, although we may wish we could measure the fitness of an individual directly, in an essentially dimensionless system, we have already seen that fitness must always be assessed relative to some environment, which means that we must have both a way to measure fitness and a way to characterize the environment. Moreover, time as well as space is involved because of the need to measure fitness in terms of the progeny produced in that environment.

Natural selection acts directly on the phenotypes of individuals. To the extent that these phenotypes are controlled or influenced by genes or genotypes, natural selection will act indirectly on genes and genotypes. If the genotypes of individuals with different phenotypes differ, those genes and genotypes of the individuals favored by natural selection will tend to increase in relative frequency. In this way natural selection brings about evolutionary change. Because population geneticists are preoccupied with changes in gene frequency and with evolutionary change, they frequently refer to selection favoring a particular gene or genotype. However, it should be kept in mind that this effect is indirect, that selection in fact is always phenotypic.

We have already seen that adaptation and fitness are meaningful only relative to a given set of environmental conditions. It is also true that the fitness of one phenotype (or gene or genotype) is usually measured relative to that of other phenotypes (or genes or genotypes). Thus fitness is essentially a relative concept. Efforts have been made to measure absolute fitness by using estimates of the intrinsic rate of natural increase, but the value of such estimates is questionable. For one thing, estimates of the intrinsic rate of natural increase are very difficult to make. For another, they are still relative in the sense that they pertain only to a particular environment and may not predict the adaptability of the individual or population over a range of environments. Furthermore, the intrinsic rate may differ significantly from the actual rate of increase in numbers so that the actual relative reproductive rate is of far more interest than the potential reproductive rate.

This preliminary discussion of the meaning of adaptation, fitness, and natural selection is intended to indicate that, despite their widespread use and the fundamental nature of the concepts involved, these words have not been rigorously defined and do not have universally-agreed-upon meanings. Instead, we shall use operational definitions that are of practical use in the study of populations. Nonetheless, the concept of adaptation is of the greatest importance. To quote Lewontin (1968), "The study of adaptation is the nexus of population ecology, population genetics and development."

One of the hazards of discussing adaptation is the tendency to slip into teleological modes of thought or speech, to refer to the "needs" of the organism or the behavior of the organism as if it were aware of its needs or behaved as it does with conscious purpose or design. A special morphological condition or physiological state or behavior pattern may well fulfill the needs of the organism in having an easily recognizable adaptive function, but this does not necessarily mean that the organism itself recognizes its own needs or acts with any conscious purpose. Nevertheless, wings are for flying and eyes for seeing. Attempts to avoid even the appearance of teleology sometimes lead into a morass of circumlocutions. Hence, even though at times our discussion of adaptation may seem teleological, the phrasing is chosen for simplicity and not for any teleological overtones.

The words in the previous paragraph *easily recognizable adaptive function* call attention to another hazard in the study of adaptation, the tendency to identify adaptive significance by inspection, or introspection, rather than by experiment. As suggested above, a test of adaptive value involves the demonstration that individuals of one phenotype (and if it is to have evolutionary significance, genotype) contribute relatively more sexually mature progeny to the next generation than individuals with other phenotypes (and genotypes) under a given set of environmental conditions. Such tests are not easily made. Otherwise respectable research papers often conclude with a flourish of speculation about the adaptive significance of the experimental results, completely unsupported by experimental evidence. A good example of this is my own statement above that tanning serves as an adaptive protection against sunburn. The fact of the matter is that a number of theories have been proposed for the adaptive function of human skin pigmentation, but the actual evidence is quite sketchy. If the study of adaptation and natural selection is to progress beyond armchair speculation, experimental rigor will have to be introduced into such studies; intuition is not enough.

To exist in a given area, for example an island, all organisms face essentially the same basic problems. First, they must get there. The options are relatively simple: migration, evolution, or spontaneous generation. Until a century or so ago, spontaneous generation was widely accepted, though more recently for a narrowing range of species types. Today migration or the evolution of the observed group from an earlier migrant population are thought to account for the existing distribution of island populations. Next, they must be able to survive in the existing physical and biological conditions. They must also be able to obtain an adequate supply of food, or, if they are plants, to synthesize it. Finally, if the group is to persist, it must be capable of reproduction. For a bird population, all the other conditions for existence might be met, but if appropriate nesting sites were not available, the population could become extinct if it could not adapt its nesting behavior to the existing conditions.

In biology, we usually tend to stress the differences between organisms and groups of organisms, to point out what distinguishes one phylum, class, order, family, genus, species, or individual from another. The truth is that all living things are very similar to one another. For example, all organisms share a similar chemical and elemental composition made up of carbohydrates, fats, proteins, nucleic acids, salts, and water. Of the nearly 100 elements available on the earth, only about one-third are actually found in protoplasm where their relative proportions differ greatly from their relative abundance on earth.

Both of these facts indicate a selective utilization of the elements in the formation of protoplasm so that some sort of selective process seems to be operating even at this level. In fact, three elements alone—carbon, hydrogen, and oxygen—account for 98.5% of living matter by weight, and the remainder for the other 1.5%, with many present only as trace elements.

Not just the chemical composition but the way the compounds are used is very similar. The food chains that exist from plants to top carnivores indicate the interconvertibility of the compounds from one species to another. More significantly, the metabolic pathways and the enzyme systems controlling these pathways are very similar from bacteria to man, a major difference being that biochemically bacteria are more versatile than humans.

At the cellular level, the higher plants and animals (the eukaryotes), despite their seeming differences, are quite similar. They are composed of cells that contain nuclei, and have similar genetic systems based on nucleic acids organized into chromosomes. Both plants and animals reproduce sexually with a life cycle that involves meiosis, mitosis, gamete formation, and fertilization. The sequence of these four phases may vary, but they are all variations on the same theme; there is an alternation of haploid and diploid generations or stages in animals and an alternation of haploid gametophyte and diploid sporophyte stages in plants. Thus, it is possible to suggest that the fundamental similarities among higher plants and animals, in fact among all living things, are greater than their differences. In reality, the differences seem to reflect the evolutionary adaptations that have occurred as this basic genetic system has evolved different modes of existence.

ADAPTIVE COLORATION

Before going further, we may do well to consider some adaptations in detail, because, although some evolutionary adaptive changes seem very simple and comprehensible, others seem so complex that it is difficult to imagine how they could have evolved by natural selection. The difficulty, however, may lie more with our imaginations than with the evolutionary process.

A classic example of an observed evolutionary adaptive shift occurred in *Biston betularia*, the peppered moth, as well as in other moth species. These moths rested in the open on lichen-covered trees, and their light cryptic coloration blended in with the lichens so that predators seldom discerned them. During the Industrial Revolution, soot from innumerable chimneys polluted the landscape, killed the in the cryptic background color of the two species may ̤erely coincidental. Further refinements of the disruptive ̤ of the frog are the eye mask, a dark line through the eye, ̤lignment at right angles to the leg of the spots on the upper ̤ parts of the leg, both serving to break up the outline of rather prominent anatomical features. All of this coloration ̤d to the dorsal surface of the frog; the ventral surface is a ̤hite. This coloration pattern, also seen in fish and other ̤nimals, is known as *countershading*, and the presumption is ̤ from below against the bright sky, these animals are less ̤en by predators. It should be added that the underparts of ̤quirrel and the gray squirrel are also white, but the adaptive ̤ce of countershading in these species, if any, has never been ̤ed, to my knowledge.

̤ood frog, *Rana sylvatica*, is widely distributed over northern ̤nerica, particularly in wooded areas. Its eye mask is even ̤nounced than the leopard frog's. There are two color phases, ̤reddish-brown, that seem to occur with different frequencies ̤nt parts of the range. Another variation is the presence or ̤f a distinct, light mid-dorsal stripe, which also varies in fre- ̤n different populations. The mid-dorsal stripe is produced ̤le dominant gene, but the mode of inheritance of the color ̤unknown. Relatively little is known about the distribution ̤equency of these traits so that it is impossible to speculate ̤eir possible adaptive significance. Interestingly enough, in ̤d grouse, whose distribution is rather similar to that of the ̤g, red and gray color phases are also known. Here, too, the ̤inheritance is not known, but is probably simple. Again, the ̤ies of the two phases are known to differ in different parts ̤nge, but as yet the adaptive significance of the two types is ̤understood.

̤book *The Descent of Man and Selection in Relation to Sex*, ̤Darwin spent much more time on sexual selection than he ̤he descent of man. He was drawn to sexual selection in an ̤to explain the sexual dimorphism observed in so many differ- ̤ies. The male red-winged blackbird, in his striking black ̤ and red and gold epaulets, differs greatly from the rather ̤yptically colored female. The male goldfinch, in his jaunty ̤nd black summer plumage, is also far more brightly colored ̤mate. In winter male goldfinches molt into a subdued olive- ̤mage so that even though flocks of goldfinches may winter ̤orth, identification is difficult at first because the male's win-

lichens, and darkened the tree trunks. With
the phenotypes of more than 90% of the m
became dark to match their new backgrou
due to physiological adaptation by individu
nearly all cases by a single dominant gene.
increased the frequency of these genes in 1
clearly demonstrated by Kettlewell (1961,
With improved pollution control in recent y
reversal of the selection pressure to which
ready begun to respond (Cook, Askew, a1
and Cook, 1975). This example of adaptive
natural selection is one to which we shall
provides us with a simple, well-studied, comp

In the case of the gray squirrel, *Sciurus*
sonable to speculate that their gray coat is
blends in with the gray bark of the decide
normally rest. Rare albino gray squirrels a
populations exist, for example around the
Ottawa, but their fitness has never been 1
squirrel, *Tamiasciurus hudsonicus*, commc
areas, presumably blends in with the reddis
and other conifers found in its habitat. In
cousins, the 13-lined ground squirrel, *Citell*
known as the Minnesota gopher, is an inha
and has a golden tan color broken up by dar
which presumably matches the light and sha

The leopard frog, *Rana pipiens*, is some
because it too is an inhabitant of grasslands
fies extremely well some of the concepts c
lined by Cott (1940), a foremost student
animals. The frog's green to greenish-brow
vides *cryptic* coloration amidst the grassy
usually are found in summer. Expansion a:
anin within the melanophores permit the
their skin color by individual adaptation to
even more precisely. The spots that give the
referred to as *disruptive* coloration because
globular green form to match the pattern
grassy habitat. Presumably, the stripes an
gopher serve a similar function. Leopard
found in the moist greener areas in grass
squirrels, which tend to live in the drier an

diffe
not
colo:
and
and
othe1
is co
crean
aqua
that,
readi
the r
signif
invest

Th
North
more
gray a
in dif
absen
quenc
by a s
phase
or the
about
the ru
wood
mode
freque
of the
not w

In I
Charle
did o1
attemp
ent sp
pluma:
drab,
yellow
than h
drab p
in the

ter and summer plumages are so different. Darwin's work on sexual selection was an attempt to explain the adaptive significance and the origin of the differences between males and females in their secondary sexual characteristics as well as in their primary sexual characteristics. The sexes often differ not just in their external appearance but in their physiology and behavior. In the American turkey, the male is larger and more striking than the female and engages in a dramatic courtship display toward the females. Darwin argued that the male's plumage and courtship ritual were adaptations that helped to ensure the success of the male in securing a mate. In these cases of sexual dimorphism, the adaptive function of the coloration of the male appears not to be cryptic but for display. The disadvantage to the males in the loss of cryptic coloration as protection against predators is apparently outweighed by the advantage gained in leaving progeny. It can be argued that this sort of evidence confirms the earlier conclusion about the importance of reproductive success rather than survival as the ultimate measure of fitness.

Colors are used not only for protection (*cryptic*) and for courtship (*epigamic*) but also for warning (*aposematic*). The function of aposematic coloration seems to be to announce the presence of the individual and to render it as conspicuous and distinctive as possible so that there will be no mistaking its presence or identity. Once while collecting frogs, I moved into shoulder-high grass because the frogs were escaping there from a more open field. I collect with a long-handled, rather strongly built net and in such tall grass had to slam the net down wherever I detected movement of the grass, without ever really seeing what caused the movement. Once, I detected movement and started to slam the net down when I suddenly became aware of the striking black and white pattern of a skunk looming up at me. Checking my swing in mid-air, I threw myself backward, but the skunk continued on his way unperturbed. If he had been cryptically colored, it could have been an embarrassing experience for both of us. The skunk is a notable example of aposematic coloration.

Some species have evolved special warning devices, such as the rattle of the rattlesnake or the white tail of the Virginia deer. These devices may serve for interspecific or intraspecific communication. In many species, the males not only engage in courtship displays to the females but engage in threat or warning displays toward one another, and their external appearance thus serves a dual function. Some species of moths appear brightly colored in flight, but the bright colors are limited to the hind wings. When the moths alight, the hind wings are covered by the cryptically colored fore wings, and the moth seems to disappear. The predator, with a search image for a conspicuous object, is confused by this combination of aposematic and cryptic coloration. Other moth species, when approached, suddenly expose the large "eye spots" on their hind wings, seemingly as a form of warning or threat to predators (Wickler, 1968). Such eye-like patterns are fairly widely distributed in the animal kingdom. The origin and evolution of these traits raise some intriguing evolutionary questions.

Many different species of animals are unpalatable, poisonous, or malodorous to potential predators. Many of these species with distasteful qualities have bright colors arranged in conspicuous patterns. Tests of palatability of cryptically and conspicuously colored members of related groups of species have revealed that predators find a much higher proportion of the cryptically colored species to be palatable. Such tests have been made with species of amphibia and insects (Cott, 1940). One may conclude that conspicuous coloration serves as an advertisement and a warning to potential predators. The presumption is that potential predators are capable of learning to associate the distinctive color pattern with the distasteful properties of their prey, and that, once having experienced this, they will subsequently refrain from further attacks. Evidence for this behavior has been obtained experimentally with members of every major group of vertebrates (Cott, 1940). The learning process in the predators obviously requires the loss of some of the protected individuals. It is clearly to the advantage of the prey species to lose as few members as possible in the process of educating predators. Therefore, selection should favor the evolution of distinctive, conspicuous color patterns in the prey to reduce the number of errors by the predators.

If each malodorous, unpalatable, or poisonous species had a different color pattern, it clearly could tax the memory of a predator, especially if the predator preyed on insects and could be confronted by hundreds of different species. Therefore, it would be to the advantage of these prey species if they were to resemble one another in both hue and color pattern. The predator then would have to learn to recognize fewer color schemes and could thus reduce the number of errors it made in attacking unpalatable individuals. An assembly of similar species would benefit from their resemblance to one another because the losses incurred during the learning process by predators would be shared among them, the losses to any one species thereby being minimized. Thus, it is not surprising to find that there are groups of species, alike in warning coloration and in being distasteful even though they may be rather distantly related. Convergence in warning coloration among a number of noxious species is known as

Müllerian mimicry. However, as Wickler (1968) pointed out, genuine Müllerian mimicry is not mimicry at all under the usual definition of the word because there is no model and no mimic and there is no deception involved. Nonetheless, the phenomenon is widely referred to as Müllerian mimicry and is exemplified by the black and yellow color pattern common to many distantly related species of bees and wasps.

Batesian mimicry, on the other hand, does involve an unpalatable model and an unrelated palatable mimic. The mimic benefits because the predators, in learning to avoid the model, also learn to avoid the mimic. Numerous examples of Batesian mimicry are given in Grant (1963) and Wickler (1968). Ants and lycid beetles are just two examples of distasteful groups that have been widely mimicked by other, more innocuous groups of insects. The numerical relations between model and mimic are significant because if the mimic becomes too frequent, the predators will learn the wrong lesson. Mimicry provides some of the most bizarre and fascinating examples of evolutionary adaptation known, with genetical and ecological ramifications of considerable interest, to which we shall return later.

MIMICRY IN PLANTS

Mimicry also occurs in plants, and some of the best studied cases are mimetic weeds. Weeds have been variously defined as plants growing in habitats disturbed by man or domesticated animals, or among crops where they are not wanted, or sometimes as alien species, or as all of these. Mimetic weeds are "useless" plants growing among crop plants. Man carries out crop selection for better yielding and hardier varieties, and also weeds them to eliminate the useless interlopers, which constitutes very strong selection pressure against them. In crops like flax and wheat, it is impractical to weed the fields so that separation must be attempted with the harvested seeds.

Camelina sativa, (in the mustard family Brassicaceae), a mimetic weed in flax fields, is a good case study. Flax, *Linum usitatissimum,* belongs to a different family, the Linaceae. Unlike the other species of *Camelina,* which are winter annuals, *C. sativa* is a summer annual whose vegetative and seed characters resemble those of flax. *C. sativa* seeds are harvested along with the flax seeds. Efforts to separate the seeds by winnowing have exerted further selection pressure on the seed characteristics of *Camelina.* Curiously enough, *Camelina* seeds have retained their distinctive appearance, but have converged only in their winnowing characteristics since selection is not for phenotypic identity but rather for seeds blown the same distance by the winnowing machines. Flax was originally cultivated for fibers to make linen, but more recently it has been grown for seeds to make linseed oil, and two somewhat different varieties have developed. Consequently, two different subspecies of *C. sativa* have evolved: *C.s. linicola,* which grows with the flax raised for fiber and *C.s. crepitans,* which grows with flax raised for oil. Thus, the evolution in cultivated flax has been paralleled by evolutionary changes in *C. sativa.* (See Stebbins, 1950, for further details and references.) One consequence of this selection is that where flax is grown for oil, flax and *Camelina* are seeded, cultivated, and harvested together and both are processed to make oil.

When wheat was first cultivated in southwest Asia, rye was thought to be a weed growing with it. When the cultivation of wheat spread to northern Europe, rye accompanied it. In this harsher environment, rye thrived better than wheat because it grew better on poor light soils and tolerated higher soil acidity and lower winter temperatures. As a result, rye became intentionally cultivated in the more northern regions, and changed from a weed to a crop plant.

A most unusual type of mimicry has evolved in many species of orchids that are pollinated owing to their resemblance to females of various insect species. The male, deceived by both the appearance and the odor of the flower, attempts to copulate with it. In the process the males carry the pollen sacs from one flower to another and thus effect pollination. The male gets no reward for his activity, as for example bees taking pollen or nectar. Instead, the orchid seems to be exploiting the males's sexual drive for its own ends. Another aspect of this relation is that each orchid species seems to be specialized to resemble the females of a particular insect species and appears to be dependent on the males of that species for cross-pollination. The evolutionary history of the development of the relationship between orchids and insects raises some mind-stretching questions about the mechanism of evolution.

OTHER ADAPTATIONS

Equally intriguing are the questions raised by brood parasitism in birds, found in a number of bird families, where a female of one species lays her eggs in the nests of other species to be reared by foster parents. A number of adaptations are associated with brood parasitism, ranging from egg mimicry to mimicry of the species-specific gaping signal for feeding or of the juvenile plumage of the host's offspring. In some cases the young intruder, instead of mimicking its nestmates, disposes of them by forcing them out of the nest. The evolution of such behavior requires that, at some point, reproductive success for the brood parasite was greater when its young were reared by others than when it attempted to rear them itself. Means of recognition of the appropriate host species, especially where mimicry is involved, had to evolve. Moreover, any form of imprinting by the young parasites on their foster parents had to be suppressed or they would seek their mates among the wrong species.

The evolution of social insects such as termites, ants, and bees with their various castes and complex social and reproductive systems has resulted in some of the most complex adaptive systems known. The story is further complicated by ant mimics, ant warfare and slavery, ant fungus gardeners, and social parasitism in ants (Wheeler 1923, 1928; Wilson, 1971).

Not all adaptations are as dramatic or obvious as some that we have been discussing. For example, the metamorphosis of a tadpole into a leopard frog results in a number of morphological transformations that change an aquatic, gill-breathing herbivore with fins into a terrestrial, lung-breathing, carnivorous tetrapod. Less noticeable but equally important biochemical adaptations occur. The visual pigment in the eye changes from porphyropsin, typical of freshwater vertebrates, to rhodopsin, typical of terrestrial vertebrates. Nitrogen excretion, primarily in the form of ammonia in the tadpole, shifts primarily to urea in the frog. Ammonia is a moderately poisonous compound that can be safely excreted if a copious supply of fresh water is available, as it is to the tadpole. Urea is a less harmful compound that can be more safely retained in the body and excreted with a smaller amount of water than ammonia. Therefore urea excretion is an adaptation for terrestrial life. Even the type of hemoglobin changes. In the adult frog, hemoglobin has a low affinity for oxygen and shows the Bohr effect, a decreasing affinity for oxygen as acidity increases. In the tadpole, the hemoglobin has a high oxygen affinity and little Bohr effect. Tadpole hemoglobin is clearly adapted for oxygen transport in an oxygen-poor aqueous environment while adult hemoglobin is adapted for oxygen transport in an oxygen-rich terrestrial environment. There is an interesting parallel to this transition in humans. Fetal hemoglobin, or hemoglobin F, has a higher oxygen affinity than the normal adult hemoglobin, hemoglobin A. Hemoglobin F is the predominant type of hemoglobin in the fetus, but its production starts to decline at birth and it has virtually disappeared in six-month-old infants. The production of adult hemoglobin rises sharply after birth, and by six months hemoglobin A has become the predominant type of hemoglobin present. There can be little doubt that hemoglobin F is adapted to the essentially aquatic life of the fetus and hemoglobin A to the terrestrial life that starts after birth.

Adaptation occurs at many levels—biochemical, physiological, morphological, behavioral, and social. Each species is adapted to its physical and biological environment, and the adaptations range from the relatively simple, such as industrial melanism, to the complex, like brood parasitism in birds. No adaptation is apt to be perfect. Since organisms carry on many functions, the adaptations tend to represent compromises among these diverse functions. Furthermore, physical and biological conditions are never static but are constantly changing, and the adaptations are continuously subjected to fine tuning by natural selection.

CHAPTER 3

Biological Variation

If someone were to ask you about variation in wild populations of gray squirrels, or robins, or cottontail rabbits, you would probably respond that, to you, one squirrel, robin, or rabbit looks pretty much like another. Superficially, this appears to be true, for marked individual variation seems to be rare in wild populations. However, in humans we can readily recognize individual differences between people. Closer observation of wild populations of other species would undoubtedly enable us to begin to recognize individual differences even though it is true that natural populations of species are usually quite uniform phenotypically. It is probably for this reason that the type concept dominated the thinking of taxonomists and other biologists for so long. The type concept is that one phenotype, the "wild type," characterizes an entire species.

Nevertheless, occasional visible variants turn up in natural populations of almost every species, albinism being a fairly common example. Albino deer are reported shot almost every hunting season, and I have seen albino leopard frogs, tiger salamanders, and even a baleful albino snapping turtle. All of these, as well as the white bluebell growing in the same rock crevice with a blue bluebell, are probably inherited in a simple Mendelian fashion; indeed some were shown to be recessives by experimental test. Not only albino but melanistic gray squirrels are occasionally seen, and ventral melanism in place of the usual white ventral surface has also been observed in otherwise gray gray squirrels. Sometimes, partial albinos have been seen, for instance in

English sparrows, but whether these white wing patches are due to germinal mutation, somatic mutation, or some developmental anomaly must remain speculative in the absence of experimental evidence.

METHODS OF DETECTION

Random observations of this sort, however, give very little idea of the amount of variation in natural populations. Many years ago Dubinin and his collaborators (1934, 1937) systematically examined thousands of wild fruit flies, *Drosophila melanogaster,* for phenotypic variation. Apart from trident, a variable and widespread morphological variant of the dorsal thorax difficult to analyze genetically, they detected visible variations in 2.08% of the 129,582 flies examined. Among them were variations in the number, size, and shape of the bristles, in the size, shape, and color of the eyes, in wing venation, in body color, and in wing and leg shape. Many of these variants were nonheritable phenocopies; less than half were attributable to single gene substitutions similar to the well-known laboratory mutants in *D. melanogaster.* Although careful scrutiny may reveal some variation among individuals, much of this variation is environmental in origin; furthermore, direct examination gives only an inkling of the extent of genetic variation in wild populations. Much of this variation is in the form of recessive genes in the heterozygous condition revealed only by special methods.

One of these methods is the *specific locus technique.* For example, if one wished to estimate the frequency of the allele for ebony body color in wild populations, one could collect a large number of males from wild populations and cross them individually to virgin homozygous ebony females. If any male is heterozygous for ebony, half his F_1 progeny in this cross would be ebony; otherwise all of the F_1 progeny will be wild type. An advantage of this method is that only one generation is required. A major disadvantage is the small amount of information derived from each cross. The amount of information gained per cross can be increased by making the test stock homozygous for more than one recessive locus. In Russell's massive study (Russell et al., 1959; Russell, 1962) of spontaneous and induced mutation in mice, he used a test stock homozygous for seven recessives and thus increased information output. However, each recessive tends to reduce the viability or fertility of a test stock so that an upper limit is soon reached as to the number of loci that can be incorporated into a given test stock. The number of loci tested in a given cross is, therefore, only a small and selected sample of the thousands of gene loci that could be tested.

Another method used to reveal hidden recessives is *inbreeding.* In *Drosophila,* for example, a number of inseminated wild females are captured and isolated individually in vials where their F_1 progeny develop. The F_1 progeny of a given female are then crossed (either by mass matings or in pairs) and the F_2 examined for homozygous recessive aberrant individuals. Analysis of several hundred females in this way will give an estimate of the types and frequencies of the recessives that may be present in the wild population. Although this approach screens for more loci than the specific locus technique, it requires an additional generation, it screens only for visible traits, and to some extent it depends on the perceptiveness of the investigator.

Still another method for the study of genetic variability in natural populations is a *chromosome assay technique* that involves extracting a single intact chromosome from a wild population and making it homozygous by first out-crossing it to a balanced marker stock carrying inversions and dominant marker genes. This technique is, in essence, a modification of the well-known *C1B* technique devised by Muller (1927, 1963) for the detection of sex-linked lethals. Although it has been used primarily for the detection of lethals, it will also reveal other types of recessive genes. Since this reliable technique has been widely used, a series of crosses to exemplify the method is shown in Figure 3-1.

Suppose we wish to determine what proportion of second chromosomes in a wild population of *D. melanogaster* carry a lethal gene somewhere along their length. First, a single male from the wild population is crossed to virgin females of the *Cy L/Pm* stock. In this stock, one second chromosome is marked by the dominant visible marker genes Curly wings *(Cy)* and Lobe eye shape *(L)* and carries complex chromosome inversions that effectively prevent crossing over in the females. The other second chromosome is marked by the dominant visible marker gene Plum eye color *(Pm)* and also carries inversions to prevent crossing over in females. It should be noted that there is no crossing over in *Drosophila* males. This stock is called a balanced marker stock because both *Cy* and *Pm* act as lethals when homozygous, and the stock will breed true even though heterozygous because these chromosomes form a balanced lethal system.

The F_1 progeny from this cross will be either *Cy L* or *Pm* and heterozygous for one or the other of the second chromosomes from the wild male parent. To ensure that only one of the original male's

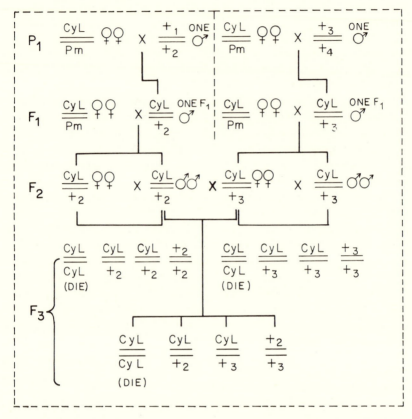

Figure 3-1. Diagram of a method for isolating a single intact second chromosome in the homozygous condition in order to detect recessive genetic variants in *Drosophila melanogaster*. See text for further details.

two second chromosomes is tested, a single F_1 male must be crossed to virgin females from the dominant marker stock, as shown in Figure 3-1, where a single $Cy\ L$ male is used. In the F_2 generation, the homozygous $Cy\ L/Cy\ L$ flies die, and the three surviving classes are distinguishable from one another phenotypically because of the different dominant markers. All of the unmarked chromosomes in the F_2 have a common origin from the single chromosome from the wild population carried by the F_1 male parent. Therefore, to make this chromosome homozygous, $Cy\ L/+$ males and females from the F_2 are mated with one another. In the F_3, the $Cy\ L$ homozygotes die, and the expected ratio among the surviving progeny is 2 $Cy\ L$ heterozygotes for each individual homozygous for the chromosome extracted from the wild population. It is worth noting that this extracted chromosome passes intact without crossing over through all of the crosses because of the lack of crossing over in *Drosophila* males and the presence of inversions in the females heterozygous for the dominant marker chromosome. In this way the genetic contents of a second chromosome from the wild population are held intact until in the F_3 they are made homozygous.

In the F_3, if no lethals are present on the extracted chromosome, the expected 2 to 1 ratio will be observed. If a recessive lethal is present at a locus anywhere along the length of the extracted chromosome, all the non-*Cy L* flies will be homozygous for the lethal, and the only surviving progeny will be *Cy L* in phenotype. It should be added that if more than one lethal is present on a given chromosome, it cannot be detected by this technique. The presence of more than one lethal can, however, be detected by suitable linkage tests to locate the loci involved, but such tests are not ordinarily run. Instead, it is usually assumed that lethals are distributed independently at different loci and that the frequency of chromosomes bearing 0, 1, 2, 3,... lethal genes will fall into a Poisson distribution with mean $\bar{x}$, the average number of lethal genes per chromsome. If L is the proportion of lethal-bearing chromosomes in the sample, then $1 - L$ is the proportion of chromosomes free of lethals. In the Poisson distribution the frequency of 0 lethals or $1 - L = e^{-\bar{x}}$, whence $\bar{x} = -ln(1 - L)$. From this estimate of the average number of lethals per chromosome, it is then possible to estimate the number of chromosomes with 1 lethal $(\bar{x}\,e^{-\bar{x}})$, 2 lethals $(\frac{\bar{x}^2}{2}\,e^{-\bar{x}})$, 3 lethals $(\frac{\bar{x}^3}{6}\,e^{-\bar{x}})$, and so on.

If a recessive visible mutant is present, all of the non-*Cy L* F_3 flies will be homozygous for this visible, which makes it somewhat easier to detect. If a deleterious recessive gene is present, reducing viability, fewer than the expected 33 1/3 % non-*Cy L* wild type flies will be observed. The following viability classes are generally recognized:

> Lethal = 0% of expected number.
> Semilethal = 0-50% of expected
> Subvital = 50-90% of expected
> Normal = 90-110% of expected
> Supervital = 110% or more of expected

If this procedure is repeated with a number of individual males from the wild population, it is possible to estimate the proportion of lethal-bearing second chromosomes in the wild population. The detection of lethals is relatively simple because scoring for lethals is simple, but with additional effort the method can be extended to the detection of semilethals, visibles, and genes affecting development or fertility. Moreover, with the use of suitable marker stocks for the other chromosomes, the entire genome can be analyzed for the frequency of lethals or other recessive variants.

This technique has other advantages. Even though the homozygotes die, the lethals can be preserved indefinitely in the heterozygous condition, balanced against the marker chromosome. Crosses among these different lethal heterozygotes will reveal how many of the lethals are allelic, and also how many different loci are involved. Moreover, if lethal chromosome 1 is allelic to lethal chromosomes 2 and 3, but 2 and 3 are not allelic to each other, it is presumptive evidence that chromosome 1 carries two lethals.

This technique provides a fairly complete picture of lethal frequency and distribution patterns in wild populations. One drawback is that the working unit is an entire chromosome rather than individual loci as in the specific locus technique. The effects observed in the F_3 are those of an entire chromosome made homozygous for its entire gene complement, in other words, the effects of many gene loci in the homozygous condition. Although lethal effects can ordinarily be traced to a single locus and even mapped if necessary, this has rarely been done, and the inference that the results from whole chromosome analyses can be extended to the effects of single loci is seldom tested.

When the tested chromosome is not lethal when homozygous, but instead reduces viability, another problem emerges. When we wish to determine whether a particular chromosome has semilethal, subvital, normal, or supervital effects when homozygous, we measure its viability relative to some standard. In the crosses outlined above, the standard was the heterozygous Cy $L/+$ flies growing in the same culture bottles with the homozygotes. However, as mentioned previously, almost any mutant has a deleterious effect on viability. Therefore, flies heterozygous for marker chromosomes, which carry one or more dominant genes, must be presumed to have a viability somewhat lower than they would if these genes were absent. As a consequence, the estimate of the viability of the homozygotes is probably somewhat inflated compared to what it would be if measured relative to the viability of flies not carrying the dominant marker genes.

To circumvent this problem, crosses are made between a number of heterozygotes carrying different chromosomes from the wild population in the balanced condition, that is, Cy $L/+_i$ x Cy $L/+_j$ (also see Figure 3-1). Then a comparison is possible between the $+_i/+_j$ heterozygotes and the $+_i/+_i$ and the $+_j/+_j$ homozygotes. This method is an improvement over the use of the heterozygotes with the dominant markers as the standard, but it too presents some problems. For one thing, the flies being compared are no longer being grown together in the same culture so that the comparison is no longer direct but indirect. Moreover, $+_i/+_j$, $+_i/+_i$, and $+_j/+_j$ are still being compared to the $Cy\ L/+$ flies in their own cultures. The presumption is that the $Cy\ L/+$ flies in each of the three types of cultures are equivalent, and hence the three types are being compared to a common standard.

However, if the wild type chromosomes have different dominant effects when in combination with the $Cy\ L$ chromosome, then the $Cy\ L/+$ flies in different types of cultures will not be equivalent. It is known, for instance, that recessive lethals usually reduce viability about 2 to 4% in heterozygotes as compared to homozygotes free of lethals. Therefore, selection of chromosomes used to generate "normal" wild type flies heterozygous for two chromosomes from the wild population $(+_i/+_j)$ raises some questions. For example, should only chromosomes free of lethals and semilethals be used in generating the $+_i/+_j$ heterozygotes to be used as a standard, or should all types of chromosomes be used because they are more representative of the situation in wild populations? Because there is no simple answer to this question, it must be kept in mind when reviewing the results of such chromosome analyses, especially when a major difference in interpretation rests on a difference of a few percent in the viability of the flies being compared.

One other consequence of this discussion should be the realization of the difficulties inherent in the type concept. There is no one "best" or "normal" genotype that can serve as a standard against which all other genotypes can be compared. In theory, one can hypothesize an individual homozygous at each locus for the most frequent allele in the population as the normal wild type individual, but as we shall now see, such individuals do not exist.

VARIATION IN DROSOPHILA

Voluminous literature exists on chromosome analysis from wild populations. I shall give a sampling of the kinds of information that have been obtained (see Crumpacker, 1967, for a more complete survey). Table 3-1 gives a summary of early data from Dubinin (1946) on the frequency of lethal second chromosomes in Russian populations of *D. melanogaster* in different months. It suggests a seasonal increase in the frequency of lethal second chromosomes, but it is inconclusive because the chromosomes came from different populations in different places and in different years. More important than the

Table 3 - 1. Frequencies of Lethal Second Chromosomes in Different Months in Different Populations of *Drosophila melanogaster* in Russia

Month	Chromosomes Tested (Samples)	Percentage Lethal	Range
July	3,506 (12)	14.0	8.8-21.6
September	2,111 (2)	25.1	22.6-29.8
October	1,766 (3)	26.1	24.1-29.2
November	480 (1)	38.9	—

Source: Dubinin, 1946.

Table 3 - 2. Frequency of Lethal Second Chromosomes in DDT-Resistant Populations and Their Controls in *Drosophila melanogaster*

Stock	N	Percentage Lethal	Percentage Allelism	Interpopulation Percentage Allelism
731 C	116	7.8	25.0	
731 R	133	6.8	25.0	0.0
91 C	138	18.8	9.7	
91 R	119	20.2	7.5	0.0
OR C	134	14.9	13.1	
OR R	137	20.4	7.2	0.0

Source: Reprinted from D. J. Merrell, Lethal frequency and allelism in DDT-resistant populations and their controls. *American Naturalist* 1965, 99:411-417 by permission of The University of Chicago Press.

possible seasonal effect, these data provide an indication of the range of frequencies of lethal second chromosomes that have been observed in natural populations, from 8.8% to 38.9%.

These data can be compared with the data in Table 3-2, which give the frequency of lethal second chromosomes in three DDT resistant stocks of *D. melanogaster* and their corresponding controls (Merrell, 1965a). The resistant populations had been separated from their controls and exposed to DDT for almost 10 years at the time these tests for lethals were made. Several points should be noted. First, the frequency of lethals in the laboratory populations was somewhat lower than in Dubinin's wild populations but still averaged 15%. The percent allelism for intrapopulation lethals is quite high. The low frequency of lethals and the high rate of allelism among them are both

Table 3 - 3. Frequency of Lethal and Semilethal Second Chromosomes
in *Drosophila melanogaster* from Different Geographic Areas in the United States

Area	Date	N	Percentage Lethals and Semilethals	Percentage Allelism-Lethals
Belfast, Maine	1938	115	51.3	
Pullman, Wash.	1951	138	39.1	
Amherst, Mass.	1938	151	45.0	0.43
	1941	108	59.3	
	1947-1952	3163	34.9	0.92
New York State (Syosset + Monroe)	1950	527	32.3	
Cannonsburg, Pa.	1952	117	28.2	
Massillon, Ohio	1941	177	49.7	
Wooster, Ohio	1951	166	36.2	
Lincoln, Neb.	1952	133	25.6	
Blacksburg, Va.	1950-1952	805	43.0	
Gallup, N.M.	1941	203	62.1	
Austin, Texas	1951	98	41.8	
Winter Park, Fla.	1940	227	67.0	} 0.44
	1942	110	61.8	
	1951	131	51.1	0.38

Source: Ives, 1954.

indicative of a relatively small population size for both exposed and control populations in the population bottles in which they were maintained. The absence of allelism between lethals from resistant populations and their controls suggests that in the decade since their separation, there has been a turnover in the lethals in one or both of the populations. Finally, there is no significant difference in lethal frequency between the DDT resistant and control populations.

Table 3-3, drawn from Ives (1945, 1954), shows the combined frequency of lethal and semilethal second chromosomes in populations of *D. melanogaster* from different geographic areas in the United States. The population data have been arranged from north to south because higher lethal and semilethal frequencies have been postulated in the southern populations, which are presumed to be larger and somewhat more stable in numbers. However, even though this trend is indicated, it is far from consistent. The frequencies are probably more directly related to the immediate past history of the populations than to some generalized area effect. Since the populations are more apt to go through bottlenecks in the more northern areas, *D. melanogaster* being essentially a tropical species, lower frequencies of lethals

Table 3 - 4. Frequency of Lethal and Semilethal Second and Third Chromosomes
in a South Amherst, Massachusetts Population of *Drosophila melanogaster*

Collection	Chromosome	N	Percentage Lethal	Percentage Semilethal	Percentage Lethal + Semilethal	Percentage Allelism (Lethals)
September 1958	II	273	20.5	15.4	35.9 ± 2.9	0.63
	III	252	20.6	19.4	40.1 ± 3.1	0.48
October 1958	II	151	21.9	4.0	25.8 ± 3.6	0.27
	III	124	23.4	2.4	25.8 ± 3.6	0.00
July 1959	II	118	21.2	8.5	29.7 ± 4.2	0.87
	III	67	31.3	9.0	40.3 ± 6.0	0.00

Source: Band and Ives, 1963. Reprinted with permission of the Genetics Society of Canada.

and semilethals and higher rates of allelism are more apt to occur in the more northern parts of its range. The allelism rates for lethals in the Florida samples are comparable to the 1938 Amherst sample, but in the 1947-1952 Amherst samples the allelism rate is double that of the others. However, all these allelism rates are less than 1%, in contrast to the rates in the laboratory populations in Table 3-2, which ranged from 7.2% to 25.0%. However, Watanabe (1969) reported considerably higher rates of allelism for second chromosome lethals in natural populations of *D. melanogaster* in Japan where the average allelic rate was 3.48% with the range from about 2% to 7%.

Table 3-4, from Band and Ives (1963), shows lethal and semilethal frequencies for not only the second but the third chromosome in wild populations of *D. melanogaster* in Massachusetts. The lethal frequencies are comparable to those found by Dubinin, and a significant percentage of the chromosomes was found to have semilethal effects. The similarity in the frequency of lethal and semilethal effects between the second and third chromosomes suggests that the observed results are not attributable to some peculiarity of the second chromosome. Moreover, the low rates of allelism again indicate a much larger size in these natural populations than in the laboratory populations in Table 3-2.

Table 3-5 (Pavan et al., 1951) gives a more complete breakdown of the effects on viability of the second and third chromosomes from a different species, *D. willistoni.* This bimodal frequency distribution of viability effects is quite typical of such analyses. One peak is at the lethal end of the viability scale; the other peak is for the group of chromosomes nearly normal when homozygous. Figure 3-2 shows

Table 3 - 5. Frequencies of Chromosomes with Different Genetic Effects
When Homozygous in *Drosophila willistoni*

Chromosome	N	Percentage Lethal	Percentage Semilethal	Percentage Subvital	Percentage Normal	Percentage Supervital
II	2004	28.6	12.6	31.9	25.7	1.2
III	1166	19.7	12.4	32.9	34.2	0.8

Chromosome	N	Percentage Sterile	N	Percentage Visible	N	Developmental Rate Percentage Slow	Percentage Fast
II	928	31.0	270	15.9	1060	31.8	13.7
III	819	27.7	477	16.1	753	35.7	2.8

Source: Pavan et al., 1951.

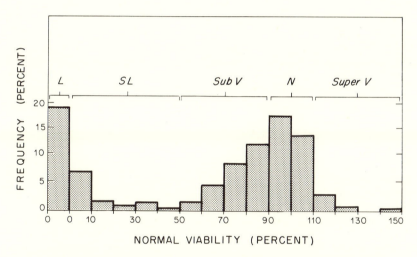

Figure 3-2. The effect on viability of homozygosity for second chromo-
somes from natural populations of *Drosophila prosaltans.* L = lethal;
SL = semilethal; Sub V = subvital; N = normal; Super V = supervital.
(Source: Dobzhansky and Spassky, 1954).

such a bimodal distribution based on the data of Dobzhansky and
Spassky (1954) for the second chromosome of *D. prosaltans.*

In addition to the viability tests, Pavan et al. (1951) tested the
quasinormal chromosomes for their effects on sterility, developmental
rate, and the visible phenotype, with the results shown in Table 3-5.
Because of the way the data are presented, these results cannot be

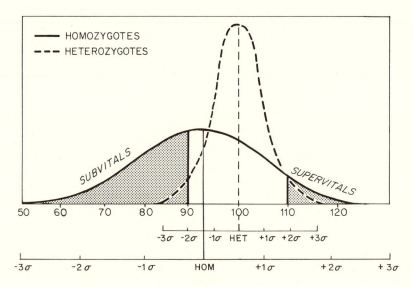

Figure 3-3. The convention used to distinguish among second chromosomes in *Drosophila prosaltans* that are subvital, "normal," and supervital when homozygous. Flies heterozygous for two different second chromosomes from the same population are used as the standard of reference. (Source: Dobzhansky and Spassky, 1954).

directly added to the viability data. However, on the assumption the effects are traceable to individual loci, careful scrutiny of Table 3-5 leads to the conclusion that the great majority of autosomes must bear at least one deleterious mutant of some kind. This in turn leads to the conclusion that the proportion of individuals in the population entirely free of lethal, semilethal, or other deleterious genes must be very small indeed. For example, if there are three large autosomes, and we reasonably assume that 95% of them carry one or another of the various kinds of deleterious genes shown in Table 3-5, then the probability that a diploid fly will be free of deleterious genes on all six of its autosomes is 0.05^6, or approximately 1 in 67 million.

The problem of finding a suitable standard of comparison in tests of the viability of chromosome homozygotes was discussed previously. Figure 3-3 (Dobzhansky and Spassky, 1954) illustrates the convention adopted to distinguish among chromosomes that are subvital, "normal," and supervital when homozygous. Under this system, normal viability is defined as the average viability of heterozygotes for pairs of chromosomes taken at random from a natural population. This mean viability is assigned an arbitrary value of 100. The analysis is based on a comparison of the residual (or genotypic) arrays of viabilities of homozygous and heterozygous individuals after the sampling and experimental ("culture") variances have been removed. A subvital homozygote is defined as one with a viability at least two standard deviations below the mean on the frequency distribution of heterozygote viabilities; a supervital homozygote is defined as having a viability at least two standard deviations above the mean on that same distribution.

One factor apparently not considered in this method is that pairs of chromosomes drawn at random from the same natural population, especially if the population is small, have a distinct possibility of carrying the same detrimental allele. If so, then a small proportion of the "heterozygotes" will be homozygous for one or another of the detrimental mutants in the natural population, and the average viability of the heterozygotes used as the standard of comparison will be lowered. Dobzhansky et al. (1963) apparently observed such an effect in their heterozygous standard for an isolated marginal population in Bogota, Colombia. They found a semilethal heterozygous combination plus two others with less than two-thirds the expected viability. Even with the semilethal heterozygote omitted from the standard, the mean viability of the remaining 209 heterozygous combinations was only 32.92% as compared to the expected 33.33%. Since the other 67.08% of the flies were *Ba gl/+,* and the dominant marker may well have had an adverse effect on the viability of those flies, as suggested previously, the viability of the heterozygous standard may be even worse than it appears. If this situation prevails, it may help to account for the apparent appearance of "supervital" homozygotes. A more appropriate standard, which would minimize this problem, could be obtained by drawing pairs of chromosomes at random from two different wild populations to construct the heterozygous standard (Merrell, 1965a).

Table 3-6 gives a summary of data on lethal and semilethal chromosome frequencies in five species of *Drosophila* and shows a rather high frequency of lethal and semilethal chromosomes in all cases. In addition the table gives estimates of the number of lethals and semilethals per genome, based on the Poisson correction and the proportion of the total genome accounted for by the chromosome in question. It can be seen that the typical value appears to be about one per genome. Thus the high incidence of lethal and semilethal autosomes is not a characteristic of a particular chromosome, or a particular species, but appears to be the norm for the genus. Moreover, reviews such as Crumpacker (1967) and Dobzhansky (1970) make clear that

Table 3 - 6. Frequencies of Lethal and Semilethal Chromosomes
in Several Species of *Drosophila*

Species	Chromo-some	Region	N	Percentage Lethal and Semilethal	Average Lethals and Semilethals per Genome
D. prosaltans	II	Brazil	304	32.6 ± 2.7	0.99
	III	Brazil	284	9.5 ± 1.7	0.50
D. willistoni	II	Brazil	2004	41.2 ± 1.1	1.33
	III	Brazil	1166	32.1 ± 1.4	1.94
D. melanogaster	II	Amherst, Mass.	3549	36.3 ± 0.8	1.13
D. pseudoobscura	II	California Yosemite	109	33.0 ± 4.5	2.00
D. persimilis	II	California Yosemite	106	25.5 ± 4.2	1.47

Source: Dobzhansky and Spassky, 1954.

this situation is by no means unique to *Drosophila* but occurs widely in cross-fertilizing species of both plants and animals, including man.

One of the problems with these studies is that the effects observed are attributable to homozygosity for an intact chromosome rather than just one locus. A further step in the study of these chromosomes was first taken by Dobzhansky (1946) when he permitted recombination to occur between chromosomes isolated from wild populations, recovered the recombinant chromosomes in the homozygous condition, and compared their effects when homozygous with the homozygous effects of the chromosomes from which they were derived. Among the recombinant chromosomes, he recovered some that were lethal when homozygous even though neither of the parental chromosomes was lethal when homozygous. He called these chromosomes "synthetic lethals." A number of studies were done following Dobzhansky's initial report (see Allen, 1966 for references), but the results were ambiguous, for some authors also generated synthetic lethals while others were unable to do so. The reasons for this difference have never been completely resolved, but it appears that the more rigorously the chromosomes are chosen for "normal" viability, the less likely is the recovery of synthetic lethals.

The primary conclusion from the numerous studies of the homozygous effects of intact chromosomes from natural populations is that they carry a considerable variety of deleterious recessive genetic material. The harmful effects are not ordinarily observed in wild populations because the chromosomes are usually heterozygous. The presence of this deleterious variation poses such questions as what is its origin, why does it persist in the populations at such seemingly high frequencies, and what is its function, if any? Even though phenotypically most individuals in natural populations appear rather similar, in reality these similar phenotypes must be produced by a wide range of genotypes.

One difficulty with chromosome analysis is that the effects of homozygous chromosomes are observed rather than the effects of homozygosity at individual gene loci. Unless time-consuming tests are run, there is no way to know whether an observed lethal or other effect is the result of homozygosity at one locus or more. Another difficulty is that the variation detected is, for the most part, in some way deleterious. It can be argued that if the environment is changed, the fitness of genotypes will also change, and that some genotypes formerly low in fitness may be the most fit in the new environment. However, many genetic effects, especially the more severely detrimental types such as those studied by chromosome analysis, are apt to be expressed in just about any environment. For example if a lethal effect is due to a gap in the intestinal tract of an animal, that defect is apt to develop under just about any set of environmental conditions. Since a primary objective of studies of genetic variation in natural populations is to extend our understanding of the evolutionary process, it may well be questioned whether the type of variation revealed by chromosome analysis plays a significant role in evolution.

BIOCHEMICAL VARIATION

In the early days of Mendelian genetics, the inheritance of visible traits was investigated including characters such as flower color, seed shape and color in peas, coat color in mice, and eye and body color, wing shape and venation in fruit flies. In each case alternative states had to exist, red and white flower color, for instance, before the mode of inheritance could be studied. Mendelian methods then revealed that, for such simple traits, alternative alleles on homologous chromosomes were usually responsible for the observed variants. If the entire population was homozygous for one of the alleles, the Mendelian approach could not detect the presence of the gene locus controlling the trait at all. In other words, Mendelian genetics was and is dependent on the presence of segregating alleles. Many of the traits studied were rather extreme variants, often of low viability. As the study of genetics progressed, other classes of mutants were studied; but attention tended to be focused on the more extreme variants because they were generally easier to work with. However, the more drastic their effects, the less important the alleles are apt to be in evolution. Thus, the more subtle gene substitutions may well play a far more significant evolutionary role than lethals and semilethals. However, these more cryptic genes are also more difficult to study by the usual Mendelian methods. Moreover, if loci can be identified only when alternative alleles have been found, an unknown number of invariant loci may exist, and we will have no idea of what proportion of the genome consists of invariant loci or, conversely, what proportion of loci are variable.

The advances in molecular genetics that opened the door to an entirely new approach to the study of variation in wild populations were pioneered by the 1966 studies of Lewontin and Hubby, and F. M. Johnson et al. in *Drosophila* and Harris in man. It is safe to say that in the last decade most population geneticists have been pre-occupied with accumulating data with the new techniques available and trying to interpret the meaning of their results (Manwell and Baker, 1970; Lewontin, 1974; Ayala, 1976).

Studies in molecular genetics revealed that DNA is the hereditary material and that through messenger RNA it controls the synthesis of proteins. The substitution, addition, or deletion of a base in the DNA genetic triplet code for protein synthesis will usually result in a corresponding change in the sequence of amino acids being assembled into a polypeptide chain. Such changes are discrete, clearcut, and qualitative unlike many other types of variation, where the discrete effects of allelic substitutions are blurred by environmental effects. Therefore, molecular genetics made it possible to study genetic variation in natural populations at the level of primary gene action rather than at a phenotypic level far removed from the primary action of the genes. Ideally, the way to study this variation would be through amino acid sequencing, now possible for proteins, but because it is so time-consuming and costly, gel electrophoresis is commonly used.

Both starch and acrylamide gel electrophoresis are widely used. In essence, proteins, such as enzymes or structural proteins, are placed in an electric field, where, because some of their constituent amino acids have either a positive or a negative charge, each type of protein molecule will have a characteristic net charge and configuration and, therefore, a characteristic rate and direction of migration. If a mutation produces a change in the amino acid composition of the protein, the mobility of the protein will change if there is a change in its charge or configuration. Not all variants will be detected by this technique because not all amino-acid substitutions will cause a change in net charge. Therefore, the amount of genetic variation will be underestimated by this technique.

After proteins have migrated through the gel in the electric field, they are made visible with a general protein stain for highly concentrated proteins. If the protein is an enzyme of low concentration, a suitable enzyme substrate plus a colorless dye is added. A color develops when, for example, the substrate is converted from its oxidized to its reduced form, thereby revealing the location of the enzyme in the gel.

A major advantage of this technique over previous methods of analyzing variation is that the proteins can be detected whether segregating alleles are present or not. For the first time, it became possible to estimate what proportion of gene loci were variable and what proportion were invariant.

This new approach to the study of genetic variation led to a great burst of research activity over the past decade, for, like children with a new toy, scientists applied the technique to the study of variation in a wide range of species (Manwell and Baker, 1970; Lewontin, 1974; Ayala, 1975a, 1976; Powell, 1975a). Instead of counting bristles, or studying wing venation patterns, or estimating the frequency of lethals and semilethals, they looked at bands of proteins in gels.

When interpreting these data, one should keep several points in mind. If two members of the same species have homologous proteins that migrate at different rates during electrophoresis, it is assumed that they carry different alleles at the locus coding for that protein. To confirm this assumption, it is necessary to make crosses to demonstrate that the differences observed are due to segregating alleles at the locus of the structural gene for that protein. Such crosses have seldom been made. However, when the genetic tests have been run, mostly with species whose genetics are well known like *Drosophila* and mice, the assumption that the electrophoretic variants are controlled by alleles at a single locus has been confirmed in nearly all cases. Therefore, in other species less well known genetically, the inference that a similar genetic mechanism controls electrophoretic variants seems reasonable though unconfirmed. Another point is that since the analysis deals only with proteins, only the structural gene portion of the entire genome is being analyzed. The amount of variation in the DNA with regulatory or other functions is not open to study with this technique.

A further difficulty is that not all of the variation at these loci can be detected. If a base substitution does not change the electrophoretic mobility of the protein molecule, it will not be detectable. Lewontin

(1974), for example, estimated that he and Hubby had detected only about a third of the total sequence variation actually present in their original study of *D. pseudoobscura*. Other, more sensitive techniques such as isoelectric focusing have been used to combat this problem. In this technique, a *pH* gradient is set up in the gel, and the proteins migrate to their isoelectric point, that is, the point along the *pH* gradient where the protein has no charge. This technique has enabled researchers to identify additional alleles, but because it is not as easy to use as gel electrophoresis, it has not been widely adopted.

A final question is whether the loci studied form a random sample or a representative sample of all of the structural genes in the genome. The proteins studied are enzymes for which a specific enzyme dye is available or proteins with a high enough concentration to be detected with a general protein stain. To the latter belong albumins, hemoglobins, and transferrins as well as proteins of unknown function. Thus, even though gel electrophoresis represents a considerable advance over the use of lethal and semilethal chromosomes in the study of genetic variation in natural populations, only a portion of the variability in the structural gene component of the genome is actually being studied.

Table 3-7 gives a summary of data more or less typical of the kinds of results that have been obtained from studies of electrophoretic variation in natural populations of a variety of different species. Such surveys produce certain basic information: estimates of genotypic or allelic frequencies for each locus tested in a given population, an estimate of the incidence of heterozygosity in the population, and an estimate of the proportion of polymorphic loci in the population.

The electrophoretic techniques permit an estimate of the number of alleles and their frequency in the population. Since the usual methods sometimes fail to distinguish between alleles generating proteins with similar mobilities, these must be regarded as minimum estimates. Usually the number of alleles and the allelic frequencies are given rather than the genotype frequencies because for a given number of alleles, the possible number of genotypes is always greater than the number of alleles. If n is the number of alleles at a locus, the possible number of genotypes is $(n^2 + n)/2$.

If the population is mating at random, the expected frequency of heterozygotes (H) at a locus can be calculated from the allelic frequencies. If there are n alleles with frequencies $q_1, q_2, q_3, \ldots q_n$, the expected frequency of homozygotes, with random mating, is $q_1^2 + q_2^2 + q_3^2, \ldots + q_n^2$, and the expected frequency of heterozygotes is $H = 1 - (q_1^2 + q_2^2 + q_3^2 \ldots + q_n^2)$. The expected frequencies of

Table 3 - 7. Genic Variation in Plants and in Several Groups of Animals

Group	Number of Species	Mean Number of Loci per Species	Mean Proportion of Loci	
			Polymorphic per Population	Heterozygous per Individual
Insects				
Drosophila	28	24	0.529 ± 0.030	0.150 ± 0.010
Others	4	18	0.531	0.151
Wasps	6	15	0.243 ± 0.039	0.062 ± 0.007
Marine				
Invertebrates	9	26	0.587 ± 0.084	0.147 ± 0.019
Snails				
Land	5	18	0.437	0.150
Marine	5	17	0.175	0.083
Vertebrates				
Fish	14	21	0.306 ± 0.047	0.078 ± 0.012
Amphibia	11	22	0.336 ± 0.034	0.082 ± 0.008
Reptiles	9	21	0.231 ± 0.032	0.047 ± 0.008
Birds	4	19	0.145	0.042
Rodents	26	26	0.202 ± 0.015	0.054 ± 0.005
Large mammals	4	40	0.233	0.037
Plants	8	8	0.464 ± 0.064	0.170 ± 0.031

Source: Selander, 1976.

heterozygotes are sometimes compared with the observed frequencies to test the assumption of random mating. However, as discussed in Chapter 6, such comparisons are not reliable unless the population is known to be in equilibrium, and that information is seldom available in studies of this type.

The average frequency of heterozygotes per locus $(\bar{H})$ is obtained by averaging H for all of the loci tested. $\bar{H}$ is a frequently used indicator of the amount of genetic variation in a population. $\bar{H}$ together with its standard error gives not only an estimate of the amount of genetic variation, but also an idea of the amount of heterogeneity in heterozygosity among the loci tested. The heterozygosity in a population is sometimes expressed as the average frequency of heterozygous loci per individual rather than the average frequency of heterozygotes per locus. This form of $\bar{H}$ is obtained by averaging, over all individuals, the proportion of heterozygous loci found in each individual. The numerical values for the two measures are the same, but their variances and standard errors usually differ. The variance of $\bar{H}$ over loci is generally larger than the variance of $\bar{H}$ among individuals, reflecting the large amount of heterogeneity among loci in this respect.

The effective number of alleles, n_e, equals the reciprocal of the frequency of the homozygotes, and also may be used to reflect the amount of genetic variation in a population. If $q_1 = 0.5$ and $q_2 = 0.5$, then $n_e = 2$. However, if $q_1 = 0.95$ and $q_2 = 0.05$, then $n_e = 1.1$. Thus, the method of calculation tends to minimize the importance of rare alleles as sources of variation.

Another measure of genetic variation is P, the proportion of polymorphic loci in a population. We shall define a polymorphic locus as one at which the frequency of the most common allele does not exceed 99%. A more conservative definition sets this frequency no greater than 95%. Both figures are arbitrary, but I prefer the 99% definition because experience with the dominant Burnsi gene in wild populations of the leopard frog, *Rana pipiens,* (Merrell, 1969b) suggested that this was in fact a polymorphism, yet the frequency of the Burnsi allele was nearly always less than 5%. The original definition of Ford (1940b, 1975) was "Genetic polymorphism is the occurrence together in the same locality of two or more discontinuous forms of a species in such proportions that the rarest of them cannot be maintained merely by recurrent mutation." Because of the difficulties in determining mutation rates and selection coefficients in natural populations, the arbitrary definition is far more practical and does not do any great injustice to the concept.

If a number of different populations of a species are analyzed, the average proportion of polymorphic loci per population ($\bar{P}$) is simply the average of P for all of the populations studied.

P is less useful than $\bar{H}$ not just because it is defined arbitrarily but also because it tells only whether or not a locus is polymorphic and nothing about the degree of polymorphism. It makes no distinction, for example, between a locus with two alleles with frequencies of 0.98 and 0.02 and another locus with five alleles each with a frequency of 0.2 even though the latter is more variable. (In the first case, $H = 0.04$; in the second, $H = 0.80$.) Although $\bar{H}$, the average frequency of heterozygous individuals per locus in a population, is more informative than P, the proportion of polymorphic loci in the population, both estimates are often given. (See Lewontin, 1974, and Nei, 1975, for further discussion.)

As can be seen in Table 3-7, the proportion of loci polymorphic in a population ranges from about 15 to 50%, and the proportion of loci heterozygous per individual ranges from roughly 5 to 15%. Thus it can be generally said that in a sexually reproducing species about a third of the loci are polymorphic, and, on the average, individuals are heterozygous for about 10% of their loci. This amount of polymorphism and heterozygosity represents a tremendous store of variability in wild populations, especially when it is recalled that these are minimum estimates based on electrophoretically detectable variation and may represent only a third to a half of the actual values. If the sample of loci studied is not random or representative, then perhaps only the more variable loci have been selected for study; but as more and more loci in more and more species have been analyzed, similar results have continued to roll in. Some patterns of difference have begun to emerge in the level of polymorphism, which we shall discuss later, but still the wealth of biochemical polymorphism continues to appear so great that it has been considered almost too much of a good thing. For example, if it is assumed that one allele when homozygous is more fit under existing conditions than the others, then with so many loci polymorphic, large numbers of homozygotes with low fitness would be generated. The loss of fitness in this population as compared to a population consisting entirely of homozygotes for the most advantageous allele has been called the *genetic load*. Alternatively, it can be assumed that the heterozygotes at a given locus are more fit than the homozygotes and the genetic load arises from both homozygotes, which will continue to be generated because both alleles must be present to produce the most fit genotype, the heterozygote. With such a high level of polymorphism and heterozygosity, it appeared that under assumptions such as the above, an intolerable genetic load would exist. One consequence of this reasoning was the suggestion that this biochemical variation must be adaptively and selectively neutral. Thus, by a simple change in the assumptions, the problem was solved. The trouble was that there was virtually no evidence concerning the possible effects on fitness of the biochemical variation that was being detected. Moreover, in many instances, the function of the enzymes and proteins that were being studied was unknown. Nevertheless, the result has been that over the past decade a considerable debate has arisen over the concept of genetic load, and more recently, over the neutralist or selectionist interpretation of these data. But before dealing further with these ideas, we need to develop some other concepts.

This review of some of the data on the study of biochemical variation should amply demonstrate that in sexually reproducing species every individual must be genetically unique. The superficially similar phenotypes of individuals in natural populations are generated by an array of different genotypes and conceal a wealth of genetic variation. Our problem is to interpret these observations in evolutionary terms.

THE CASTLE-HARDY-WEINBERG EQUILIBRIUM

Evolution may be defined as descent with modification, which, in modern terms, may be restated as a change in the kinds or frequencies of genes in populations. Therefore, the basic problem becomes to determine what causes changes in gene frequency or the substitution of one allele for another in natural populations. Thus far, four independent factors have been identified that can produce gene frequency change: mutation, natural selection, migration or gene flow, and random genetic drift. Evolution is a complex phenomenon ordinarily resulting from the combined effects of these factors operating simultaneously in the same population. However, in order to comprehend the more complex situation, it is helpful to study it first in the simplest possible terms. For this reason and to establish certain principles, we shall first consider each factor separately in relation to its effects on a single gene locus. Following this approach, we shall be better prepared to consider how these factors may interact with one another in the more complex and realistic situation. The first question is, What happens in a population where none of these factors operates, in which there is no mutation, no selection, no gene flow, and no random genetic drift? This situation was first examined by Castle (1903) and later by Hardy (1908) and Weinberg (1908) independently.

At this point we must introduce a few concepts, the first of which is *gene frequency*. Let us consider a single autosomal locus at which there are just two alleles, *A* and *a*. Each diploid individual will then have two "*A*" genes, one on each of the homologous chromosomes carrying the "*A*" locus. The possible genotypes are three: *AA*, *Aa*, and *aa*. We shall give the number or proportion of *AA* as *D*, of *Aa* as *H*, and of *aa* as *R*, so that $D + H + R = N$, or if given as proportions or relative frequencies, $D + H + R = 1$ (where D = dominants, H = heterozygotes, and R = recessives). Since 100 diploid individuals have 200 chromosomes carrying the "*A*" locus, they carry 200 "*A*" genes of some sort, and N individuals may be said to carry $2N$ "*A*" genes. The gene frequency of the dominant allele *A* is simply the proportion of dominant alleles present among the total of all "*A*" alleles of any kind. In other words, if the frequency of *A* equals p, then

$$f(A) = p = \frac{2D + H}{2(D + H + R)} = \frac{2D + H}{2N} = \frac{D + \frac{1}{2}H}{N}$$

or if $D + H + R = 1$

$$p = D + \frac{1}{2}H.$$

Similarly,

$$f(a) = q = \frac{2R + H}{2N} = \frac{R + \frac{1}{2}H}{N}$$

or if $D + H + R = 1$

$$q = R + \frac{1}{2}H.$$

This sort of analysis simply enumerates the genes in the population and finds the proportions for each type of allele to determine its frequency. Suppose in a certain population, the number of AA individuals is 4; Aa, 8; and aa, 28. Then

	A alleles	a alleles
$D = 4$	8	0
$H = 8$	8	8
$R = 28$	0	56
$N = 40$	16	64

$$2N = 80$$

and

$$p = \frac{2D + H}{2N} = \frac{8 + 8}{80} = \frac{D + \frac{1}{2}H}{N} = \frac{4 + 4}{40} = 0.2$$

$$q = \frac{2R + H}{2N} = \frac{56 + 8}{80} = \frac{R + \frac{1}{2}H}{N} = \frac{28 + 4}{40} = 0.8.$$

Also, given as proportions,

$$D = \frac{4}{40} = 0.1$$

$$H = \frac{8}{40} = 0.2$$

$$R = \frac{28}{40} = 0.7$$

and

$$p = D + \frac{1}{2}H = 0.1 + 0.1 = 0.2$$

$$q = R + \frac{1}{2}H = 0.7 + 0.1 = 0.8.$$

It must be emphasized that the gene and genotype frequencies obtained in these calculations are not equilibrium frequencies, but are simply the frequencies obtained by determining the proportion of each type of allele and genotype in the population.

Another concept we must consider is *random mating* or *panmixia*. In a population with separate sexes, mating is random if any individual of one sex is equally likely to mate with any individual of the opposite sex. In a hermaphroditic population, mating is random if any individual is equally likely to mate with any other individual, including itself.

As an example, let us suppose that we introduce 500 homozygous gray (AA) rats, 250 of each sex, and 500 homozygous black (*aa*) rats, also 250 of each sex, onto an island. If the matings are at random, the gray will be as likely to mate with black as with gray, and vice versa. The possible matings are

It can readily be seen that the progeny will occur in a 1*AA* : 2*Aa* : 1aa ratio. This ratio is not an ordinary Mendelian 3 : 1 ratio, for these are F_1 progeny, not the F_2, and it has resulted from random mating among all types present rather than controlled matings of the sort that generate a 3 : 1 F_2 Mendelian ratio.

If this F_1 generation now mates at random, there will be nine possible kinds of matings, as shown below.

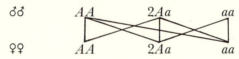

This sort of approach is cumbersome, but it can be shown, by determining the kinds and frequency of the progeny from each of the nine possible kinds of matings, that the F_2 ratio among the progeny will again be 1*AA* : 2*Aa* : 1aa.

However, if mating, segregation, and fertilization are all random processes in the population, gene frequencies can be used rather than genotype frequencies, and the calculations are much simpler because the gametes will combine at random with one another in proportion to their frequencies in the population. In the above case the frequency of the two kinds of gametes will be equal, $p = q = 0.5$ in both males and females. With random mating, the frequency of a particular zygotic combination will simply be the product of the frequencies of the two gametes that contribute to it. Thus, $f(AA) = p^2 = 0.25$ and $f(aa) = q^2 = 0.25$. The heterozygotes can be formed in two different ways because either parent may contribute the recessive and the other the dominant allele. Hence $f(Aa) = 2pq$. This relationship can be better visualized from the following:

♀ gam \ ♂ gam	p A	q a
p A	p^2 AA	pq Aa
q a	pq aA	q^2 aa

Therefore, the frequency of the various genotypes can be obtained by the expansion of the binomial $(p + q)^2$ where

$p + q = 1$

$p = f(A) = 1 - q$

$q = f(a) = 1 - p$

$p^2(AA) + 2pq(Aa) + q^2(aa) = 1.$

If more than two "A" alleles are present, the equation becomes $(p + q + r + . . .)^2$, the expansion of which will give the expected frequencies of the various possible types of homozygotes and heterozygotes.

Let us suppose that the initial population introduced on the island consisted of 300 gray AA and 700 black aa rats, again with equal numbers of males and females for each genotype. In this case, the gene frequency, the genotype frequency, and the phenotype frequency will all be the same. Initially, $p = 0.3$ and $q = 0.7$. After one generation of random mating, heterozygotes as well as homozygotes will be present in the island population in the following proportions:

$p^2(AA) + 2pq(Aa) + q^2(aa) = 1$

$(0.3)^2 + 2(.3)(.7) + (.7)^2 = 1$

$0.09(AA) + 0.42(Aa) + 0.49(aa) = 1.$

If the population size remains constant at 1000, the genotypic ratio will now be 90AA: 420Aa: 490aa, and the phenotypic ratio 510A-: 490aa, quite different from the initial 300 AA: 700 aa. However, the gene frequencies will be unchanged, for

$p = D + \frac{1}{2}H = 0.09 + 0.21 = 0.3$

$q = R + \frac{1}{2}H = 0.49 + 0.21 = 0.7.$

Moreover, from this generation on, with random mating and no evolutionary forces acting, not only the gene frequencies but the genotypic and phenotypic frequencies will also remain constant, and the population is in the Castle-Hardy-Weinberg equilibrium. The relationship has been shown above algebraically, but it can also be put into words. In a large, random-mating population, in the absence of selection, mutation, and migration, the relative frequency of the genes in the population will remain constant.

One of the most troubling problems to Charles Darwin in his development of the theory of natural selection was his belief that, because of blending inheritance, variability was being rapidly lost every generation. He felt that a high level of variability was essential for natural selection to work effectively, but he knew of no mechanism that could replenish this lost variation rapidly enough. The discovery by Mendel that inheritance was particulate and not blending, and the realization by Castle, Hardy, and Weinberg that, rather than a constant erosion of variability each generation, the variability would remain constant in the absence of evolutionary forces, resolved Darwin's dilemma completely. Unless acted on in some way by one or more of the factors mentioned previously, the store of variability discussed earlier will continue to persist unchanged in the populations.

The Castle-Hardy-Weinberg equilibrium is usually the starting point for the study of the genetics of populations. For example, if the frequency of one homozygous type is known in a population, which is assumed to be in C-H-W equilibrium, it is possible to estimate the gene and genotype frequencies. For example, suppose that $f(aa) = 0.04$ and then assume that $0.04 = q^2$ of the C-H-W equation,

then $q = \sqrt{0.04} = 0.2$

   $p = 1 - q = 0.8$

and $f(AA) = p^2 = 0.64$

   $f(Aa) = 2\,pq = 0.32$

   $f(aa) = q^2 = 0.04.$

Recall that in our initial discussion of the calculation of gene frequencies, the numbers of the different genotypes were $AA = 4; Aa = 8; aa = 28$, and $p = 0.2$ and $q = 0.8$. Calculation of the expected genotype frequencies from the gene frequencies will show that the observed proportions are not very close to the expected values although the difference may not be significant because the numbers are small. However,

	Expected C-H-W	Observed
$f(AA)$	0.04	0.10
$f(Aa)$	0.32	0.20
$f(aa)$	0.64	0.70
	1.00	1.00

the difference between observed and expected is great enough to raise suspicions about whether the population is in equilibrium.

For sex-linked genes, the situation is somewhat different than for autosomal loci. If the population is in equilibrium, the gene frequencies in males and females will be the same, that is, $p_\delta = p_\varphi$. However, in a species with heterogametic (XY) males, each male carries only one X chromosome but each female has two X chromosomes (XX). The males then are hemizygous for genes on the X, and can have only two possible genotypes, but the females can have three possible genotypes, as shown below. If the population is not in equilibrium,

Male (XY)		Female (XX)		
A	a	AA	Aa	aa
p	q	p^2	$2pq$	q^2

that is, $p_\delta \neq p_\varphi$, and the sex ratio equals one, the equilibrium gene frequency can be obtained from the equation

$$q = 1/3(q_\delta + 2q_\varphi).$$

The equation takes this form because if the numbers of males and females are equal, the females (XX) in the population will have two-thirds of the X chromosomes in the population and the males (XY) will have only one-third.

Rather than reaching equilibrium with a single generation of random mating as autosomal loci do, sex-linked gene frequencies oscillate about the equilibrium value, with the deviation from equilibrium halved for each generation of random mating, with the sign reversed, as shown in Figure 3-4. In this case, it should be noted that the gene frequencies in the males in one generation are simply those of the females of the preceding generation since the males receive their single X chromosome from their mothers. The gene frequency of the females equals the mean of the gene frequencies of their male and female parents since they receive one X chromosome from each parent. Therefore, when the gene frequencies for sex-linked genes differ in males and females, a plot of the gene frequencies separately by sex with random mating shows the curve for the males identical to that for the females, but lagging a generation behind it. The difference between the sexes is halved each generation and the deviation from the equilibrium value is also halved each generation. However, even though the distribution of alleles between males and females oscillates during the approach to equilibrium, the overall gene frequencies based on all the X chromosomes remain constant throughout.

The significance to evolution of the Castle-Hardy-Weinberg law is that any population tends to conserve at the existing level whatever genetic variability may be present, and this variation is not lost through crossing. Therefore, the C-H-W equilibrium is conservative,

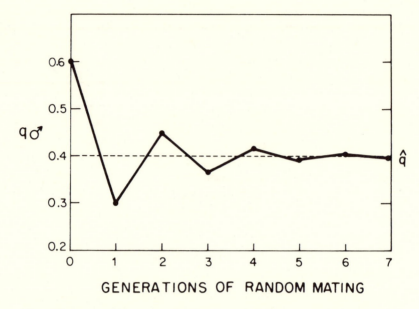

Figure 3-4. Approach to the Castle-Hardy-Weinberg equilibrium for a sex-
linked allele with random mating. Only the male frequencies are shown, for
these are the same as those of the females of the preceding generation.
The approach is oscillatory about the equilibrium value of 0.4.
For further details, see text.

tending to maintain the status quo, and evolution may be redefined
as a shift in the Castle-Hardy-Weinberg equilibrium.

CHAPTER 4

Mutation

The modern concept of evolution stems from several sources. Mendel's discovery of the particulate nature of inheritance (1866) formed the basis for the Castle-Hardy-Weinberg equilibrium (Castle, 1903; Hardy, 1908; Weinberg, 1908). Darwin, in his *Origin of Species* (1859), developed the theory of natural selection as the mechanism for adaptation. In 1901 de Vries postulated that mutations also played a role in the evolutionary process. These separate strands were woven into a single pattern by the work of R. A. Fisher (1930) in *The Genetical Theory of Natural Selection,* Sewall Wright (1931) in "Evolution in Mendelian Populations," and J. B. S. Haldane (1932b) in *The Causes of Evolution.*

The modern concept postulates that evolutionary change is brought about by the combined effects of natural selection, gene flow, and random genetic drift on the genetic variation in natural populations. The ultimate source of genetic variation is mutation, which, in the broad sense, may be defined as any hereditary change not caused solely by Mendelian recombination.

In the narrow sense, mutation refers to gene or point mutations, mutations at a single gene locus in which there is a change in the information contained in the DNA of the chromosome. As understanding of the nature and function of DNA, the hereditary material, has increased, various kinds of genes have been recognized. Structural genes code for messenger RNA, which, in turn, mediates protein synthesis; other genes code for transfer RNA, still others for ribosomal RNA. A considerable portion of the DNA, especially in higher forms of life, is thought to have various regulatory functions. Presumably, mutations can occur affecting any of these types of genes, but the mutations most commonly studied have been in structural genes. DNA consists of the familiar Watson-Crick double helix, a double-stranded chain of nucleotides held together by hydrogen bonding between the bases. The nucleotides consist of deoxyribose, phosphate, and a purine (adenine, guanine) or a pyrimidine (cytosine, thymine) base, and the specificity of the DNA resides in the sequence of bases. The genetic code in structural genes is a three-letter code in which three bases (a triplet) code for a single amino acid, and a series of triplets code for a polypeptide chain. (For further details, consult Merrell, 1975a, or any introductory genetics text.)

Mutation in a structural gene may involve *base-pair substitution,* the substitution of one type of purine or pyrimidine base for another, or *frameshift mutations,* the addition or deletion of one or a few base pairs. Base-pair substitution of one of the three bases in a triplet codon may lead to a change in a single amino acid in the protein controlled by this gene. The code is ordinarily non-overlapping and is read from a fixed starting point so that the frameshift mutations will alter the reading of the code beyond the point where the addition or deletion occurs. The so-called *macrolesions* consist of deletions, duplications, or rearrangements (inversions, translocations, insertions) of small segments of DNA. Chromosome mutations may involve rearrangements of chromosome fragments (duplications, deletions, translocations, insertions, and inversions) or change in number of chromosomes (aneuploidy, polyploidy). Position effect is difficult to categorize in this context; but it is sometimes observed following rearrangement of genes into new associations and may cause some confusion in mutation research if not properly identified. Since DNA has been identified in plastids and mitochondria, cytoplasmic mutations may also occur, but they do not follow Mendelian patterns of inheritance.

Although this scheme of classification is useful, the degree of experimental sophistication varies considerably among different species. Therefore, in some species it may be possible to distinguish between base-pair substitutions and frameshift mutations, whereas in others, it may be difficult even to distinguish between point mutations and chromosome mutations.

MUTATIONS IN POPULATIONS

The effect of mutation is to increase the amount of variability in a population. Our treatment of mutation in populations will deal with gene mutations, but it could also be applied to chromosome mutations, although they are less apt to be recurrent. It can also be used for systems of multiple alleles, but in such cases, attention is often concentrated on one allele and all the others are lumped together and treated as a single allele to simplify matters.

Consider first the fate of a single autosomal recessive *a* mutation in an otherwise homozygous *AA* population. If one of the dominant alleles mutates to a recessive allele, then a single *Aa* individual will be present in the population. The *Aa* heterozygote must mate with an *AA* homozygote since this is the only possible type of mate present. Therefore, the cross will be *Aa* x *AA*. If just one offspring is produced from this mating, it has a 50% chance of being *Aa* and a 50% chance of being *AA*. Thus, with only one offspring, there is a 50% chance of loss of the new mutation from the population in the next generation.

Suppose two offspring are produced from the cross *Aa* x *AA*. The possible types of offspring and the probability of each combination are as follows:

1. 1st *AA*: 2nd *AA* = 25%
2. 1st *AA*: 2nd *Aa* = 25%
3. 1st *Aa*: 2nd *AA* = 25%
4. 1st *Aa*: 2nd *Aa* = 25%.

Thus, with two offspring there is still a 25% chance that the recessive mutation will be lost, there is a 50% chance that one or the other of the two will receive *a* and the population will still have one *a* allele in the next generation, and there is a 25% chance that both will receive the recessive allele and the number of *a* genes in the population will increase from one to two. All that is involved here is pure chance; no selection is involved.

Family size, of course, is variable. However, if family size is equal to k, then the probability of loss of this new recessive mutant is obviously $(\frac{1}{2})^k$. If a constant population size is assumed, then average family size, $\bar{k}$, equals 2. If A and a are assumed to be selectively neutral, and if the distribution of family sizes (k) fits a Poisson distribution (that is, the mean and the variance are approximately equal and in this case equal 2), then the probability of loss of the single mutant a is shown by Table 4-1. The values in the table are obtained as follows:

Family size (k)	0	1	2	3	. . .	n
Frequency of family of given size from Poisson distribution (F_n)	e^{-2}	$2e^{-2}$	$\dfrac{2^2}{2!}e^{-2}$	$\dfrac{2^3}{3!}e^{-2}$	. . .	$\dfrac{2^n}{n!}e^{-2}$
Probability of loss of a in family of given size (l_n)	1	$\dfrac{1}{2}$	$\dfrac{1}{4}$	$\dfrac{1}{8}$	. . .	$\left(\dfrac{1}{2}\right)^n$.

Table 4 - 1. Probability of Extinction or Loss
of a Single Mutant Gene from a Population

Generation (N)	Neutral Mutation		Mutation with 1% Selective Advantage	
	Extinction (l_n)	Survival $(1-l_n)$	Extinction (l_n)	Survival $(1-l_n)$
1	0.3679	0.6321	0.3642	0.6358
2	0.5315	0.4685	0.5262	0.4738
3	0.6259	0.3741	0.6197	0.3803
5	0.7319	0.2681	0.7246	0.2754
7	0.7905	0.2095	0.7825	0.2175
15	0.8873	0.1127	0.8783	0.1217
31	0.9411	0.0589	0.9313	0.0687
63	0.9698	0.0302	0.9591	0.0409
127	0.9847	0.0153	0.9729	0.0271
.				
.				
.				
Limit	1.0000	0.0000	0.9803	0.0197

Source: Fisher, 1930.

Then the chance of loss values (l_n) are multiplied by the frequency of occurrence of the corresponding family size (F_n) and summed to provide the values in the table.

From the table it can be seen that even after just two generations, the chances are better than 50 : 50 that a new, selectively neutral mutation will have been lost by chance from the population; after 63 generations, the chances are only 3 in 100 that the mutation will still persist in the population. Even more surprising, perhaps, is the fact that these probabilities are not greatly changed even if the mutation has a 1% selective advantage over the existing allele. Thus, it appears that the fate of most new mutations in a population will be loss due to chance and only a few "lucky" ones will become established. Even mutations with a slight selective advantage are apt to be lost; however, the greater the selective advantage, the better the chances that the mutation will become established. From these considerations, it should be clear that as new mutations occur, they are not necessarily soaked up by a population like a sponge. Quite apart from any selective pressures against them if they are detrimental, any particular mutation that occurs also has a distinct probability of being lost simply by chance.

Now consider the effects of recurrent mutations in populations. Suppose that dominant A mutates repeatedly to recessive a. One can then say that mutation pressure is at work to increase the frequency of a in the population. If this trend continues unchecked, in due time all of the dominant A alleles could be transformed into a by mutation. By definition, such a transformation is an evolutionary change, and thus mutation alone is an independent evolutionary force.

Next let us address the question, How long will it take recurrent mutation to effect a given change in gene frequency? As before, let $p = f(A)$ and $q = f(a)$, and designate the mutation rate from A to a as u. The change in the frequency of a is called Δq, and

$$\Delta q = q_1 - q_0$$

or

$$q_1 = q_0 + \Delta q.$$

Keep in mind that for any increase in q, there is a corresponding decrease in p, and vice versa, so that

$$\Delta q = - \Delta p.$$

In this case

$$\Delta q = u p_0,$$

that is, the mutation rate times the frequency of the dominant allele available to mutate. Therefore

$$q_1 = q_0 + u p_0$$

and since

$$p_0 = 1 - q_0,$$

this becomes

$$q_1 = q_0 + u(1 - q_0) = q_0 + u - u q_0$$

or

$$q_1 = u + (1 - u)q_0.$$

Similarly, you can find q_2 in terms of q_1, and hence of q_0, q_3 in terms of q_0, and so on. When this is done, the general equation takes the form

$$q_n = u + (1 - u)u + (1 - u)^2 u + \ldots + (1 - u)^n q_0.$$

The first n terms of this expression (excluding the last term) form a geometric series with the common ratio $(1 - u)$. (A geometric series is a sequence of numbers in which the same quotient is obtained by dividing any term by the preceding term. This quotient is called the common ratio.) The sum of these n terms equals $1 - (1 - u)^n$. Hence, the general expression becomes

$$q_n = 1 - (1 - u)^n + (1 - u)^n q_o$$

and, on factoring, it becomes

$$q_n = 1 - (1 - u)^n (1 - q_o).$$

This equation can be rearranged as follows:

$$1 - q_n = (1 - u)^n (1 - q_o)$$

$$(1 - u)^n = \frac{1 - q_n}{1 - q_o} = \frac{p_n}{p_o}$$

$$p_n = p_o (1 - u)^n.$$

Therefore, from this equation, p_n, the frequency of the dominant allele, can be calculated following n generations of mutation from dominant to recessive at mutation rate u, or else n, the number of generations required for a given change in p, can be estimated.

For example, if $p_o = 1.0$, $u = 1 \times 10^{-5}$, and $n = 100$, p_n after 100 generations of mutation at that rate will equal approximately 0.9991. To change from $p_o = 1.0$ to $p_n = 0.9$ at that rate of mutation would require about 11,500 generations; to change from $p_o = 0.6$ to $p_n = 0.5$, about 20,000 generations; and from $p_o = 0.2$ to $p_o = 0.1$, about 75,000 generations. Obviously, mutation as an independent evolutionary force can act only very slowly to bring about evolutionary change.

Moreover, recurrent reverse mutations may also occur from the recessive a back to the dominant A. Call the rate from a to A equal to v. Now there will be opposing mutation rates, one tending to decrease, the other to increase the frequency of the dominant A allele. The decrease in A equals up; the increase in A, vq, and the net change in A will equal

$$\Delta p = vq - up.$$

If u equals 1×10^{-5} and v, the reverse mutation rate, equals 5×10^{-6} and $p = 0.3$ and $q = 0.7$ then

$$\Delta p = 5 \times 10^{-6} (0.7) - 1 \times 10^{-5} (0.3)$$

$$\Delta p = 5 \times 10^{-7}.$$

Then the frequency of dominant gene A will increase by 5 per 10 million genes, despite the fact that the mutation rate from dominant A to recessive a is twice the reverse mutation rate. The increase, of course, results from a larger proportion of recessive alleles available to mutate. A change in frequency of 5×10^{-7} is very small and raises the question whether an equilibrium is possible. An equilibrium will exist when Δp equals zero. This equilibrium will be not static, but dynamic, for it will exist when the absolute or actual number of mutations in each direction is equal. Thus, at equilibrium

$$\Delta p = 0 = vq - up$$

$$up = vq.$$

Substituting $(1 - p)$ for q, and rearranging, we obtain

$$\hat{p} = \frac{v}{u + v}; \hat{q} = \frac{u}{u + v}.$$

From the equation, it can be seen that the equilibrium value of p is independent of the initial values of p and q and depends only on the mutation rates. It will be reached from any value of p from zero to one, and is a dynamic but stable equilibrium. For the mutation rates given above,

$$\hat{p} = \frac{v}{u + v} = \frac{5 \times 10^{-6}}{10 \times 10^{-6} + 5 \times 10^{-6}} = 0.333$$

$$\hat{q} = 0.667.$$

Therefore, at equilibrium, there will be twice as many recessives mutating at a rate half as great as the rate from dominant to recessive resulting in the same absolute number of mutations in each direction.

A typical mutation rate is often given as 1×10^{-5} or 1×10^{-6}, which seems rather low. However, if the total mutation rate per gamete is calculated, one gets a very different perspective on the frequency of mutations. Assume that there are 50,000 gene loci each mutating at an average mutation rate of 1×10^{-6} (both quite reasonable assumptions), then $5 \times 10^4 \times 1 \times 10^{-6}$ or 5% of all the gametes formed will carry a new mutation. Changes in the assumptions will modify the value obtained, but any reasonable assumptions about mutation rates and numbers of loci lead to the conclusion that a fairly substantial fraction of the gametes will carry a new mutation. Selander (1976), for example, estimates that less than 10% of the DNA in the eukaryotic genome contains about 40,000 genes of the type transcribed and translated to produce enzymes and other proteins. Hinegardner (1976) estimates that the number of genes in eukaryotes range from 10,000 to 400,000 in the "primary DNA," which he calls the "selectively constrained DNA" and which includes more than just the structural genes, but still accounts for only 0.6 to 24% of the haploid DNA. Thus, the estimate that 5% of the gametes formed will carry a new mutation seems conservative.

MUTATION RATES

Mutation rates are usually expressed as the number of mutations observed per locus per generation. Studies of mutation rates are fraught with pitfalls. For example, a gene or a gene locus can be considered as a DNA segment that codes for a particular polypeptide. Two loci may have different mutation rates simply because one gene consists of a larger DNA segment than the other. It would seem that the ideal way to study mutation rates would be to find the rate at which one particular base-pair is substituted for another at a given site in a particular codon of the gene in question. The frequency of exactly the same base substitution must be extremely small. In most studies of mutation rates, however, the different mutations have not been identified at the level of individual base-pair substitutions. Instead, they have been identified at the level of the cistron, and identified by complementation tests, at best. Therefore, the observed mutations used to estimate the mutation rate at a locus may consist of a potpourri of base-pair substitutions, frameshift mutations, and macrolesions, or even chromosome rearrangements if the methodology for the detection and separation of such changes from intragenic changes is crude.

Furthermore, the mutation rate observed depends not only on the rate of occurrence of the primary lesion, but on the probability of repair of this lesion by DNA repair systems. An observed difference in rate thus could be due not to a difference in the incidence of primary lesions, but to the frequency of repair. It will also depend on the probability that the mutation will produce a recognizably altered phenotype. It has already been pointed out, for example, that gel electrophoresis fails to pick up a proportion of the variants. Thus, successfully detecting new mutations depends on whether the mutation causes a phenotypic change that is detectable by the screening technique in use. In addition, it should not be forgotten that the mutation rates reported may be derived from a nonrandom group of loci, probably mostly structural genes, and there is as yet very little knowledge of mutation rates in those portions of the DNA that do not code for proteins. Other troublesome problems are phenocopies, the occurrence of rare recombinants that appear to be new mutations, and the occurrence of suppressor mutations at loci other than the one under study. Loss of the suppressor then may permit the trait to be expressed as if caused by a new mutation. In other cases, mutations at two or more distinct loci may give rise to very similar phenotypes. Another problem is the labor involved in screening large numbers of cells or organisms to detect the mutations, especially the "spontaneous" mutations of most interest to students of natural populations. This problem has been reduced in some cases by the use of what Auerbach and Kilbey (1971) have called the "suicide principle" whereby all the cells or organisms die except those with new mutations. However, spontaneous mutations occur at much lower rates than mutations induced by radiation or chemical mutagens, and thus larger numbers of progeny must be scored in order to study spontaneous mutation rates.

Mutation rates have been estimated in various ways and with various kinds of material. Estimation in haploids, in which all new mutations will usually be expressed immediately, is simpler than estimation in diploids. Even in haploids, however, suppressor mutations and mutations at different loci with similar phenotypic effects may complicate matters.

In diploids, direct estimates of mutation rates can be made for dominant mutations, for example, by estimating the frequency of achondroplastic dwarfs (a dominant trait) born to normal parents. Here, too, one must beware lest clinically similar traits are caused by mutations at different loci, or to the loss of a suppressor, or even to a rare recombinant. Sex-linked mutation rates can also be estimated directly because the heterogametic sex is like a haploid for the sex-linked loci. Care, of course, must be taken to ensure that the chromosome is free of mutations in the homogametic sex at the outset of the study.

Direct estimates of mutation rates for autosomal recessives are also possible if the mutant is not completely recessive and its presence can be detected in heterozygotes. However, the difficulties in detecting such heterozygotes are usually so great that estimates are not made, and mutation rates for autosomal recessives are estimated indirectly. For indirect estimates, it is assumed that the population is in equilibrium between selection and mutation, with selection against the recessive being balanced by mutation to the recessive. It is, of course, extremely difficult to demonstrate that a population is in equilibrium because inbreeding, changing selection pressures, and the like may negate the assumption of equilibrium. A very slight and, in a practical sense, undetectable heterozygote advantage can also maintain an equilibrium so that even if an equilibrium exists, it is not certain that it is the result of a balance between mutation and selection. Quite apart from these problems, the usual bugaboos of phenocopies, recombinants, suppressors, and different loci with similar effects must be guarded against. If this discussion gives the impression that the determination of mutation rates is difficult, that is its intent.

For the experimental determination of mutation rates, the specific-locus technique is probably the most reliable. In this case, large numbers of progeny from the cross $+/+ \times a/a$ are reared. If any a/a homozygotes appear among the F_1 progeny, then one of the $+$ alleles must have mutated to a. The method is very straightforward. If one a/a individual appears among, say, 100,000 F_1 $+/a$ flies, the mutation rate is 1 per 100,000 gametes per generation. However, counting 100,000 flies to find one mutation is a lot of work. Since a/a flies are being used in the cross, there is always the hazard that the presumed mutant fly is a contaminant. Moreover, care must be taken to ensure that the $+/+$ parents are free of the recessive a allele and that none among them happens to be $+/a$. The efficiency of the technique is increased, of course, by the use of test stocks that carry a number of homozygous recessive genes. Probably the best estimates for spontaneous mutation rates for higher eukaryotes have been obtained for *Drosophila* and mice (*Mus musculus*).

Spontaneous mutation-rate data (Table 4-2) come from the untreated controls in studies of induced mutation at specific loci in *Drosophila* and mice. The rates are somewhat lower than those usually given for these species, but they were chosen as representative of the rates observed in carefully conducted large studies. As is usually reported, the rate for the mouse is somewhat higher than that for *Drosophila*. The data on the mouse suggest a somewhat higher mutation rate for the same loci in males than in females, and also show a four-fold higher rate for forward mutations as compared to reverse mutations. More extensive data, from Schlager and Dickie (1971), are shown in Table 4-3, where it can also be seen that some loci appear to be more mutable than others. This difference in mutability is even more dramatically demonstrated in Table 4-4 from Stadler's (1942, 1948) observations on mutation rates for seed characters in maize. In addition, the same gene placed in different genetic backgrounds (Cornell vs. Carrion), may have a very different mutation rate, as shown at the bottom of the table.

The mutation rates reported in bacteria, viruses, and unicellular organisms are usually somewhat lower than those reported in higher organisms. However, these mutation rates are given as the rate of mutation per gene per cell division in contrast to the higher forms where it is given as the rate per gene per generation, which may account for some of the difference. Mutation rate studies are much easier to carry out with these lower forms because it is so much easier to screen large numbers of organisms for mutations. Table 4-5 shows the large numbers screened to establish that the mutation rate was

Table 4 - 2. Specific Locus Mutation Rates in *Drosophila* and the House Mouse

Species	Number of Loci Tested	Total Locus Tests	Mutations	Mutation Rate x 10^{-6}
Drosophila melanogaster				
Bonnier and Lüning (1949)	2	307,158	1	3.2
Mickey (1954)	8	150,416	0	-
Alexander, M.L. (1960)	8	797,624	0*	-
Frye (1961)	3	1,088,572	1**	0.9
		2,343,770	2	0.9
Mus musculus				
Russell (1962) ♂♂	7	3,720,500	28	7.5
Russell et al. (1959) ♀♀	7	327,341	0	-
Lyon and Morris (1966)	5	8,570	0	-
Schlager and Dickie (1966)				
Forward mutation	5	721,597	8	11.1
Reverse mutation	5	4,473,786	12	2.7
		9,251,794	48	5.2

* 1 questionable "slight ebony allele."

** 4 additional "mutations" were due to chromosomal rearrangements.

independent of cell number for reverse mutations at the histidine locus. The other rates given range from about 10^{-6} to 10^{-9}, which seems to be typical of the findings with these lower forms. Ryan (1963) pointed out, however, that even these rates should be divided by four to put them on a per gene basis because the *E. coli* cell normally contains four bacterial chromosomes, each with a histidine locus.

Finally, Table 4-6 contains estimates of mutation rates for autosomal dominants in humans. Some of these rates are comparable to those found in the mouse, but others are much higher. However, most estimates of human mutation rates are subject to errors, which, rather than tending to cancel out one another, all act in the same direction to lead to overestimates of the mutation rates. In the first place, there is usually a biased sample of loci because only loci known to mutate are apt to be included in the survey. Moreover, somatic mutations, phenocopies, and recombinants may be scored as new mutations unless steps are taken to detect them. Further, if several loci produce a similar phenotypic effect, as appears to be the case for chondrodystrophy, then the mutation rate will be the sum of the rates at several loci rather than just one. Finally, incomplete penetrance

Table 4 - 3. Forward and Reverse Mutation Rates[*]
at Specific Loci in the House Mouse

Locus	Number of Gametes Tested	Number of Mutations	Mutation Rate x 10^{-6}
Forward			
a^+	67,395	3	44.5
b^+	919,699	3	3.3
c^+	150,391	5	33.2
d^+	839,447	10	11.9
ln^+	243,444	4	16.4
	2,220,376	25	11.2
Reverse			
a	8,167,854	34	4.2
b	3,092,806	0	-
c	3,423,724	0	-
d	2,286,472	9	3.9
ln	266,122	0	-
	17,236,978	43	2.5

Source: Schlager and Dickie, 1971.

[*]Actually, an array of mutations was recovered rather than just forward or reverse mutations at a given locus.

Table 4 - 4. Specific Locus Mutation Rates
for Seed Characters in Corn, *Zea mays*

Mutation	Gametes Tested	Mutations	Mutation Rate x 10^{-6}
$Wx \to wx$	1,503,744	0	
$Sh \to sh$	2,469,285	3	1.2
$Y \to y$	1,745,280	4	2.2
$C \to c$	426,923	1	2.3
$Su \to su$	1,678,736	4	2.4
$Pr \to pr$	647,102	7	11.0
$I \to i$	265,391	28	106.0
$R \to r$	554,786	273	492.0
$R^r \to r^r$ (Cornell)	109,904	94	860.0 (Limits - 780-960)
$R^r \to r^r$ (Carrion)	25,911	0	(Limits - 0-140)

Source: Stadler, 1942, 1948.

Table 4 - 5. Specific Locus Mutation Rates
in the Bacterium, *Escherichia coli*

Mutation	Total Number of Bacteria Tested	Rate of Mutation per Bacterium per Generation
$his^- \to his^+$	9.6×10^5	1.5×10^{-8}
	2.2×10^6	5.0×10^{-8}
	5.6×10^6	2.8×10^{-8}
	1.3×10^7	2.9×10^{-8}
	2.4×10^7	1.5×10^{-8}
	3.2×10^7	2.0×10^{-8}
	6.3×10^7	3.0×10^{-8}
	Mean	2.7×10^{-8}
$lac^- \to lac^+$	-	2×10^{-7}
$T1-s \to T1-r$	-	2×10^{-8}
$his^- \to his^+$	-	4×10^{-8}
$his^+ \to his^-$	-	2×10^{-6}
$str-s \to str-d$	-	1×10^{-9}
$str-d \to str-s$	-	1×10^{-8}

Source: Ryan, 1963.

Table 4 - 6. Estimated Mutation Rates
for Autosomal Dominant Mutations in Man

Trait	Mutation Rate Per Gamete x 10^{-6}
Epiloia	8
Aniridia	5
Microphthalmus	5
Waardenburg's syndrome	4
Facioscapular muscular dystrophy	5 (Direct)
	< 0.5 (Indirect)
Pelger anomaly	9
Amyotrophic lateral sclerosis	30
Myotonia dystrophica	16
Myotonia congenita	4
Huntington's chorea	2
Retinoblastoma	4 (Bilateral cases corrected for phenocopies)
Chondrodystrophy	60 (probably more than one locus)
Neurofibromatosis	130 − 250 (Direct)
	80 − 100 (Indirect)
Deafness	47 (all loci causing dominant deafness)

Source: Reprinted by permission from J. F. Crow. Mutation in man. In *Progress in Medical Genetics*, Vol. 1, A. G. Steinberg, ed., Grune and Stratton, New York.

of the dominant in the parent may lead one to score an affected child as a new mutant, when in fact he or she is not.

The average of the autosomal dominant mutation rates in Table 4-6 is about 3×10^{-5}. However, as Cavalli-Sforza and Bodmer (1971) argue quite cogently and persuasively from the systematic survey of mutation rates in X-linked genes by Stevenson and Kerr (1967), this estimate is far too high. The selection of only the more mutable loci was minimized in this study, and the mean mutation rate was then estimated to be about 4×10^{-6} for these X-linked mutations, a whole order of magnitude lower than the estimate of 3×10^{-5}. However, the median mutation rate, on the assumption of a lognormal distribution of mutation rates, is more readily obtained than the mean from the data and is estimated to be only 1.6×10^{-7}, a far cry from the figures usually used for human mutation rates.

Another approach to the study of mutation rates involves the estimation of whole chromosome mutation rates. In this case, a chromosome known to be free of lethals and semilethals is crossed to a balanced marker-stock and then maintained intact in the balanced condition for a number of generations to permit the accumulation of recessive mutations. Each generation the chromosome runs the risk of acquiring recessive mutants, which will accumulate since they are not expressed in the heterozygotes. Then the number of chromosome lines tested multipled by the number of generations will give the number of chromosome generations of exposure to the risk of mutation. To score the chromosomes, they must be put in the homozygous condition in a manner comparable to that used for the analysis of genetic variation in wild populations. Table 4-7 shows the results of such an analysis in homologous chromosomes from four species of *Drosophila*. It can be seen that the expectation of a new lethal or semilethal mutation somewhere along the length of this chromosome is approximately 1% each generation. The lethals appear to be several times more frequent than the semilethals.

Mutation rates are known to be increased by the effects of such environmental agents as ionizing and non-ionizing radiation, chemical mutagens, and thermal shock. Less familiar, perhaps, are the *mutator* genes, which enhance the mutation rates of genes at other loci. The dominant Dotted (*Dt*) allele in maize is an example (Rhoades, 1941). In the presence of the recessive *dt* allele, the recessive a_1 allele is stable, but when *Dt* is present, a_1 becomes highly mutable, reverting to the dominant A_1 allele. The *hi* gene in *D. melanogaster* (Ives, 1950) behaves somewhat differently from *Dt*, which is quite specific in effect, for it increases mutation rates at many loci rather than just one, although different loci may respond differently. In addition, *hi*

Table 4 - 7. Mutations in Homologous Chromosomes of Four Species of *Drosophila*

Species	Chromo-some	Chromo-somes Tested	Chromo-some Generations	Mutations L	SL	V	Corrected L + SL	L + SL Frequency %
D. pseudoobscura	II	206	4082	23	14	1	40.78	0.999
D. persimilis	II	191	3820	50	8	2	68.12	1.783
D. willistoni	III	323	5715	35	12	8	51.12	0.894
D. prosaltans	III	391	6547	31	6	2	41.80	0.638

Source: Dobzhansky, Spassky and Spassky, 1952, 1954.
    L = lethal; SL = semilethal; V = visible

also causes an increase in the incidence of inversions. These genes are rather difficult to work with because they are only known to affect the mutation rate. Mutator genes have also been identified in several bacterial species, yeasts, and viruses (Drake, 1969); a more thorough analysis of their nature is possible than in the higher forms. Furthermore, some antimutator genes that suppress mutation have also been identified. Among the mutators, some have been found to affect the frequency of the primary lesions and others to affect the frequency of repair of the lesions. The evolutionary significance of mutator and antimutator genes is to provide a mechanism through which mutation rates can be controlled by natural selection. In other words, by favoring or eliminating the mutators, natural selection can control the rate at which new variability is fed into the gene pool of a population.

One type of evidence that mutation rates are not solely dependent on the effects of external environmental agents comes from a comparison of the mutation rates per generation in bacteria (generation length ≅ 30 minutes), *Drosophila* (≅ 10−12 days), mice (≅ several months), maize (≅ 1 year), and humans (≅ 25 years). Despite the great difference in generation length, and hence the great difference in time of exposure to external mutagenic agents, the mutation rates per generation in all of these species appear to be surprisingly similar, as shown earlier. Since the exposure to external mutagenic agents must be comparable, the species with longer generation lengths would be expected to manifest higher mutation rates. That the rates are similar suggests some form of control. Along these same lines, the mutation rates from the smaller prokaryotes through the lower fungi are relatively constant, about 0.5% per genome per duplication, despite the fact that genome size in these organisms varies well over 1000-fold. Again, these observations argue for some form of control over the mutation rates in these species.

CHAPTER 5

Natural Selection

The concept of natural selection won wide acceptance following the publication of Darwin's *Origin of Species* in 1859. Darwinian natural selection was based on several observations and the logical conclusions to be drawn from them. First, Darwin observed that the reproductive potential of any species was far greater than needed to replace the parents in the previous generation. The reproductive capacity is geometric or logarithmic. Even for elephants, which he assumed to produce, between the ages of 30 and 90, only six progeny, he estimated that one pair would have 19 million living descendants after just 750 years. Second, despite this reproductive potential, population size does not increase geometrically but rather fluctuates about some average number from year to year. The obvious conclusion from these observations is that many of the fertilized eggs produced must die before reaching maturity. His third observation was that no two individuals are ever identical, that biological variation is universal. Hence, some individuals must be better adapted than others to survive and reproduce under the existing environmental conditions. Even though many of the deaths may be random, if a differential rate exists for the remainder, it will still be effective, though more difficult to detect. To the extent that the differences among individuals are hereditary, the next generation will then contain a greater proportion of the favored types. Expressions such as "survival of the fittest" and the "struggle for existence" were commonly used, and the effects of competition and predation loomed large as factors in natural selection. Darwinian selection thus emerged as a concept of "nature, red in tooth and claw."

Darwin realized that this picture was incomplete and attempted to remedy this deficiency with his theory of sexual selection, published in 1871 in *The Descent of Man and Selection in Relation to Sex.* With this theory he attempted to explain the origin of the differences in secondary sexual traits of males and females, of courtship displays, gaudy colors, male weapons, and the like. He postulated that two factors were involved in sexual selection: male competition and female choice. Although Darwin appeared to regard sexual selection as equal in importance to natural selection, sexual selection never drew as much attention as did natural selection, perhaps because it came to be regarded as an integral part of natural selection rather than separate from it.

The concept of natural selection has undergone a change since Darwin's time. Darwinian natural selection seemed to emphasize differential survival by different types. Today the emphasis is placed on differential reproduction. Any factors leading to differential reproduction, that is, the production of more progeny by one type in proportion to its numbers than by other types, are factors in natural selection if the differences are to any extent hereditary. As an evolutionary force, natural selection includes all systematic changes in gene frequency not caused by mutation or immigration. The factors at work may include differences in survival and longevity, fecundity and fertility, mating success, competition or cooperation, resistance to parasites and disease, developmental rate, food requirements, physiological tolerances, behavior patterns or color patterns, and so forth. The crux of natural selection is that some individuals in the population are better able to survive and reproduce than others. Moreover, survival is not enough—as suggested previously, the ultimate test of evolutionary fitness is reproductive success. If there are differential rates of reproduction by individuals with different genotypes, the genes carried by the more successful individuals will tend to increase in frequency in the population, and the population as a whole will tend to become better adapted to the existing conditions. Natural selection, then, is the mechanism of adaptation. It may produce improved adaptation to relatively stable environmental conditions, or it may permit the population to adapt to changing environmental conditions; but adaptation is always related to existing conditions.

As mentioned in Chapter 2, natural selection acts on phenotypes. Even though we shall be treating selection in this chapter as if it were acting on genotypes or on single gene loci, always keep in mind that selection is phenotypic. To the extent that the phenotypic differences reflect genotypic differences, selection will also act on genotypes, but its action on these genotypic differences is indirect. In other words, a crippled animal will be easy prey to a predator regardless of whether its crippling is caused by an environmental accident or by a genetic defect. Thus, natural selection acts to eliminate the less well-adapted individuals from the population. If the differences between the well-adapted and the poorly adapted individuals are to any degree influenced by heredity, those genes conferring greater fitness will tend to increase in frequency in the population as the result of natural selection. Thus, only those cases involving genetic differences are of evolutionary significance, and it is on these that we shall focus attention. However, it is well to remember that just about any phenotypic trait has both an environmental and a genetic component. Even a jackrabbit crippled by a truck may have suffered, not just from misfortune, but also from a reduced genetic capacity for agility or cunning, compared to its fellows, in evading trucks.

Note, also, that natural selection may operate within a breeding population among the phenotypes and genotypes of the individuals in that population, or, conceivably, between different breeding populations if the same selection pressures bear on individuals in both populations. In this chapter, we shall deal only with intragroup selection, deferring consideration of possible intergroup relations until later.

Some people find natural selection disturbing because it seems to give rise to such a brutal concept of nature and seems a threat to any higher view of nature or humankind. But predation and parasitism do occur, and members of the same species may compete for food, space, mates, light, and other essentials. Every time you eat, some plant or animal is being sacrificed to your needs. Hence, natural selection does involve a struggle for existence. However, cooperative behavior, parental care, social behavior, intelligence, and even altruism may also contribute to reproductive fitness and, therefore, be enhanced by natural selection. Thus, even though you may impose your own value judgments on the consequences of natural selection, natural selection itself can hardly be judged as either good or evil. Those factors, whatever their nature, that increase reproductive fitness will be favored by natural selection, and those decreasing it will be eliminated.

GAMETIC AND ZYGOTIC SELECTION

Natural selection differs from mutation and migration in being absent if $p = 0$ or if $p = 1$. In other words, there can be no selection if there are no alternatives. To study selection in its simplest form, first consider genic (or gametic or haploid) selection. If, for example, for every 100 A gametes formed, 100 are transmitted to the next generation while only 99 of 100 a gametes are transmitted, the fitness, W_A, of the dominant allele equals 1.00, but the fitness of the recessive allele, W_a, equals 0.99. By definition, the relationship between fitness (W) and the selection coefficient (s) is given by the equation,

$$W = 1 - s.$$

Thus, in this case,

$$W_a = 1 - s$$

and

$$s = 1 - 0.99 = 0.01.$$

Therefore, s is a measure of the selective disadvantage of a as compared to A. Owing to the selection against the recessive allele, the frequency of A will increase. The amount of increase can be calculated as follows:

Allele	Initial Frequency F	Relative Fitness W	Frequency after Selection FW
A	p_o	1	p_o
a	q_o	$1 - s$	$q_o(1-s)$
Total	1.0		$1 - sq_o$

Therefore,

$$p_1 = \frac{p_o}{1 - sq_o}$$

and

$$\Delta p = p_1 - p_o = \frac{p_o}{1 - sq_o} - p_o$$

$$\Delta p = \frac{sp_o q_o}{1 - sq_o}.$$

If s, q, or both are small, then the product sq_o is small and $1 - sq_o \cong 1$ so that

$$\Delta p \cong sp_o q_o.$$

If

$$s = 0.01, p_o = 0.3,$$

and

$$q_0 = 0.7,$$

then

$$\Delta p = (0.01)(0.3)(0.7) = + 0.0021$$

and

$$p_1 = 0.3 + 0.0021 = 0.3021.$$

The increase in the frequency of A results then from natural selection.

A rather interesting relation exists for genic selection between the average time in generations that a harmful allele persists in a population $(\bar{n})$ and the selection coefficient (s), which is simply

$$\bar{n} = \frac{1}{s}.$$

Thus the more deleterious the allele, the more rapidly it will be lost. For a complete lethal, $s = 1$ and $\bar{n} = 1$ generation. If $s = 0.01$, on the other hand, $\bar{n} = 100$ generations. Both types of alleles cause a genetic death sooner or later, but the question may be raised, which has the more serious effect, the lethal, which is eliminated at once from the population, or the slightly harmful allele, which persists in the population for 100 generations on the average before being finally eliminated.

Next consider zygotic selection or selection in diploids. In this case, the effects of selection can be estimated as follows:

Genotype	Initial Frequency F	Relative Fitness W	Frequency after Selection FW
AA	p_0^2	1.0	p_0^2
Aa	$2p_0q_0$	$1 - hs$	$2p_0q_0(1 - hs)$
aa	q_0^2	$1 - s$	$q_0^2(1-s)$
Total	1.0		$\bar{W}$

In this case, hs is the selection coefficient for the Aa heterozygotes. Here, h is a coefficient with no separate biological meaning that, when multiplied by s, the selection coefficient for aa, gives the selection coefficient for Aa. If either Aa or aa is more fit than AA, appropriate changes in sign of h or s can make the fitness values greater than 1. $\bar{W}$ indicates the average fitness for the entire population, and it equals the sum of the proportions of individuals of the three genotypes to be expected in the next generation. That is

$$\overline{W} = p_o{}^2 + 2p_oq_o(1 - hs) + q_o{}^2(1 - s)$$
$$= 1 - 2 hsp_oq_o - sq_o{}^2.$$

The frequency of A after one generation of selection is

$$p_1 = \frac{D_1 + \frac{1}{2}H_1}{D_1 + H_1 + R_1}$$

and

$$\Delta p = p_1 - p_o.$$

Therefore

$$\Delta p = \frac{p_o{}^2 + p_oq_o(1 - hs)}{1 - 2hsp_oq_o - sq_o{}^2} - p_o$$

which, with considerable algebraic juggling, converts to the general equation

$$\Delta p = \frac{sp_oq_o[h(2p_o - 1) + q_o]}{1 - 2hsp_oq_o - sq_o{}^2}.$$

If $2hsp_oq_o$ and $sq_o{}^2$ are small, as is often the case, then $\overline{W} \cong 1$ and the equation simplifies to

$$\Delta p \cong spq[h(2p - 1) + q]$$

or

$$\Delta p \cong spq^2 + spq[h(2p - 1)]$$

where spq^2 is the portion of Δp resulting from selection against the homozygous recessives (aa) and $spq[h(2p - 1)]$ is the part of Δp resulting from selection against the heterozygotes (Aa).

In the special and rather common case of complete dominance where $W_{AA} = W_{Aa} = 1.0$, then $hs = 0$, and h must equal zero, and the equation

$$\Delta p = \frac{spq[h(2p - 1) + q]}{1 - 2hspq - sq^2}$$

simplifies to

$$\Delta p = \frac{spq^2}{1 - sq^2},$$

and if sq^2 is small, to

$$\Delta p \cong spq^2.$$

This equation represents an important special case because of the prevalence of deleterious recessive genes in natural populations. Examination of the equation reveals some interesting properties. First $\Delta p = 0$ if $s = 0$, $p = 0$, or $q = 0$. Moreover, the magnitude of Δp depends on the p and q values as well as on s itself. The most rapid changes in gene frequency, the largest Δp values, will occur at intermediate gene frequencies. If either p or q is very close to zero, Δp will be very small and selection is therefore relatively ineffective at low gene frequencies.

The discussion of gametic and zygotic selection is general enough to illustrate how natural selection brings about gene-frequency changes in populations. In these cases, *directional selection* prevails, for one allele is favored over the other, and the less-favored allele is driven toward extinction. Now we shall consider other possible ways in which selection may operate. For example, the habits and behavior of males and females often differ, and accordingly the selection pressures acting on males and females may also differ. If this is the case, and the numbers of males and females are equal, the selection coefficient acting on the population as a whole is simply the average of $s_{\male}$ and $s_{\female}$, that is,

$$\bar{s} = \tfrac{1}{2}(s_{\male} + s_{\female}).$$

This $\bar{s}$ can then be used in the equations derived above. If the loci of interest are X-linked and the sex ratio equals one, then because the females carry two-thirds of the X chromosomes,

$$\bar{s} = \frac{2}{3}s_{\female} + \frac{1}{3}s_{\male}$$

Thus far, we have assumed that the selection coefficient is a constant, but this may not always be so. For example, the intensity and even the direction of selection may vary with gene frequency — that is, the value of s may vary for different values of p and q. Frequency-dependent selection of this kind has been reported in which the rarer allele is always favored by selection. This sort of selection will generate a stable equilibrium that is of considerable interest because, at the equilibrium point, selection will favor the continued presence of both alleles (in the simplest case) in the population, but no genetic load will be generated. It is also possible that the selection coefficient varies with gene frequency, but the rarer it becomes, the stronger the selection against it, in which case its rate of elimination from a population would be hastened. Actually, there is no reason to expect selection coefficients to remain constant under changing conditions;

however, they are usually assumed to be constant in order to simplify the mathematical treatment of selection.

If the heterozygote is less fit than either homozygote, an unstable equilibrium may exist but will be very ephemeral because as soon as one allele becomes less frequent than the other allelomorph, selection will tend to drive it to extinction. On the other hand, if the heterozygote is more fit than either homozygote, a stable equilibrium will develop. The nature of this equilibrium is demonstrated below:

Genotype	Initial Frequency F	Relative Fitness W	Frequency after Selection FW
A_1A_1	p_0^2	$1 - s_1$	$p_0^2(1 - s_1)$
A_1A_2	$2p_0q_0$	1.0	$2p_0q_0$
A_2A_2	q_0^2	$1 - s_2$	$q_0^2(1 - s_2)$
Total	1.0		$1 - s_1p_0^2 - s_2q_0^2$

Here

$$\Delta p = \frac{p_0^2(1 - s_1) + p_0q_0}{1 - s_1p_0^2 - s_2q_0^2} - p_0$$

which can be rearranged to give

$$\Delta p = \frac{p_0q_0(s_2q_0 - s_1p_0)}{1 - s_1p_0^2 - s_2q_0^2}.$$

At equilibrium $\Delta p = 0$, and in the equation above, the equilibrium point of interest will exist when

$$s_2q - s_1p = 0.$$

Hence, at equilibrium

$$s_2\hat{q} = s_1\hat{p}$$

from which

$$\frac{s_2}{s_1} = \frac{\hat{p}}{\hat{q}}$$

and

$$\hat{p} = \frac{s_2}{s_1 + s_2}$$

and

$$\hat{q} = \frac{s_1}{s_1 + s_2}.$$

Therefore, the equilibrium values for p and q are dependent on the relative fitnesses of the two homozygotes.

It is worth noting that at equilibrium the mean fitness of the population is $\overline{W} = 1 - s_1 p^2 - s_2 q^2$. Substituting the equilibrium values of p and q above, we obtain

$$\overline{W} = 1 - s_1 \left(\frac{s_2}{s_1 + s_2}\right)^2 - s_2 \left(\frac{s_1}{s_1 + s_2}\right)^2$$

which simplifies to

$$\overline{W} = 1 - \frac{s_1 s_2}{s_1 + s_2}.$$

The maximum average fitness of this population exists at the equilibrium gene frequencies, $\hat{p}$ and $\hat{q}$, yet even so, compared to the optimal value for fitness of 1.0 assigned to the heterozygotes, the fitness of the population is lowered by $s_1 s_2/(s_1 + s_2)$. This value is called the segregational load of the population and may be of considerable magnitude.

Now consider the case of selection against a homozygous recessive lethal—that is, where $W_{AA} = W_{Aa} = 1.0$ and $W_{aa} = 0$, and hence $s = 1$. In this case, the following relations hold:

Genotype	Initial Frequency	Frequency after Selection
AA	p_o^2	p_o^2
Aa	$2p_o q_o$	$2p_o q_o$
aa	q_o^2	0
	1.0	$p_o^2 + 2p_o q_o$

Here

$$q_1 = \frac{\frac{1}{2}H}{D + H} = \frac{p_o q_o}{p_o^2 + 2p_o q_o} = \frac{q_o}{1 + q_o}$$

and

$$p_1 = 1 - \frac{q_o}{1 + q_o} = \frac{1}{1 + q_o}.$$

In the next generation before selection:

Genotype	Frequency before Selection
AA	$\dfrac{1}{(1+q_0)^2}$
Aa	$\dfrac{1}{(1+q_0)^2}$
aa	$\dfrac{q_0{}^2}{(1+q_0)^2}$

The relation between q_0 and q_1 is

$$q_1 = \frac{q_0}{1+q_0},$$

which is the sequence equation in an harmonic series, for which the general expression for q_n after n generations is

$$q_n = \frac{q_0}{1+nq_0}$$

which transforms into

$$n = \frac{1}{q_n} - \frac{1}{q_0}.$$

Thus, for example, with complete selection against a homozygous recessive lethal, a change in gene frequency from $q_0 = 0.01$ to $q_n = 0.001$ will require 900 generations. If any two values are given or known in the equation, the third can be calculated. If you wish to calculate how long it will take for selection to halve the frequency of the recessive allele, let $nq_0 = 1$; then

$$q_n = \frac{q_0}{1+nq_0} = \frac{q_0}{2}.$$

Obviously, the product of many combinations of n and q_0 will equal one (2 X ½; 10 X 1/10; 50 X 1/50; etc.). In general, the number of generations needed to halve the gene frequency can be obtained from $n = 1/q_0$. If $q_0 = 0.5$, $n = 2$; if $q_0 = 0.05$, $n = 20$; if $q_0 = 0.005$, $n = 200$, and the efficiency of selection obviously declines as q decreases.

One additional point worth mentioning is that when the gene frequency is halved, the frequency of the homozygous recessive individuals is reduced to one-fourth its previous frequency. For instance, if $q = 1/50$, $q^2 = 1/2500$ but for $q = 1/100$, $q^2 = 1/10,000$.

MUTATION AND SELECTION

Now consider what happens in a population in which both mutation and selection are affecting the same deleterious autosomal recessive allele. Mutations result from changes in the sequence of nucleotides in the Watson-Crick DNA double helix. Selection acts upon the phenotypes and genotypes of individuals. The utility of the approach we are using lies in the fact that the effects of these quite different biological phenomena can be studied simultaneously through their effect on gene-frequency change. From the preceding discussion, it should be clear that

$$\Delta p = vq - up + \frac{spq^2}{1 - sq^2}.$$

Ordinarily vq is small relative to up and can be ignored—or else up can be regarded as the net mutation rate. Moreover, sq^2 is so small that $1 - sq^2 \cong 1$ so that the equation simplifies to

$$\Delta p = spq^2 - up.$$

At equilibrium

$$\Delta p = spq^2 - up = 0$$

or

$$spq^2 = up.$$

Hence,

$$\hat{q}^2 = \frac{u}{s} \text{ and } \hat{q} = \sqrt{\frac{u}{s}}.$$

For a homozygous recessive lethal,

$$s = 1 \text{ and } \hat{q}^2 = u.$$

These simple relationships have often been used to estimate mutation rates in populations. However, in weighing the validity of mutation rates so obtained, one should keep the assumptions in mind. An equilibrium is assumed, but it is extremely difficult, if not impossible, to prove. Unless the gene is lethal, an estimate of s is required, but a reliable estimate is difficult to obtain.

However, if the assumptions are correct, and, for example, 1/90,000 of the individuals born dies of a recessive lethal condition, the mutation rate can then be estimated directly to be 1/90,000. At first, or perhaps at second glance, this appears to present a paradox, for mutation rates are expressed as mutations per locus per gamete, but the deaths are expressed in terms of individuals. However, for each homozygous recessive individual who dies, two recessive alleles are lost, which must be replaced to maintain the equilibrium. But 90,000 diploid individuals carry 180,000 genes at this locus so that the rate of elimination is actually 2/180,000 rather than 2/90,000, and the mutation rate needed to replace this loss is, in fact, 1/90,000. Even though this homozygous recessive condition is rare, 1/90,000, it can easily be shown that the heterozygous carriers are much more frequent, for approximately 1 in 150 individuals will be heterozygous for this lethal. Even if it could be established that an equilibrium exists, that does not necessarily prove that the equilibrium is being maintained by a balance between mutation and selection. For example, it is possible to calculate how great an advantage the heterozygote would need for these frequencies to be maintained by heterozygous advantage. In this case

$$W$$

AA	$1 - s_1$	$\hat{q}^2 = \dfrac{1}{90,000}$
Aa	1.0	$\hat{q} = \dfrac{1}{300}$
aa	$0 \; (s_2 = 1)$	$\hat{p} = \dfrac{299}{300}.$

Therefore

$$\hat{p} = \frac{s_2}{s_1 + s_2}$$

$$\frac{299}{300} = \frac{1}{s_1 + 1}.$$

Solving, we find $s_1 = 0.003$. Hence, a difference in fitness between $AA = 0.997$ and $Aa = 1.000$ would be sufficient to maintain this equilibrium. It would also be extremely difficult to detect such a slight difference experimentally. Therefore, even though in theory $\hat{q}^2$ may equal u/s, the assumptions underlying this relation should always be kept in mind, and it should probably be treated as a first approximation.

The relation above may also be written $u = s\hat{q}^2$, where $s\hat{q}^2$ is the frequency of genetic deaths. This means that at equilibrium the frequency of genetic deaths is determined by the mutation rate rather than by the harmfulness of the gene, another seeming paradox. This is resolved by the fact that the more harmful genes reach equilibrium at lower frequencies. For example, if $s = 0.1$ and $\hat{q}^2 = 1/10,000$, then $u = 1 \times 10^{-5}$. If $s = 1.0$, then u would have to be 1×10^{-4} to maintain $\hat{q}^2 = 1/10,000$, an improbably high value for a mutation rate. If the mutation rate for the lethal were also 1×10^{-5}, then $\hat{q}^2$ would be $1/100,000$. In other words, for a more harmful gene to exist in a population at the same frequency as a less harmful gene, a higher mutation rate is required.

The equation $\hat{q}^2 = u/s$ applies to cases of *incomplete or partial selection* against a homozygous recessive condition. If there is no dominance, in other words, the heterozygote Aa is intermediate in expression between AA and aa then

$$\Delta p = \tfrac{1}{2}spq - up = 0$$

and

$$\hat{q} = \frac{2u}{s}.$$

For gametic selection,

$$\Delta p = spq - up = 0$$

and

$$\hat{q} = \frac{u}{s}.$$

If selection acts against the dominant A allele,

$$\Delta p = vg - spq^2 = 0$$

and

$$v = spq$$

$$pq = \frac{v}{s}.$$

Since

$$H = 2pq$$

$$\hat{H} = \frac{2v}{s} \text{ or } v = \tfrac{1}{2}\hat{H}s.$$

Therefore, if one in 100,000 people is dying owing to a dominant lethal condition, the mutation rate to the dominant lethal is $v = \tfrac{1}{2}\hat{H}s$. Since s is 1.0 and $H = 1/100,000$, all the affected being heterozygotes, then $v = 1/200,000$. Again, recall that 100,000 people represent 200,000 genes, of which only one, in this case, is a dominant lethal. Therefore, the mutation rate needed to replace the lost dominant lethal is $1/200,000$. If, on the other hand, the mutation rate to a dominant lethal is 1/100,000 gametes per generation, the number of affected persons will be twice as great, or 1 in 50,000 since $\hat{H} = 2v/s$.

Selection against sex-linked recessive alleles is still another type of selection. Here, selection in the XY males (or ZW females) is like selection in haploids, and selection in the XX females (or ZZ males) is like selection in diploids. For example, consider hemophilia in humans. The sex ratio is about 1:1 so that one-third of the X chromosomes, which bear the hemophilia locus, are carried by the males, and two-thirds of the X chromosomes are carried by the females. Since the fitness of the hemophilic males is reduced, the frequency of homozygous females, whose fathers normally would be hemophilic males and who would be uncommon even if the Castle-Hardy-Weinberg equilibrium prevailed, is negligible. Therefore, to all intents and purposes, selection against the gene for hemophilia acts only in males. The selection coefficient for the population as a whole then is $s_{\delta}/3$, and the appropriate equation to measure the effects of selection is

$$\Delta p = \frac{s_{\delta} pq}{3},$$

since selection in the males resembles haploid selection.

The net fertility or fitness of male hemophiliacs has been estimated to be only about one-quarter that of non-hemophilic, normal individuals, and the frequency of male hemophiliacs is about 4×10^{-5}, which is a direct estimate of the frequency of the gene for hemophilia. Thus, despite a selection coefficient of $s = 0.75$ (in effect, the gene is semilethal in males) and haploid selection against it, the gene for hemophilia has a frequency of 1 in 25,000. If this frequency is being maintained by mutation, the relevant equation is

$$\Delta p = \frac{s_{\delta} pq}{3} - up = 0$$

or

$$u = \frac{s_{\delta} q}{3} = \frac{3}{4} \cdot \frac{4 \times 10^{-5}}{3} = 1 \times 10^{-5}.$$

As pointed out previously, selection is most effective in bringing about gene frequency change at intermediate gene frequencies. It has been calculated, for instance, that selection against a slightly deleterious recessive gene ($s = 0.01$) would require 900,230 generations to reduce the frequency of this deleterious recessive from $q = 0.001$ to $q = 0.0001$. With selection so ineffective, mutation becomes of greater importance as a cause of gene frequency change. If, for example, the recessive allele a has a slight selective advantage (that is,

s against dominant A = 0.001), it has been calculated that 321,444 generations would be required for the recessive a to increase in frequency from $q = 1 \times 10^{-6}$ to 2×10^{-6}. However, if the mutation rate is 1×10^{-6}, the same change could occur in a single generation in a large population.

In general, mutation and selection can be thought of as opposing forces, with mutation generating deleterious alleles when compared with the existing wild type (most frequent) alleles, and with selection winnowing them out and eliminating them.

TRANSIENT POLYMORPHISM

The type of selection involved in the balance between mutation and selection is often referred to as *normalizing selection* because, by the elimination of harmful mutants, the adaptive norm of the population or the integrity of its gene pool is protected from deterioration. Normalizing selection thus is conservative, tending to maintain the status quo, and does not produce evolutionary change. *Directional selection* is the expression sometimes used for selection that causes evolution to occur, a type of selection also called *transient polymorphism*. It is observed during the period when a previously favored allele (the most frequent or wild-type allele) is being replaced by another allele that is currently favored by natural selection and is on its way to becoming the new wild type (Figure 5-1). In this case, either a new mutation occurs that is more favorable under the existing environmental conditions than the wild-type allele, or else the environmental conditions may change so that the existing wild-type allele is no longer favored by selection as much as some other allele that produces better adapted phenotypes in the new environment. Thus, transient polymorphism exists during the period of transition while the newly favored allele is increasing in frequency and spreading through the population. Evolution can again be redefined—this time as transient polymorphism.

Ordinarily, evolution is thought of as a long-drawn-out, very gradual affair, but that is not necessarily the case. Agriculture was invented only about 10,000 years ago so that, geologically speaking, the domestication of plants and animals has taken place during a relatively short time. Striking differences now exist between domesticated species and their wild ancestors, which must have evolved very rapidly. Domestication involves artificial selection—that is, the intervention or control by humans of the reproduction of crops or animals. Thus, domesticated species provide a laboratory for the study of evolution, particularly for the efficacy of selection (Merrell, 1975a). Even though

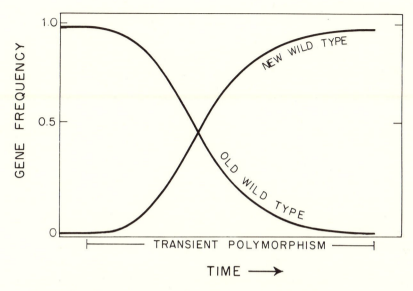

Figure 5-1. Selection curve during transient polymorphism when a new
wild-type allele replaces the old wild-type allele.

it may be argued that these are only microevolutionary changes, in
many cases the changes are so great that it is clear that evolutionary
change can be quite rapid under strong selection pressures. Moreover,
the changes in domesticated species are continuing, perhaps at an
accelerated pace, because of our better understanding of the underlying genetic mechanisms. It is not always realized that today's breeds
of animals are all less than 200 years old. The oldest breed is the
thoroughbred horse for which the herdbook was begun in 1791.
"Purebred" breeds of sheep, cattle, pigs, and dogs were first established
in the nineteenth century. The history of the Santa Gertrudis breed
is well documented because it was developed by the Klebergs on the
King Ranch in a deliberate effort to develop beef animals that
thrived in the hot subtropical climate of southern Texas. The breed
resulted from crosses between Shorthorn beef cattle and Brahma
cattle of India followed by selection among the progeny for favorable
beef qualities and heat, disease, and tick resistance. This breed, developed in the twentieth century in a relatively short time, is eloquent
testimony to the possibilities for rapid evolutionary change.

In Santa Gertrudis cattle, it is estimated that approximately seven-
eighths of their genotype is derived from Shorthorn cattle and only
one-eighth from Brahma cattle. More recently, an even wider cross

between beef cattle and the American bison or buffalo has produced the beefalo, whose ancestry is about five-eighths beef cattle and three-eighths buffalo.

Under controlled conditions, artificial selection for increased or decreased size in mice and other species has been dramatically successful. Selection in a single base population of mice has produced large strains that are two to three times heavier than the downward-selected small strains (Falconer, 1960). This difference is not nearly as impressive, however, as the 75-fold difference in weight between the largest and smallest breeds of dogs (Sierts-Roth, 1953). The degree of success of experimental selection in mice is probably limited by the small number of animals, the relatively short time period, and the fact that a closed population is used. Associated with the changes in size are changes in fertility and behavior. The large mice tend to have larger litters and rear more young in the laboratory than the small mice; however, they are also more lethargic and less reactive to disturbance than the small mice. Hence, even though in the laboratory they may be more fit than the small mice, in nature their greater fertility must be offset by their tendency to fall easy prey to predators, and the fittest mice would be intermediate in size.

In pigs selection has generally been directed toward producing large animals with a rapid rate of weight gain. When a smaller pig was desired as an experimental research animal, Drs. L. M. Winter and Wm. Rempel soon developed a line of miniature pigs. Like the mice, these little pigs differ considerably in temperament from larger breeds, for they are alert and quick moving in contrast to their porcine relatives. Equally prepossessing are the ponies, about the size of a Golden Retriever dog, that have been produced by artificial selection in horses.

Similar success stories are known for selection in plants. A shortage of sugar in France during the Napoleonic wars led to efforts to increase the sugar content of common garden beets, *Beta maritima*. Selection was so effective that today sugar beets contain about 25% sugar in contrast to about 5% in their garden beet ancestors. In corn, *Zea mays*, 50 generations of upward and downward selection in a variety that initially contained 10.9% protein and 4.7% oil led to lines with a low of 4.9% protein and a high of 19.4% protein and a low of 1.0% oil and a high of 15.4% oil.

The introduction of chemotherapeutic agents such as the sulfa drugs and of antibiotics such as streptomycin and aureomycin was soon followed by reports of resistant bacterial strains. Although initially puzzling, the resistance was soon demonstrated to result from the survival and reproduction of rare resistant mutant cells in the presence of the agent, which killed off the susceptible cells. This story has been repeated recently with the discovery of a penicillin-resistant strain of gonococcus, the organism responsible for gonorrhea. In the case of streptomycin, a streptomycin-dependent mutant strain of *Escherichia coli* has been discovered, which thrives better in the presence of the antibiotic than in its absence. If such mutants occur often, drug or antibiotic therapy may do a patient more harm than good.

The insecticide DDT was first used during World War II, about the same time that the sulfa drugs and penicillin were first put into use. Here, too, reports of resistance soon appeared. Resistance to DDT in houseflies was first reported in 1946, and by 1965, resistance to DDT had been reported in 165 pest species (Crow, 1966). When American troops moved into Korea, they brought DDT into that country for the first time. In about a year, reports of resistance began to appear for such species as the human body louse.

Although the earlier examples involved artificial selection, these last few examples involve natural selection. Natural populations of bacteria and insects, confronted by new and previously unknown environmental agents, were able to adapt very rapidly. All of these examples make clear that evolutionary change does not necessarily have to be slow and gradual, but may be extremely rapid.

Although, at first, physiological adaptation was thought by some to be the mechanism of resistance to DDT by insects, the evidence for genetic adaptive changes in the exposed populations soon became overwhelming. The genes for resistance apparently already existed in the populations and were not induced by exposure to DDT nor did they occur as spontaneous mutants subsequent to exposure to DDT. The reason for this statement is that 2½ years of selection in inbred strains of *D. melanogaster* produced no increase in resistance to DDT (Merrell and Underhill, 1956), so that DDT apparently did not act as a mutagenic agent to cause mutations to DDT resistance in these inbred lines nor did spontaneous mutations to resistance occur often enough during this period to permit an increase in resistance to develop. In contrast, an increase in resistance was observed in less than six months in non-inbred strains derived from wild populations. Since flies from the resistant lines were resistant to DDT when tested, even when they had never previously been exposed to DDT, the basis for the resistance appeared to be genetic and not physiological adaptation.

Studies of DDT resistance in a number of different insect species revealed that a variety of genetic mechanisms was involved. The genes responsible might be dominant or recessive, autosomal or sex-linked, and in some cases just one locus was involved while in others inheritance was polygenic. It appeared that whatever genetic variability was present in the population would be utilized. The mechanism for resistance appeared to involve dominant genes quite often, apparently because selection for rare favorable dominants is more efficient than selection for rare favorable recessives or for favorable combinations of polygenes. Furthermore, a number of mechanisms for resistance were found. When the walls of houses were sprayed with DDT to control malaria, the mosquitoes landing there were killed, and a behavioral change evolved such that the surviving population no longer tended to land on the walls. In other cases, the cuticle of resistant insects was less penetrable to DDT; or the fat bodies, which accumulate fat-soluble DDT, increased in size; or the nerves were found to be less sensitive to DDT; or an enzyme that breaks down DDT to a relatively harmless compound was present. Sometimes different mechanisms of resistance were found in different populations of the same species. The variety of mechanisms of resistance is testimony to the opportunism of evolution. In the presence of a lethal agent like DDT, any mechanism, behavioral to biochemical, that enhances the insect's chances of survival will be favored by selection, and the genes responsible will increase in frequency. Little wonder that such a diversity of mechanisms has evolved.

Industrial melanism, mentioned in Chapter 2, is probably the best-studied example of transient polymorphism. This shift from light-colored to melanic populations of moths in industrial areas has been found in England, continental Europe, and in the United States, and has been observed in over 100 species of moths in England alone. The facts are well known. The transition from light-colored to melanic populations occurred rapidly, in less than a century following the Industrial Revolution. In nearly all cases, dominant genes for melanism have been favored by selection in industrial areas. Differential predation by birds, eliminating dark forms in unpolluted areas and light forms in polluted areas, has been demonstrated both by direct observation and by release and recapture experiments with marked individuals. Pollution-control measures in industrial areas in England have already begun to be reflected in the populations, in which a decline in frequency of the genes for melanism has been noted.

One thing common to all of these examples of selection at work is the short time required for significant evolutionary changes to occur. It is not necessary to turn to the fossil record in order to study evolution, for there are many examples of evolutionay change in living species available for study. Accidental or purposeful introductions of species into new areas, disruption of the habitat by pollution or agriculture, the introduction of herbicides or pesticides into the environment—all may provide material for the study of evolution.

THE ORIGIN OF DOMINANCE

It is most interesting that in over 90% of the species of moths in which the genetics of industrial melanism has been studied, the melanistic phase is controlled by a single gene that is dominant to the allele for the lighter form that it has replaced (Kettlewell, 1973, p. 86). This fact raises fundamental questions about the process of evolution. Why should a similar genetic mechanism be observed in so many different species, and why should this mechanism involve a gene at a single locus, dominant to the previous wild-type allele?

Dominance has been a matter of interest to geneticists ever since Mendel first noted it in his crosses with garden peas. It was also the subject of a long-standing controversy between R. A. Fisher and Sewall Wright, two of the founding fathers of population genetics, who differed over the origin of dominance. Mendel observed that the hybrid offspring of parents homozygous for different alleles at the same locus tended to resemble one or the other of the parents rather than being intermediate between them. The trait that appeared in the hybrids or heterozygotes was referred to as the dominant trait, that not shown was recessive. These terms have sometimes been transposed to the alleles themselves, but such usage may lead to ambiguity. For instance, in garden peas, the round shape of the pea is dominant to wrinkled in hybrid progeny from homozygous round and wrinkled parents. However, the starch grains in these same hybrids are intermediate in size and shape to those found in homozygous round and wrinkled peas so that dominance for this trait is lacking or incomplete. Dominance, then, is used to refer to the phenotype. When a gene has pleiotropic effects, as in this case, if one of the traits is dominant, it does not necessarily follow that the other will be dominant too.

Furthermore, dominance is not necessarily fixed even for a particular phenotypic trait. In the cross between homozygous red and homozygous ivory snapdragons, the color of the flowers of the F_1 hybrids, all of the same genotype, will depend on the external environmental conditions. If reared at high temperatures in the shade, the F_1 flowers will be ivory; if reared at lower temperatures in full sun, they will be red. The internal environment may also be a factor in the expression of dominance. Baldness is inherited in humans, but is much more common among men than women because the trait is sex-influenced: the hormonal differences between the sexes cause the gene to be expressed as a dominant in men, but not in women. The genetic background of a gene may also influence its expression. The short tail or brachyury mutant in mice is expressed as a dominant in the heterozygous condition in the European house mouse, *Mus musculus*, but in the Asiatic house mouse, *M. bactrianus*, it is recessive in heterozygotes. With such a spectrum and range of effects, it should be clear that no hard and fast laws can be drawn up about dominance. Dominance is not a property of the gene per se but rather depends on the nature of the allelic interactions at a given locus and on the influence of the rest of the genotype and the environment.

Even though there are no laws of dominance, some generalizations are possible. If there are two or more alleles at a locus, usually one is dominant. This dominant allele is also usually the most frequent allele at that locus; in other words, the wild-type allele is usually dominant. Conversely, the less frequent, and more deleterious, alleles are recessive. Moreover, when the wild-type allele mutates, the mutants recovered are usually recessive and deleterious compared to the wild type.

A number of theories have been proposed to explain the origin of dominance, most of which have been framed so as to explain the dominance of the wild-type allele as well. These are evolutionary theories, for dominance is usually thought to be a product of evolution. However, two distinct questions can be posed. First, Why is one allele dominant over another? and second, Why is the wild-type allele usually dominant? These questions are usually treated as parts of the same problem, but, in reality, are separable because the theories explaining the origin of dominance of the wild type simply cannot account for all cases of dominance. For example, in humans, the five-fingered (pentadactyl) condition is far more common than polydactyly, and hence clearly qualifies as the wild type. Nevertheless, polydactyly is inherited as a simple dominant, and a separate theory seems necessary to explain the dominance of such rare traits.

Bateson and Punnett (Punnett, 1911) were the first to propose a theory of dominance, the presence-absence hypothesis; the wild type represented the presence of a gene, and the mutant type represented its absence, with loss of the corresponding trait. Such a loss is apt to be harmful, as most mutants are, and recessive because in the heterozygote one dose of something (the wild type) will be more like two doses of something than it will be like two doses of nothing. The Bateson-Punnett theory soon encountered difficulties. In a system of multiple alleles, the intermediate alleles posed a problem; also, known deficiencies like Haplo-IV and Notch, which should have been recessive under this theory, were dominant. The discovery of reverse mutations from recessive to dominant meant that nothing was capable of mutating to something. Finally, if evolution occurs by the gradual substitution of mutants for existing wild-type genes, and the mutants represent losses, then evolution must occur as a series of losses from the ancestral genotypes, which seems highly improbable.

R. A. Fisher (1928a, 1928b, 1931) proposed that dominance of the wild type evolved as the consequence of selection for modifying factors in heterozygotes. He assumed that the first time a deleterious mutation ever occurred in a species, dominance would be lacking, and the heterozygote would be intermediate between the homozygous types. Since the heterozygote would be less fit than the existing homozygous wild type, it would be selected against. Any modifiers at other loci tending to make the heterozygote more like the homozygous wild type would be favored by natural selection. If the deleterious mutations are recurrent, in time, selection would build up a system of modifying factors that would make the wild-type allele dominant to this recurring deleterious mutant.

Sewall Wright (1929a, 1929b, 1934) criticized Fisher's theory of the origin of dominance. First, the heterozygotes for the deleterious mutants would be so rare that the selection pressure for the modifiers of dominance in the heterozygotes would be too weak to be effective. Furthermore, these modifying factors will have their own primary effects so that their frequency will depend on the nature of the selection for these primary effects rather than for their secondary effects on dominance at some other locus.

One further criticism should be added. Fisher (and more recently Wallace, 1968a, p. 329) assumed that the heterozygote for a completely new mutation at its first occurrence will be intermediate in expression to the corresponding homozygotes. However, this assumption is untestable and seems unjustified because it concerns the very matter about which proof is being sought. The level of dominance of a new mutation is inevitably related to the previous history of the population in which it occurs since it will be expressed within the framework of the existing gene complex, which is the product of evolutionary forces. Its effect in the heterozygous condition will depend on its ability to function in relation to its allele, to the rest of the genotype, and to the environment. Therefore, a completely new mutation might show any degree of dominance initially. Its level of dominance may become modified owing to changes in the gene complex or in the environment. However, there is no a priori reason to suppose that new mutations will always be intermediate in expression in heterozygotes.

Wright (1929a, 1934) proposed an alternative "physiological" theory of dominance that is a more sophisticated version of the Bateson-Punnett theory. He suggested that dominant genes are functional, but the recessive alleles are partially or completely inactive, and dominance thus is related to the level of activity of alleles. If there is an upper limit to the expression of a trait and one dose of a given allele suffices to reach that limit, then it will act as a dominant, for two doses of the same allele will have no greater effect than a single dose. In some cases neither allele is dominant to the other, and each is expressed independently of the other. In the ABO human blood groups, for example, the A/A^B heterozygotes produce both antigen A and antigen B in their red cells. Similarly, heterozygotes for the gene causing sickle-cell anemia and for its normal allele (Hb^S/Hb^A) produce almost equal amounts of hemoglobin S and normal adult hemoglobin, hemoglobin A, so that the synthesis of the two types of protein molecules goes on independently. Such alleles are said to be codominant. Thus, these results show that different alleles may produce qualitative differences as well as quantitative differences, and also that alleles may act quite independently of one another.

J. B. S. Haldane (1930b, 1939) proposed still another theory for the origin of dominance of the wild type, the selection of more efficient wild-type alleles. For example, if wild-type allele A_1 produces, in combination with the deleterious gene a, a heterozygote intermediate in fitness between A_1A_1 and aa, then it will be replaced by a new wild-type allele A_2 if the A_2a heterozygote is closer to the well adapted wild-type phenotype (A_2A_2, A_1A_1) than the A_1a heterozygote. A_2 could then be replaced by a more efficient allele A_3, and so on.

C. R. Plunkett (1932, 1933) and H. J. Muller (1932) independently proposed a theory of modifiers different from Fisher's to account for the origin of dominance of the wild type. This theory stemmed from the observation that the wild-type phenotype is usually modified less by various stresses during development than are mutant phenotypes. They argued that selection tends to build a factor of safety into the wild phenotype by the accumulation of modifiers that stabilize it in the presence of environmental and genetic stresses. Thus, selection of the modifiers occurs in the frequent wild-type homozygotes rather than in the rare heterozygotes as Fisher suggested. More recently, this hypothesis has been phrased in terms of the concept of canalization (Waddington, 1957b; Rendel, 1962).

Thus, the theories fall into two categories, one, the "physiological" theories stressing the primary action of the gene, and the other, the origin of dominance by the selection of modifiers. However, these theories are not mutually exclusive, and none of them alone can account for all of the observations concerning dominance. Furthermore, there is some experimental evidence in support of each. Dosage studies with recessive genes in *Drosophila*, for example, often showed the recessives to be hypomorphs or amorphs (in Muller's terminology, 1932) which supports Wright's conception of partial or complete inactivation in recessive alleles. Increased doses of the hypomorphic allele gave phenotypes closer to the wild type while increased doses of the amorph had no effect. Antimorphs, neomorphs, and hypermorphs were usually dominant to the wild type.

The demonstration of the existence of iso-alleles by Stern and Schaeffer (1943) makes Haldane's concept of the selection of more efficient wild-type alleles plausible. They demonstrated that three different wild-type alleles exist at the *cubitus interruptus* locus in *D. melanogaster*. All of these alleles gave indistinguishable wild-type phenotypes when homozygous at $25°C$, but were shown to differ when tested at low temperatures, as hemizygotes, or as heterozygotes in combination with *ci* or *ci^W* alleles.

The existence of modifying factors was first clearly demonstrated by Castle (1919) in selection experiments modifying the expression of the hooded trait in rats, which is due to a homozygous recessive. Morgan (1929) reported similar results with the eyeless trait in *D. melanogaster*, also caused by a homozygous recessive. He also noted that eyeless, like other distinct mutants in *Drosophila*, tended to become less distinct from the wild type when it was maintained in stock culture for a number of generations. Fisher (1935, 1938) reported his experiments on dominance modification in poultry through the selection of modifiers, and Ford (1940a) reported similar findings with Currant Moths, *Abraxas grossulariata*, drawn directly from wild populations. Fisher and Holt (1944) modified the expression of the short-tailed mutant in mice by the selection of modifiers for greater tail length. Thus, there appears to be adequate evidence for the existence of modifying factors for dominance, and for the modification of dominance by selection acting on these factors.

However, Fisher and Holt's interpretation of their data was questioned by Dunn and Gluecksohn-Schoenheimer (1945). The short-tailed *Sd* mutant affects not only tail length but also has profound effects on the development of the urogenital system. The homozygous *Sd/Sd* individuals are tailless, lack kidneys, and die shortly after birth. The heterozygous *Sd/+* mice have short tails and less severe urogenital malformations. Thus, the *Sd* gene acts as a recessive lethal, as a dominant in its effects on the tail, and as an incomplete dominant in its effects on the urogenital system; and in its pleiotropic effects, this mutation shows different degrees of dominance. Fisher and Holt assumed that selection of modifiers for increased tail length in heterozygotes—that is, for modifiers enhancing the dominance of the wild type allele of *Sd*—would also result in increased viability. However, Dunn and Gluecksohn-Schoenheimer found a case in which increased tail length was associated with decreased viability and another in which decreased tail length was associated with increased viability so that Fisher and Holt's assumption about a positive association between viability and tail length was obviously unwarranted. If modifiers of *Sd* are selected, they will be selected to make the urogenital effects of *Sd* recessive because of the strong association between viability and the urogenital syndrome produced by *Sd*. Length of tail has much less relation to viability than imperforate anus or defective kidneys. The same genetic constitution that suppresses the urogenital syndrome and enhances the dominance of the + allele for this trait in heterozygotes may cause shorter tails and, hence, reduce the dominance of the + allele with respect to tail length. Therefore, genes with pleiotropic effects lead to further complexities in the study of dominance.

Support for the concept of Muller and Plunkett that modifiers have been selected, not in relation to the relatively rare heterozygotes as Fisher postulated, but rather to stabilize the phenotype of the much more frequent homozygous wild type, comes from studies such as Rendel's (1959, 1962) with scute in *D. melanogaster*. Wild-type flies almost invariably have 4 bristles on the scutellum of the thorax; the wild-type phenotype is very stable. The sex-linked recessive mutant scute when homozygous in females (*sc/sc*) or hemizygous in males (*sc/Y*) leads to much greater variation in the phenotype, for the flies may have 0, 1, 2, or 3 scutellar bristles. If selection is practiced on the rare +/+ and +/Y flies with 3 or 5 bristles, it is possible, with difficulty, to establish lines with either more or less than the normal 4 bristles. Although progress is slow at first, it becomes more rapid after some selective progress away from 4 bristles has occurred. If selection is practiced on the *sc/sc* and *sc/Y* flies for increased bristle number, the response is rapid and a scute line with 4 bristles is soon produced. However, if upward selection is continued, progress to 5 or more bristles is extremely difficult. Therefore, even though considerable genetic variation affecting the number of scutellar bristles is present, modifiers have apparently been selected to stabilize or *canalize* the number of bristles at 4.

This work and similar results with other mutants indicate that the wild type is phenotypically more stable than mutant phenotypes and that modifying factors play a role in this greater stability. These modifiers have been selected for their effects in the frequent homozygous wild-type individuals, rather than as modifiers of dominance in heterozygotes. The dominance observed in heterozygotes is an incidental by-product of the canalizing selection for stability of the wild phenotype.

The controversy over the origin of dominance continues, usually over whether or not Fisher's theory is correct (Crosby, 1963; Ewens, 1965a, 1965b, 1966, 1967; Sheppard and Ford, 1966; Sved and Mayo, 1970; Wright, 1964, 1977). Most of the arguments are theoretical and mathematical. However, at least some of the differences in opinion stem from differences in the questions posed. Some geneticists seem to be concerned primarily with the origin of dominance per se, the question why one allele should be dominant over another; they express their views in terms of gene action. Others are more interested in the evolutionary origin of dominance; but here too, different questions can be posed.

Fisher's theory was designed to explain why deleterious mutations are generally recessive. This question is somewhat different from the question of why, in cases of transient polymorphism such as industrial melanism, the evolutionary change to a new wild type is mediated by an allele dominant to the existing wild type (Ford, 1953, 1971; Kettlewell, 1973). Dominance of an existing wild-type allele over a deleterious mutant and dominance of a new favorable mutation over an existing wild-type allele both involve dominance of the wild type, but the dominance may not be achieved in the same way. Although the evolution of dominance by the selection of modifiers is often postulated for both situations, this may not occur. As Ford (1971, p. 316) says, "The fact that nearly all industrial melanics are complete dominants when they begin to spread poses a considerable problem." To circumvent this problem for Fisher's theory, Kettlewell (1956, 1961, 1973) suggested that the dominance of the genes for melanism had evolved 8,000 to 10,000 years ago when most of Great Britain was covered by pine forests. Since melanic forms are found in present-day relict pine forests, he argued that melanism conferred fitness in the past during the Boreal period and that the modifiers making the genes for melanism dominant were selected during that time. Then when the Industrial Revolution came, the gene complex of these species already contained the necessary modifiers to make the alleles for melanism dominant.

Ford had earlier (1937, 1940b) proposed that the melanic forms were "hardier" or more viable than the light-colored cryptic types, but, until the coming of the Industrial Revolution, their greater viability was counterbalanced by selection against them because of their lack of protective coloration in the unpolluted countryside. Then, when the Industrial Revolution occurred, selection was strongly in their favor because of their greater vigor as well as their cryptic coloration against the dark polluted background, and the melanistic type rapidly replaced the light-colored wild type in industrial areas.

The evidence for greater viability of the melanistic types was re-examined by Merrell (1969a), who concluded that the published data do not unequivocally support the greater viability of the melanistic types. In the data published before 1940, no significant differences in viability were observed in 87 families (3,443 moths), a significant excess of melanics was found in 19 families (230 moths), and a significant excess of typicals was found in 4 families (175 moths). The data published after 1940 showed a similar mixed pattern of significant and non-significant results. A breakdown of the data revealed that where a significant excess of the melanistic type was found, it was traceable to one or a few out of a large number of broods, the rest of which gave the expected 1:1 ratio. The most likely explanation for this heterogeneity in the data is that in the aberrant broods, the homozygous typicals $(+/+)$ were also homozygous for a deleterious gene linked to the typical gene. The melanic type in these backcrosses, being heterozygous $(M/+)$, would have a much lower probability of being homozygous for such linked deleterious genes. Since there was little knowledge or control over the rest of the genotype in these moth crosses, and it is clear from the methods used that a few related individuals were ordinarily used as progenitors, this explanation seems more consistent with the data than superior viability of the melanic type compared to the typical form. The results with the dominant genes for melanism seem to parallel the results with the dominant, Lobe, in *D. melanogaster* (Merrell, 1963), in which crosses from $L/+ \times +/+$ parents gave an excess of $L/+$ progeny. Analysis showed that, rather than an excess of $L/+$, there was a deficiency of $+/+$ flies because some of the chromosomes bearing the $+$ allele of Lobe carried a linked deleterious gene in common, which reduced the number of $+/+$ progeny below expectations.

The argument that the genes for melanism were dominant at the time of the Industrial Revolution because the necessary modifiers had already been selected during the Boreal period has a definite weakness. If modifiers making the melanic alleles dominant were accumulated during the Boreal period, then selection should have led to the accumulation of modifiers that enhanced the dominance and viability of the light-colored cryptic typical forms during the subsequent thousands of years they served as the wild type. Unless it is postulated that systems of modifiers are selected only in relation to genes for melanism and that these systems would be uninfluenced by subsequent events in these populations for thousands of years, Kettlewell's explanation for the dominance of the genes for melanism seems inadequate. Furthermore, Kettlewell's experiments (1956) showed decisively that strong selection pressure by birds operated against either the melanic or the typical form, depending on the type of background. Thus, the selective advantage of the melanic form in industrial areas is clearly related to its cryptic quality against a dark background, and there is no need to postulate some other type of advantage, such as viability, to account for its rapid increase in frequency.

The origin of resistance in insects to DDT and to other pesticides has been even more closely observed than the origin of industrial melanism. In most cases, resistance has appeared within a very brief time, a matter of a few generations or a few years. With pesticides, the evolutionary change has to be recent, for the insects have never before been exposed to agents such as DDT (Brown and Pal, 1971). In a high proportion of such cases, the basis for resistance could be traced to a single dominant or semi-dominant gene (Crow, 1957; Oppenoorth, 1965; Georghiou, 1969; Brown and Pal, 1971; Plapp, 1976). The brief period of time between the first use of the insecticide and the appearance of resistance with a dominant mode of inheritance almost precludes the evolution of dominance for these genes for resistance. They must have been dominant when they were first favored by selection, yet this dominance could not have been developed by selection at some earlier date because the species had never before been exposed to DDT. 236172 .

It should be noted that the evolutionary role of dominant genes is not confined to industrial melanism or to pesticide resistance in insects, but that they appear to play an important role in a number of cases, particularly where a species is forced to adapt rapidly to strong new selection pressures (Merrell, 1969a). Among these pressures are the resistance of plants to rusts, smuts, and other plant diseases (Flor, 1971), of corn to the corn borer, and, of mammals such as mice to various infections. In humans, resistance to malarial infection has been shown to be enhanced by the genes for hemoglobin S (sickle-cell trait), hemoglobin C, thalassemia, or G6PD deficiency (glucose-6-phosphate dehydrogenase) and probably genes at other loci as well (Cavalli-Sforza and Bodmer, 1971). Many types of mimicry are also controlled at a single locus. Microbial resistance to antibiotics and to chemotherapeutic agents is generally caused by a mutant at a single locus although dominance is not involved because microorganisms are ordinarily haploid. The point, however, is that changes at individual gene loci can make a difference in evolution; the entire genotype does not have to be restructured for an adaptive shift to occur. Just as in the cases of industrial melanism and insecticide resistance, not all of the evolutionary changes cited above are mediated by dominant or semi-dominant genes, but a rather sizable proportion is. To realize why this should be so, we need to consider how a previously well-adapted population responds to a new selection pressure.

When the environment changes so that a new phenotype is better adapted than the existing wild-type phenotype, and, hence, becomes the new wild type (as in the case of melanics replacing typicals in industrial areas), this new phenotype may be approximated by various genes or combinations of genes. Any genes tending to produce the new wild type and present as part of the genetic variability of the population will be favored by natural selection. If such variability is not present, the adaptive shift cannot occur unless, or until, it becomes available either by mutation, migration, or possibly by recombination.

The new wild-type phenotype could be produced in several ways: (1) from the action of a dominant mutant; (2) from the action of a recessive mutant in the homozygous condition; or (3) from the combined effects of genes at two or more loci. In a large population all three of these possibilities could coexist in the gene pool. Melanism, for example, has been shown to be controlled by dominants, or by recessives, or in some cases, by a combination of genes at several loci (Ford, 1953). If all three types of variation were present at low frequency in the population, it is instructive to consider which one is most likely to become the genetic basis for the new wild phenotype when selection starts to favor melanism.

Since sexual reproduction will constantly tend to break up favorable combinations of genes, progress in the desired direction will be slow when the adaptive shift is controlled by genes of low frequency at two or more loci (Haldane, 1932). Progress will also be slow if selection favors a rare homozygous recessive since nearly all such individuals will mate with the more frequent dominant type, most of which will be homozygous dominant. Thus, with rare exceptions, all of the progeny of a favored homozygous recessive individual will be of the dominant rather than the favored recessive phenotype, and selection will have no chance to act on the next generation.

However, selection for any dominant or partial dominant that produces the favored phenotype will be more effective than selection for any other type of genetic control because half of the progeny of a dominant heterozygote will also show the trait and will again be subject to favorable selection in the next generation. Even if the initial frequency of the favorable dominant gene is low, in a finite breeding population it has a reasonably good chance of becoming established. Therefore, of the available genetic means of reaching a new adaptive phenotype, the most efficient is by selection of a dominant or partially dominant mutant. If both dominant and recessive favorable genes are present in a population at a low frequency, selection would favor both types of genes, but the new wild-type gene will almost invariably turn out to be the dominant because of the greater efficiency of selection in its favor (Haldane, 1932b; James, 1965).

If a gene acting as a complete dominant under the existing genetic and environmental conditions is available, this gene has a good probability of becoming the new wild-type allele. If no such dominant is available in the gene pool of the population, then a partially dominant gene may serve instead. The presence of the completely dominant *carbonia* melanic form in the moth *Biston betularia* and its allele (Clarke and Sheppard, 1964, 1966), the partially dominant melanic form *insularia*, in some parts of the range of the species (Kettlewell, 1956, 1958) indicates the opportunism of natural selection in seizing upon whatever variation may be available to effect the necessary adaptive phenotypic change.

The development of insecticide resistance in insects occurs in a similar way. In this case, not only different genetic mechanisms for the same mode of resistance may exist, but also quite different modes of resistance have evolved (Brown and Pal, 1971). Selection for insecticide resistance in natural populations seems to lead to forms of resistance governed by dominant genes more often than selection in laboratory populations. In laboratory populations, multiple-factor or polygenic modes of inheritance for the resistance are often found. The most reasonable explanation for this difference is the difference in the size of the populations exposed to the insecticide. The laboratory populations are small relative to the size of the wild populations, and their gene pools are less apt to contain rare favorable dominants than those of the much larger wild populations. Under these circumstances, natural selection seizes upon any available genetic variation for resistance in the laboratory populations, which often leads to multigenic forms of resistance. Thus, natural selection is forced to act upon the genetic material available in the population. If the population is large so that dominant, recessive, and multigenic options are all available, the greater efficiency of selection for favorable dominants dictates that they will win out almost every time. If they

are not available in the population, other genetic mechanisms of resistance will evolve. Therefore, it is not surprising that the genetic
basis for resistance has been variously reported as autosomal or sex-
linked, dominant or recessive, and unifactorial or polygenic, even
within the same species.

If both dominant and semi-dominant genes are available, as in the
case of the *carbonaria* and *insularia* forms of *B. betularia*, the dominant apparently wins out. If only a partial dominant is available, it
seems reasonable to expect that selection for modifiers enhancing the
dominance of this favorable allele will occur, or if a more dominant
favorable allele subsequently appears owing to mutation or immigration, it may replace the semi-dominant.

When dominant genes for melanism or for mimicry have been
crossed into populations other than the ones in which they naturally
occur, the dominance sometimes breaks down (Sheppard, 1959, 1961a;
Sheppard and Ford, 1966; Kettlewell, 1961, 1965). These results
have been interpreted to support the concept that the dominance of
these genes has evolved in their native gene complexes by the selection
of modifiers enhancing dominance. However, this breakdown of dominance on outcrossing does not prove that dominance originated in
the original population by the selection of modifiers, for it reveals
nothing about the initial level of dominance of these genes in their
original gene complex. As we have already seen with melanism and
insecticide resistance, in many cases the newly favored alleles appear
to have been dominant from the outset. Similarly, the many experiments demonstrating the existence of modifiers of dominance, or the
success of selection in enhancing dominance, are suggestive, but do
not prove how dominance of the wild type is actually achieved in
natural populations. Only direct observations and experiments at
the time the newly favored allele is first becoming established in the
population can determine its level of dominance at that time. The
available evidence indicates that in quite a few cases a gene already
dominant within the existing gene complex mediates the shift to a
better adapted phenotype. In other cases, where partial dominants
are involved, it seems reasonable to expect their degree of dominance
to be increased by the selection of dominance modifiers. If no dominants are available, any genes in the population modifying the phenotype in the desired direction will tend to increase in frequency, and
multigenic systems may be selected.

Perhaps the essential difference between the ideas outlined above
and previous theories of the origin of dominance of the wild type is
the change in approach to the problem. Rather than focusing attention

on possible changes related to one major locus, we have considered the possibility that similar phenotypic or adaptive shifts may be mediated in various ways within the gene complex of the population. The pressure of natural selection is directed toward producing the best adapted phenotypes rather than increasing the frequency of a gene at any particular locus. When a strong new selection pressure is applied to a population and the best adapted phenotype suddenly changes, the population must adapt quickly or become extinct. Since rapid adaptive change is mediated most effectively by dominant genes, they appear to play a more significant role in evolution than has sometimes been recognized. Darlington (1956), for example, represented a widely held view about the mode of evolution with this statement: "While marker genes, the chief legacy of classical genetics, with their pedigrees and their mutation rates, are of great importance for the study of evolution, they are of little importance in carrying it out." The evidence, particularly with respect to such well-studied cases as industrial melanism, insecticide resistance, and disease resistance in plants, suggests that such genes, especially dominant and semi-dominant genes, can and do play a significant role in evolution.

MIMICRY

In Batesian mimicry, a harmless or tasty mimic closely resembles a poisonous or noxious model even though they are not closely related. The resemblance may extend to visible external morphological structures, color patterns, and behavior, but, like beauty, it is only skin deep. The similarity between model and mimic is superficial; their convergence in appearance and behavior is far from complete. Nevertheless, it is sufficient to deceive predators (Brower, 1958, 1960; Brower and Brower, 1962, 1966; Brower, Brower, and Westcott, 1960; Brower et al., 1964), and mimicry is an adaptation for survival.

The origin and genetic basis of mimicry have also been a matter of controversy. The mimetic resemblances are surprisingly good and usually involve a number of quite different characters such as structure, color, color pattern, and behavior. Nevertheless, this combination of diverse traits usually is inherited as if controlled by an allele at a single locus. That so many seemingly unrelated traits should be controlled by a gene at a single locus and, furthermore, that all of these traits together should produce an exact mimetic resemblance poses a considerable evolutionary problem. Attempts to resolve this apparent paradox resemble the explanations that have been offered for the origin of dominance.

In 1915 Punnett suggested parallel mutations as the explanation for the close resemblance between mimic and model. In other words, homologous genes in the two species responsible for the same cluster of traits had mutated in both model and mimic. This theory runs into immediate difficulty because, as noted above, the resemblances between model and mimic are superficial whereas parallel mutations would lead to fundamental similarities between model and mimic.

Fisher (1927, 1930, 1958) and Ford (1953, 1971), on the other hand, argued that mimicry arose gradually. They stated that if a mutation occurs in an unprotected species that confers even a slight resemblance to a protected species, the mutant gene will spread in the population owing to its selective advantage, and the mimetic effects of the gene will be gradually improved by the selection of modifiers. Just how great the initial "slight resemblance" needs to be is not altogether clear, but in their writings, both Fisher and Ford seem to regard any resemblance, however slight and however small the selective advantage, as sufficient to initiate the evolution of mimicry. A further part of their hypothesis is that the various components of the mimetic resemblance were originally controlled at different independent loci, separately affecting such traits as structure, color, color pattern, and behavior, and that each was favored separately by natural selection. Then, since these independently inherited genes cooperated to produce the mimetic effect, natural selection favored rare structural rearrangements—chromosomal translocations and inversions—that would bring these genes together on the same chromosome and prevent crossing-over between them. The reduction of crossing-over can be achieved by a reduction in chiasmata formation in the region concerned, or by rearrangements that bring the loci into close proximity. In this way, a cluster of genes can be brought together to form a single *switch-mechanism* or *super-gene*, which behaves like an allele at a single locus. Alternatively, the same effect could be achieved if several of these genes were included within a single inversion, which would then also act as a super-gene. The questionable parts of this theory seem to be the assumption that any slight resemblance between mimic and model would be sufficient to trigger the evolution of mimicry, and the evolution of super-genes from independently inherited loci.

Goldschmidt (1945), however, again reviewed the mimicry question and concluded that Punnett's theory of the origin of Bateson mimicry by mutation (or saltation) agreed better with the facts than Fisher's "neo-Darwinian theory." He based his conclusion on his studies of the developmental genetics of pattern formation in the wings of Lepidoptera, arguing that different developmental potentialities for pattern existed and that a mutation could act as a simple switch mechanism to change from one developmental system to another. His theory was criticized strongly, especially by Ford (1953, 1971). The main objection is that Goldschmidt's theory deals primarily with the developmental systems controlling pattern formation and color in Lepidopteran wings, and fails to deal with mimicry involving other traits such as behavior or structure, and with groups other than the Lepidoptera.

A considerable amount of experimental work has been done on the genetics of mimicry, especially by Clarke and Sheppard (Clarke and Sheppard, 1955, 1959a, 1959b, 1960a, 1960b, 1960c, 1960d, 1962; Clarke, Sheppard, and Thornton, 1968; Sheppard, 1958, 1959, 1961a, 1961b, 1969). In *Papilio dardanus*, one of the best studied species, the genetics of 22 mimics were studied. Of the 22 forms 15 were governed by a single completely dominant allele to the non-mimetic form (Sheppard, 1959). The other 7, which were not complete dominants, were either very rare or were poor mimics. On the other hand, expression in 12 of 14 heterozygotes from crosses between individuals from allopatric populations was intermediate. Sheppard used these results to argue that the dominance of the mimetic forms must have evolved through the selection of modifiers. As pointed out earlier, this type of evidence shows that the means of achieving dominance in the two different populations must differ, but reveals nothing about the initial level of dominance of these genes within their own population.

Although Sheppard, like Fisher and Ford, believes that mimicry has evolved through the selection of modifiers, he differs from them in believing that the initial step requires the appearance of a mutant form that in some way resembles a model sufficiently well to deceive a predator from the outset (Sheppard, 1961a). Subsequently, natural selection refines and polishes the resemblance to the model by the selection of modifiers. The difference between Fisher and Ford, on the one hand, and Sheppard on the other is that they believe that the evolution of mimicry was a gradual process throughout, whereas Sheppard believes that an initial breakthrough was necessary, which was later followed by a gradual improvement in the resemblance. It is interesting that Punnett (1915) also considered this latter possibility for the origin of mimicry, but dismissed it. This theory, of course, represents a compromise between the ideas of Punnett and Goldschmidt and those of Fisher and Ford. The difficulty with accepting the ideas of Punnett and Goldschmidt lies in the improbability that the many nuances of resemblance between mimic and model would arise fully perfected at a single step. The difficulty in accepting the ideas of Fisher and Ford stems from the improbability that selection could work effectively to produce mimicry with only very slight resemblance between model and mimic as the starting point.

That the evolution of mimicry could occur through the occurrence of a genetic variant in a harmless species that in some significant way resembles a distasteful or harmful model is suggested by evidence obtained in a different context. Tinbergen (1951), working with the three-spined stickleback, *Gasterosteus aculeatus*, showed that the spring fighting of the males was directed primarily against other males in nuptial colors. In spring, males differ from other individuals in having an intense red throat and ventral surface. During spring he presented to the males a series of stickleback models, some of which were very crude representations that did not look like sticklebacks or even like fish, but had a red belly. Others were exact replicas of male sticklebacks, but lacked the red belly. The males reacted much more strongly to the models with the red belly, no matter how odd their shape, than they did to the exact models. Lack (1943) performed a similar experiment with the English robin and showed that a territorial male would threaten a bundle of red feathers more aggressively than a mounted robin complete in every detail except the red breast. These experiments on aggression suggest that a mimic need not be perfect in every detail to deceive a potential predator. There seems no reason to suppose that stimuli leading to aversive behavior would differ significantly from those leading to aggressive behavior. If some initial degree of deception is achieved, selection will undoubtedly continue to improve the resemblance through the selection of modifiers in the mimic population.

A considerable amount of work has been done on single-trial learning in mice and rats, but a footnote in Brower, Pough, and Meck (1970) suggests how effective single-trial learning could be in protecting mimics, once a predator has had a single unpleasant experience with the model. After two blue jays had each eaten a single emetic monarch butterfly, they were each offered 120 monarchs over a two-day period. One jay rejected all 120 monarchs on sight, and the second rejected the first 96, but finally attacked the 97th.

The experiments on single-trial learning and on aggressive behavior, considered together, suggest how mimicry may have originated. If a genetic variant in a harmless species causes it to resemble a noxious species enough to cause a potential predator to avoid it, that variant will tend to increase in frequency in the population. Once again, if the variant is dominant to any degree, it will have a better probability of becoming established in the population than if the mimic trait is governed by some other mode of inheritance. In this case, especially, it seems highly probable that natural selection will continue to improve the resemblance between mimic and model, either through the selection of modifiers of the original mimetic trait or perhaps through selection favoring other more or less independent traits that contribute to the overall mimetic effect. If the latter is the case, then the control of all of these mimetic traits by one or, perhaps, two dominant switch genes or *super-genes* presents a bit of a problem. The theory that a number of separate independent loci governing the various phases of mimicry have become closely linked on the same chromosome to form a *super-gene* through chromosomal rearrangements is the most plausible explanation suggested for the origin of the existing genetic control of mimicry. However, the actual evidence showing the origin of *super-genes* through chromosomal rearrangements is minimal.

Compared to Batesian mimicry, the origin of Müllerian mimicry seems much more straightforward. All Müllerian mimics are similar in appearance and are characterized by aposematic or warning colors that advertise their presence. It is generally believed that the less common species will evolve toward the appearance of the more common species (Marshall, 1908; Punnett, 1915; Sheppard, 1958) and that the more nondescript species will evolve in the direction of more distinctive and unmistakable warning colors. The less numerous noxious species will evolve to resemble the more numerous noxious species because predators are more apt to learn of the aversive nature of the common species. Then selection in the rare species will tend to favor those individuals who most resemble the noxious image that the predators have already learned. Distinctive aposematic colors will tend to evolve in all species of the Müllerian complex so that the predators' chances of mistaken identity will be minimized. However, Sheppard (1958) argued that in Müllerian mimicry, the more common species and the less common species will tend to converge toward one another in appearance, but that the less common species will undergo the greater amount of change in the direction of the more common species. He believed in convergence rather than in the evolution of the rarer type to resemble the common species because he felt that genes of "marked effect have little chance of surviving in nature," and, therefore, the evolution of Müllerian mimicry must be a gradual process of convergence involving changes in both species. Whatever the actual nature of the change—and convergence by both species certainly seems possible, especially if they are evolving toward a more conspicuous common appearance—the reason given for favoring convergence is not valid. Sickle-cell anemia is a case in point. This gene has a major marked effect, for it is lethal in the homozygous condition. Nevertheless, its frequency often reaches 30 or 40% in malarial regions where it confers protection against infection in hetero-zygotes. Thus, genes of marked detrimental effect can and do play a role in adaptation. Genes such as the sickle-cell gene may subsequently be supplanted by other genetic mechanisms that confer resistance without the associated detrimental effects, but to argue that evolutionary change must be gradual because major genes may have detrimental effects is clearly open to question. Recent discoveries of the tremendous store of genetic variability in natural populations make clear that selection has a far greater range of options than was previously suspected. Moreover, the relationship between protein polymorphism and fitness is still relatively unexplored. The dire consequences of the sickle-cell gene in the homozygous condition are traceable to a single amino acid substitution in the hemoglobin molecule. Fisher (1958), like Sheppard, argued in favor of gradual convergence of Müllerian mimics, but his attitude arose from his commitment to the idea of gradual evolutionary change to produce the evolution both of dominance and of Batesian mimicry as well as Müllerian mimicry, and from his unwillingness to entertain any possibility of evolutionary change that smacked of sudden changes or saltations or genes of major effect.

GENETIC ASSIMILATION

We have already discussed at length the remarkable precision of the adaptation of living things to their physical and biological environments. However, the environment is not constant and a phenotype well adapted in one set of environmental conditions may be less well adapted in others. A phenotype is not, however, a fixed attribute of a genotype. As mentioned previously, the genotype determines a range of possible phenotypes. The particular phenotype that develops for a given genotype will depend on the particular combination of environmental conditions influencing it during its development. One of the remarkable aspects of this relationship between genotype and phenotype is that the phenotype that develops will be better adapted to the environment that evokes it than the other phenotypes possible with that genotype. This phenomenon has been referred to by Thoday (1953) as "developmental flexibility." These adaptive changes in the course of development are often thought of as acquired traits; yet, in some instances, the acquired trait appears to have become hereditary.

For example, calluses normally appear on the body surface at points of stress. If you chop a lot of wood, or play tennis, or dig ditches, or do a lot of writing, calluses will appear on your hands in different places depending on the particular stress points. Calluses also develop on your feet, their thickness depending on the amount of walking or running that you do. However, even before birth, human infants have been shown to have thicker skin or calluses on the soles of their feet than elsewhere on their bodies. Similarly, calluses appear at the fore and aft points on the ventral surface of an ostrich (Waddington, 1957b) where the body makes contact with the ground; yet these thickenings appear before the egg hatches. In these cases, the calluses appear before the stress and must be genetically controlled independent of the external stress. These and other examples suggest Lamarckism, the inheritance of acquired characters, yet modern geneticists generally dismiss Lamarckism, primarily because of the lack of convincing evidence that the inheritance of acquired characters ever occurs. Despite this lack of evidence, Lamarckism has proved to be a remarkably hardy and long-lived theory, in part because of the emotional and political appeal of the idea that by improving the environment of individuals, one can thereby effect permanent, hereditary change, and in part because it seems to be a reasonable explanation for phenomena such as the development of calluses in ostriches and humans.

To the question, Does the inheritance of acquired characters occur? the simple answer is still no. However, the question remains, Is there any mechanism known within the framework of modern evolutionary theory by which "acquired" traits could become hereditary, as appears to have happened with calluses? In other words, is it possible to reconcile to any extent the apparent conflict between Lamarckism and the modern theory of evolution?

Studies by Waddington and his co-workers, which he has called "the genetic assimilation of an acquired character," have shed some interesting light on how adaptive shifts during development can become incorporated into the genotype so that, subsequently, they appear even in the absence of the stress that initially induced them.

First, it should be realized that any trait is to some extent inherited and to some extent acquired. A genotype is required and so is an environment; the interaction between them during development produces the individual. However, under unusual conditions, non-heritable phenotypic modifications called phenocopies sometimes appear. Phenocopies result from aberrant environmental conditions and simulate the phenotype caused by a gene mutation (Goldschmidt, 1935, 1938). In the initial experiments, Waddington (1953) subjected pupae of *D. melanogaster* to temperature shock and induced a number of different wing aberrations, from which he chose the trait crossveinless for study. This induced non-heritable phenocopy resembled the phenotype of flies homozygous for the recessive crossveinless mutant, *cv.* The stock of flies used was a wild Edinburgh strain known as *S/W* 5, which was phenotypically wild type and never lacked the crossveins in the wings. Nevertheless, when pupae from this strain were subjected to 40°C for 4 hours starting 21 to 23 hours after puparium formation, a number of crossveinless flies appeared even though none appeared in the untreated controls. Waddington then started upward selection lines in which he interbred crossveinless flies from the treated stock, and downward selection lines in which the normal flies from the treated stock were interbred. Selection in both directions was successful, especially in the upward line. By the 5th generation, 60% of the treated pupae in the upward line developed into crossveinless flies; by the 23rd generation, this had increased to 96%.

For each generation he ran parallel untreated controls for the upward selected line. For the first 13 generations, no crossveinless flies appeared among these controls. In other words, crossveinless was behaving as a phenocopy. In the 14th generation a few crossveinless individuals appeared in the control population despite the fact that they had never been subjected to the temperature shock. By the 16th generation, the frequency of crossveinless had risen to about 2%. At this point, Waddington changed his selection procedure, starting a number of selection lines with single-pair matings between crossveinless flies from the untreated controls and rearing the progeny at 25°C without heat shock. This type of selection within lines produced 4 exceptionally high lines that gave the following results even in the absence of heat shock:

Line	Percentage Crossveinless	
	25°C	18°C
H 1	68	100
H 16	87	100
H 24	67	99
H 26	95	100

Thus, from an original population that had produced no crossveinless flies except as phenocopies following heat shock, Waddington had produced populations that were virtually 100% crossveinless even without heat shock. This result was the phenomenon he called the "genetic assimilation of an acquired character" and seemed to verge on Lamarckism. Bateman (1959b) similarly achieved genetic assimilation of four venation phenocopies. Both Waddington and Bateman attempted to produce genetic assimilation of crossveinless in inbred lines, but failed. That fact makes clear that the observed phenomenon was not simple Lamarckism. Furthermore, it suggests that assimilation was dependent on the initial variation present in the populations and that new mutations were not occurring frequently enough in the populations to be useful.

The most straightforward interpretation of these results was presented by C. Stern (1958) who pointed out that in the original environment, the genotypic variation present in the population was below the threshold of expression and thus was not open to selection. In the new environment (the heat shock), these formerly subthreshold differences were expressed, and it was possible to single out those genotypes that produced the crossveinless phenotype. Once it was possible to distinguish between the different genotypes, artificial selection and interbreeding of these individuals not only led to an increase in the frequency of the trait under heat shock, but eventually, through recombination and continued selection, produced individuals with genotypes that gave the desired phenotype in either environment, with or without heat shock.

Comparably successful experiments on genetic assimilation were also run involving dumpy and bithorax (Bateman, 1959a; Waddington, 1956, 1957a, 1957b, 1959a, 1959b, 1961). In these experiments, as well as in Bateman's (1959b) work on the four venation phenocopies, genes of major effect often appeared to play a significant role. In Waddington's initial experiments with crossveinless there was little evidence for the existence of any definite switch-gene (Waddington, 1953, 1959b). However, Bateman (1959a, 1959b) reported that "the assimilation of dumpy appears to be fundamentally due to the presence of the gene $dpTP2$," a recessive lethal allele of dumpy, and that, for the four venation phenocopies, even though all three major chromosomes were involved, their individual effects were unequal. She adds, "The pronounced effect of, in most cases, a single chromosome suggests that it carries a major gene. . . . Nevertheless, the hypothesis of single-gene control is not unequivocal, . . . It is, however, clear that the assimilated characters are not controlled merely by an indefinite number of genes of individually small effect, i.e. are not 'polygenic'." Furthermore, "It is therefore suggested that assimilation has been due to the selection of penetrance-modifiers for one or more major genes."

In the initial $S/W\,5$ population, Waddington (1953) found no crossveinless flies before the experiments began, but did not examine a very large number of flies. Bateman (1959b), using a different stock and examining many more individuals in the initial population, found less than 1% of spontaneous low-grade crossveinless individuals. From these flies, she was able to develop a "nonassimilated" crossveinless stock, which she later tested against the "assimilated" crossveinless stock, and found that the two stocks appeared identical in genetic constitution. Again, these results suggest that the genes necessary to produce genetic assimilation were already present in the initial population, but at very low frequencies. For the other three venation phenocopies studied by Bateman, no individuals of the abnormal types were found in the initial populations. This lack does not necessarily mean that the genes were absent, but merely that they might have been extremely rare or that their expression was subthreshold under the existing genetic and environmental conditions.

In the experiments on the genetic assimilation of bithorax (Waddington 1956, 1957a, 1957b), a dominant bithorax gene (bx^D) with a recessive lethal effect appeared in two separate selection experiments. In still another assimilated bithorax stock, an important component in the genotype was an X-linked recessive condition that caused a maternal effect; homozygous recessive mothers laid eggs that tended to develop into bithorax phenotypes. More recently, Thompson and Thoday (1975) discovered a major factor on the second chromosome responsible for the genetic assimilation of part of the veinlet phenotype—a gap in only the L 4 vein—modified by one or more factors on the third chromosome.

Therefore, in many of the experiments on genetic assimilation, success was dependent on the presence of a major gene having the desired effect, whose expression in turn was influenced by modifiers. These genes include dumpy and the four venation patterns studied by Bateman (1959a, 1959b), two dominant bithorax alleles (Waddington, 1956) and the recessive bithorax maternal effect (Waddington, 1957a), and the L 4 veinlet gene (Thompson and Thoday, 1975). In some cases, the genes were apparently present in the initial population before selection, either at very low frequencies or below the threshold needed for expression. Failure to produce genetic assimilation of crossveinless in inbred lines (Waddington, 1953; Bateman, 1959b) suggests that favorable mutations of the desired type during the course of the experiments were too rare to be a factor. However, Waddington (1957b) states that the bx^D allele certainly arose by mutation during the course of the experiments and that the dumpy allele and the bithorax maternal effect probably did as well. Because several hundred thousand flies were involved in these experiments, it seems probable that these were merely normal random, spontaneous mutations rather than adaptive mutations induced by and in response to the treatment.

The picture that emerges from these studies is similar to the observations on the evolution of insecticide resistance. Under strong new selection pressures any suitable available genetic variation is utilized, major genes or polygenes or combinations of both types, whose inheritance may be autosomal, sex-linked, dominant, or recessive. Again, perhaps the most surprising discovery is the frequency with which genes of major effect are involved in mediating rapid evolutionary change.

Waddington (1959a) reported a different kind of assimilation experiment. In his previous studies, the traits used were not known to be adaptive in nature, and, moreover, some sort of threshold phenomenon was involved. In this experiment, natural selection rather than artificial selection operated and the trait had no threshold. For this purpose he added sufficient sodium chloride to the normal food medium to kill more than 60% of the larvae. Selection was maintained at a more or less constant level for 21 generations by increasing the salt concentration in later generations to 7%, twice the concentration of sodium chloride in seawater. The area of the larvae's anal papillae was used as a measure of their adaptation to this osmotic stress. The anal papillae were believed to be involved in osmotic regulation of the hemolymph of the larvae, and they were known to increase in size with increasing salt concentration.

After 21 generations of selection, the flies of the selected strains survived better on high salt concentrations than did the unselected controls, and the size of the anal papillae was greater in the selected than in the unselected strains at all salt concentrations. Therefore, the increased size of the anal papillae had been genetically assimilated; and the selected lines responded more than the unselected lines to increased salt concentrations, for their anal papillae increased relatively more in size than those of the unselected controls. This greater response to stress indicates that the selected lines had also improved their adaptability to this stress. The genetic mechanism mediating the genetic assimilation was not reported, but would have been of interest, for these experiments show that genetic assimilation can occur for adaptive traits as well as those that are not clearly adaptive. Nevertheless, to quote Waddington (1961), "We have, therefore, experimental justification for using the notion of genetic assimilation to explain all those evolutionary phenomena which people in the past have been tempted to attribute to the inheritance of acquired characters in the Lamarckian sense."

GENETIC HOMEOSTASIS

Cannon (1929, 1932), a physiologist, coined the word *homeostasis* to describe the relatively stable internal environment maintained in the bodies of higher organisms in the face of varying internal and external conditions. Physiological homeostasis refers to the capacity of an organism to adjust to varying conditions through self-regulating mechanisms that tend to maintain various physiological processes in a steady state. For example, body temperature in warm-blooded vertebrates is maintained at a fairly constant level despite a wide range of external temperatures, and pH, oxygen concentration, and glucose level of the blood are all maintained within rather narrow limits by biological feedback mechanisms that tend to counteract any shifts away from the optimum values.

Later, homeostasis was given new and somewhat different meanings. Physiological homeostasis is a property of individuals as is developmental homeostasis (Lerner, 1954). Developmental homeostasis, which resembles Waddington's (1942, 1957b) "canalization" or "homeorhesis" and Schmalhausen's (1949) "stabilizing selection," does not, however, refer to the maintenance of a steady state, but rather to the attainment of a particular phenotypic outcome of development despite varying conditions during development. The wild-type phenotype of *D. melanogaster*, for example, shows developmental homeostasis or canalization as compared to the phenotype of flies homozygous recessive for vestigial wings. Grown at temperatures from 17° to 30°C, the wild-type flies show much the same-size wings, but the flies homozygous for vestigial, being less well "buffered" or "canalized," show a much wider range of wing sizes, with larger wings in flies reared at higher temperatures. This greater stability in the phenotype of the wild type is said to be the result of *stabilizing selection* or *canalizing selection*.

However, physiological homeostasis and development homeostasis are not equivalent. If urodele larvae are grown under low oxygen tension, the external gills become larger and have thinner walls than they do at normal oxygen concentrations. In this way physiological homeostasis for blood oxygen is maintained by a departure from developmental homeostasis. However, the respiratory rate per unit body weight is reduced, a departure from physiological homeostasis that presumably occurs because developmental canalization is too strong to permit maintenance of the constant rate.

Thoday (1953) introduced another expression, "developmental flexibility," with two rather different connotations. On the one hand, he used it to refer to a genotype that produces different phenotypes in different environments, each phenotype better adapted to the environment that evokes it than the others. On the other hand, he used it to refer to a genotype so buffered against environmental variables that the same adaptive phenotype appears in a range of environments. In the latter sense, developmental flexibility is equivalent to developmental homeostasis or to canalization. Both physiological homeostasis and developmental homeostasis permit the individual to adapt to varying environmental conditions and to survive and reproduce in a wide range of environments. Homeostasis in these senses is closely associated with adaptation.

With the publication of *Genetic Homeostasis* in 1954, Lerner extended the use of homeostasis into a new area, the genetics of populations. In essence, he argued that the gene frequencies in natural populations are in equilibrium. If this equilibrium is disturbed in some way—for example, by artificial selection or by inbreeding—then when the constraint is relaxed, the system of gene frequencies will tend to return to its original equilibrium state as the result of natural selection. Thus, genetic homeostasis is a population phenomenon, comparable to the concept of genetic inertia introduced at about the same time by Darlington and Mather (1949). However, homeostasis seems preferable to inertia because homeostasis carries the connotation of autoregulation but inertia does not. This extended use of homeostasis introduces some complications. Physiological homeostasis and developmental homeostasis are related to individual adaptation, but genetic homeostasis is a population phenomenon. Its relation to adaptation is ambiguous, for genetic homeostasis implies a resistance to changes in gene frequency, which is essentially the antithesis to an adaptive response.

The essence of Lerner's concept of genetic homeostasis is as follows. In cross-fertilizing species, evolution has led to developmental canalization or developmental homeostasis so that the individuals of a given population display quite uniform phenotypes even though they are genetically variable. Moreover, evolution has produced genetic homeostasis in breeding populations so that they tend to retain the gene pool that confers maximum average fitness in the environments in which these populations exist. The crux of Lerner's argument is that the mechanism for both developmental homeostasis and genetic homeostasis is the superior fitness of heterozygotes as compared to the corresponding homozygotes. In other words, he postulated that heterozygotes would show a greater degree of canalization or developmental homeostasis than homozygotes and that genetic homeostasis is based on natural selection eliminating the more extreme homozygotes from the population but favoring the phenotypically intermediate heterozygotes. If so, then there should be some optimum level of heterozygosity in a breeding population. At this level developmental homeostasis will lead to uniformity of adaptive phenotypic expression with minimum reproductive waste, and yet genetic homeostasis will permit retention of sufficient genetic variability to protect the population from drastic environmental change by enabling it to respond to new selective pressures.

Lerner (1954) adduced considerable evidence in support of his views that heterozygotes showed greater canalization than homozygotes and that genetic homeostasis existed in natural populations. Others subsequently reported similar findings (summaries in Falconer, 1960, and Mayr, 1963). However, still others have shown that a decline in developmental homeostasis was not caused by increasing homozygosity, for the same decline was seen in crosses between the selected lines (e.g., Thoday, 1958b). Moreover, the precise meaning of genetic homeostasis and its relation to other forms of homeostasis have been subjected to criticism (Waddington, 1957b; Lewontin, 1957). Muller (1958) expressed his opinion of the concept in rather strong terms. "In its application to matters of genetic variation, the term homeostasis has recently been given a special meaning, whereby it denotes an essentially mystical doctrine, representing a revival from pre-Mendelian times. According to this doctrine, an organism's vigor is *per se* enhanced as a result of the hereditary elements derived from its two parents being unlike one another." This statement apparently was a reaction to the idea that genetic homeostasis involved an optimum level of heterozygosity without any particular regard to the nature of the loci that were heterozygous. Although Lerner's work on genetic homeostasis aroused considerable interest in the decade following its publication, recently it has received very little attention. With the discovery of the high levels of protein polymorphism in natural populations, for which adequate explanations are still being sought, it is perhaps surprising that there has not been a revival of interest in genetic homeostasis.

CHAPTER 6

Balanced Polymorphism

In the literal sense, polymorphic means having many forms. Thus, not all polymorphisms in natural populations are necessarily genetic. In species with several generations per year, the different environmental conditions at different seasons may cause individuals with different phenotypes to develop, even though their genotypes are essentially the same. Furthermore, individuals developing at the same time, but under different environmental regimes, may have different phenotypes. In some species juvenile and adult forms (bird plumage, for example) may be quite distinct and therefore polymorphic. These developmental and environmental polymorphisms may lead to some confusion in the study of natural populations because they may be mistaken for genetic polymorphisms.

Genetic polymorphisms may be genic, under the control of segregating alleles, as in the blood groups in humans, or they may be chromosomal, as in inversion polymorphisms in *Drosophila*. The observation of distinct forms under genetic control in a natural population is no assurance that it represents an example of genetic polymorphism in the sense of Ford's original definition (1940b, 1971, and Chapter 3). The situation may be due simply to a Castle-Hardy-Weinberg equilibrium, with all the forms having equal fitness, or it may represent a balance between mutation and selection, or it may result from a balance between migration and selection, with a gene favored in one locality but not in an adjacent area. None of these three phenomena fits Ford's definition of genetic polymorphism;

nevertheless, they may be responsible for the observed polymorphism. We dealt in Chapter 5 with transient polymorphism, which Ford did include in his definition, but this is apt to be relatively rare because of its transient nature. Most genetic polymorphisms are apt to be *balanced* and stable, maintained in some way by opposing selective forces.

Recall that in Chapter 3 we defined a polymorphic locus as one at which the frequency of the most common allele does not exceed 99% rather than using Ford's definition that "Genetic polymorphism is the occurrence together in the same locality of two or more discontinuous forms of a species in such proportions that the rarest of them cannot be maintained merely by recurrent mutation." Ford's definition could be improved by using "in the same breeding population," rather than "in the same locality," because migratory birds of the same species but from different regions may overwinter in the same locality but never interbreed. He specifically intended (1971) to exclude geographic races of the same species so that he regarded genetic polymorphism as an intra-population phenomenon maintained by some form of *balancing selection*.

Ordinarily, it is relatively simple to distinguish between developmental and environmental polymorphisms on the one hand, and genetic polymorphism in the broad sense on the other. The discovery of the tremendous amount of genetic variability in natural populations, especially through gel electrophoresis, has led to intensive efforts to discover the underlying mechanisms maintaining this variability in natural populations. In fact, much of the research in population genetics over the past decade has been directed toward this problem. It may be difficult to distinguish balanced polymorphism because of the time required to see whether gene frequencies are changing. It is also difficult to determine whether the genetic polymorphism is actually caused by some form of balancing selection or by a balance between mutation and selection, or between migration and selection, or merely represents an array of neutral genes in Castle-Hardy-Weinberg equilibrium. Therefore, even though there can no longer be doubt about the great amount of genetic variability in wild populations, there is considerable doubt about the role it plays in these populations.

HETEROZYGOUS ADVANTAGE

If balanced polymorphism is defined as a stable equilibrium of two or more genetic alternatives in a breeding population at such frequencies that the most common does not exceed 99%, the possible ways such an equilibrium might be maintained can now be explored. In many cases, there is a strong tendency to equate balanced polymorphism with heterozygous advantage. For example, Ford (1971, p. 100) wrote, "The most general basis of genetic polymorphism is a balance of opposed advantage and disadvantage such that the heterozygote is favoured compared with either homozygote." In other words, heterosis is thought to be the cause of balanced polymorphism in most cases. The evidence that heterosis is indeed the mechanism maintaining a balanced polymorphism is often not especially satisfactory. In fact, the very existence of balanced polymorphism sometimes seems to be regarded as *ipso facto* evidence for heterozygous advantage. However, a number of other possibilities exist.

FREQUENCY-DEPENDENT SELECTION

Recently, frequency-dependent selection has come to the fore as an alternative to heterozygous advantage. This theory postulates that selective values are not constant, but change as relative gene frequencies in the population change. One possibility is that a rare genotype would be selected against more strongly the rarer it becomes, which would hasten its elimination from the population. Another possibility is that the rarer genotype would always be favored by natural selection. If this were so, it would lead to a balanced polymorphism with a stable equilibrium point at which the various types will be equal in fitness. If the equilibrium is disturbed, the rarer type, whichever it may be, will be favored by selection, and will return to the equilibrium point and neutrality. This theory has considerable appeal as an explanation of balanced polymorphism because, at equilibrium, all genotypes have equal fitness and there is no genetic load comparable to that generated by the less fit homozygotes under the theory of heterozygote advantage.

Kojima and his associates have recently reported frequency-dependent selection acting on allozyme polymorphisms (Kojima and Yarbrough, 1967; Kojima and Tobari, 1969; Huang, Singh, and Kojima, 1971; Kojima, 1971). The type of data obtained is shown in Table 6-1, for the fast (F) and slow (S) alleles at the esterase-6 locus in *D. melanogaster*, when grown on preconditioned medium. However, such results have not always been observed (Lewontin and Matsuo, 1963; Yamazaki, 1971), so the generality and importance of the phenomenon are questionable.

Table 6 - 1.   Relative Viability of Genotypes of the Esterase-6 Locus
in *Drosophila melanogaster* on Preconditioned Medium.

Tested Genotypes	Preconditioning Genotypes		
	FF	FS	SS
FF	0.923 ± 0.053	1.068 ± 0.056	1.130 ± 0.058
FS	1.090 ± 0.057	1.000 ± 0.057	1.087 ± 0.057
SS	1.146 ± 0.059	1.078 ± 0.057	0.928 ± 0.053

Source: Huang, Singh, and Kojima, 1971.

Another form of frequency-dependent selection is the rare-male mating advantage (Petit and Ehrman, 1969; Spiess, 1970) shown in various species of *Drosophila*. Here, the rarer of the two male types has been shown to be relatively more successful in mating when the males were of two different mutant types, or one was mutant and the other wild type, or when they carried different gene arrangements (inversions), or they came from strains of different geographic origin, or even if they came from the same strain but were reared at different temperatures. To show that the phenomenon was dependent on olfactory cues, Ehrman and Probber (1978) placed large numbers of males, or dead males, or extracts of dead males of one inversion type in the lower chamber of a double Elens-Wattiaux observation apparatus, and an equal number of male and female pairs of two different inversion types in the upper chamber, which was separated from the lower chamber by a piece of cheesecloth. The Chiricahua (CH) and Arrowhead (AR) gene arrangements of *D. pseudoobscura* were used. If the males in the lower chamber were of the CH type, the females in the upper chamber treated them as part of the total population, for they mated significantly more often with the AR males even though the number of AR and CH males in the upper chamber was equal. If AR males were placed in the lower chamber, the females then preferred to mate with the CH males. The similar results obtained when dead males or male extracts were used eliminated the possibility that vibrational or auditory cues were involved rather than olfactory cues. This frequency-dependent type of mating should lead to a genetic equilibrium, a balanced polymorphism not dependent on heterosis.

One of the intriguing aspects of this study is that it indicates that the sensory apparatus of the females enables them in some way to distinguish between males carrying different chromosome inversions. Presumably, the genetic contents of the inversions differ in a way reflected in the phenotype and detectable by the females. However, the females must not only be able to distinguish between the male types, but they must also be able to determine which type is common and which type is rare and be more sexually receptive to the rare type of male. It is not simply a matter of the female identifying and preferring to mate with a particular male type, as in the case of positive or negative assortative mating. The female must identify the male types, somehow quantify their numbers, and then mate with the rare-male type, no matter whether the male's genotype is like her own or different.

Another way that frequency-dependent balanced polymorphism could arise is if predators tended to form a search image for the more common form in a polymorphic population and preferentially preyed upon that form until it became rare. At some point, they would start to learn a new search image for the previously rare type as it became relatively more abundant and then switch to this type of prey.

COMPULSORY CROSS-FERTILIZATION

Enforced out-crossing is still another way that balanced polymorphism may be maintained. In hermaphroditic plants, for example, self-sterility alleles prevent selfing. These systems of multiple alleles may include large numbers of alleles, say, s_1, s_2, s_3 . . . s_n. All plants will be heterozygous, as for example, s_1s_2, s_1s_3, s_2s_3. A pollen grain bearing the s_1 allele will not grow on the stigma of any plant whose genotype includes s_1. Therefore, s_1 pollen will not grow on s_1s_2 or s_1s_3 plants, but will grow on s_2s_3 plants. Therefore, no homozygotes can be formed, a balanced polymorphism will be maintained, and since the rarer an allele is in the population, the more plants it can fertilize, a form of frequency-dependent selection is operative as well.

A somewhat different genetic mechanism is found in the primrose (*Primula vulgaris*) to ensure cross-pollination. This species is polymorphic for two types of flowers called pin and thrum because of the different relative positions of the anthers and the pistil. Flower type is controlled at the P locus, with pin flowers in homozygous recessive (pp) plants and thrum flowers in the heterozygotes (Pp). Thrum pollen is of two types, P or p, both of which will grow on pin plants, but not on thrum. The pollen from pin is all p and grows well down the styles of thrum plants, but grows so slowly on pin styles that it can not compete with pollen from thrum plants. Thus, the genetic mechanism in the primrose is different from self-sterility alleles, but the effect is the same—it ensures out-crossing and a balanced polymorphism.

In species with separate sexes, negative assortative mating, that is, preferential mating between individuals of different genotypes, will also lead to balanced polymorphism. This type of mating system is also a form of frequency-dependent selection, for rare males and females have a greater number of potential mates. "One-sided mating preferences" have frequently been reported (Dobzhansky and Mayr, 1944; Dobzhansky and Streisinger, 1944; Dobzhansky, 1944; Merrell, 1950, 1960), which is a form of selective mating with one male type, say, more successful in mating with either female type than the other type of male. Positive-assortative mating gives rise to sexual isolation, of which many examples are known, of course. However, despite the folklore that opposites attract, examples of true negative-assortative mating are rare or nonexistent so that this mechanism for maintaining balanced polymorphism must be regarded as theoretical. Perhaps it should be added that the rare-male mating advantage is a form of one-sided mating preference rather than negative-assortative mating.

Finally, the separation of the sexes is itself a mechanism that prevents selfing, ensures outcrossing, and maintains a form of balanced polymorphism involving the sex chromosomes and probably many other loci as well.

OPPOSING SELECTION PRESSURES

Another theoretical possibility for the maintenance of balanced polymorphism is where different selection pressures operate on males and females (Owen, A. R. G., 1953; Kidwell, Clegg, Stewart, and Prout, 1977), the simplest case being where an allele is favored in one sex but selected against in the other. For example, conspicuous coloration in male birds might aid their success in courtship and mating, but be maladaptive in females if it made them vulnerable while nesting.

Still another theoretical possibility for maintaining balanced polymorphism would be situations in which an allele favored at one stage of development is selected against at another. Meiotic drive, perhaps, can serve as an example; but it is a rather unusual phenomenon, and other more general examples may well be found. In forms with larval and adult stages, the selection pressures operating at the different stages of the life cycle may be quite different. However, it should be added that it seems probable that in this situation as in the case of differing selection pressures in males and females above, selection will operate to resolve these problems through the suppression of gene action in one of the stages or through developing separate systems of genes to control development in the different stages or sexes.

MEIOTIC DRIVE

Meiotic drive was originally defined by Sandler and Novitski (1957) as any alteration of the normal process of meiosis such that a heterozygote produces an excess of one of the two possible types of gametes. If unchecked, this pattern of gamete formation would soon lead to the fixation of the favored or *driven* allele. However, in the cases studied, the driven allele has been found to be associated with a loss of fitness at some other stage of development, and, hence, the known cases of meiotic drive are examples of balanced polymorphism. However, more recently, the concept has been broadened to include transmission anomalies that are not caused by meiotic disturbances as such, but have similar effects in the population. Included are such phenomena as gametic selection and gametic competition. Meiotic drive has similar effects and may be difficult, in practice, to separate from them. However, the latter are selection phenomena in the gametic or haploid phase and are dependent on the gene contents of the gamete, whereas meiotic drive is independent of the gene content of the gamete in the usual sense.

The first well-studied instance of meiotic drive was the sex-ratio X chromosome reported in *D. obscura* by Gershenson (1928). He discovered that males carrying the sex-ratio X chromosome produced primarily or exclusively female progeny. Subsequently, sex-ratio (*sr*) chromosomes were found in a number of *Drosophila* species, and in all cases, the *sr* chromosomes were associated with inversions (Sturtevant and Dobzhansky, 1936). Since there was no evidence of excess zygotic mortality, the males with the *sr* chromosome were apparently producing mostly X-bearing sperm. Gershenson pointed out that, given these properties, the *sr* X chromosome would eventually replace other X chromosomes and the species could run into difficulty because of the lack of males. However, the sex-ratio condition has apparently existed for some time in *Drosophila*, for it is widespread in natural populations (Stalker, 1961). The sex-ratio example in *Drosophila* is a case of genic meiotic drive regulated by the *sr* factor on the X chromosome.

The mechanism of *sr* is unclear. Sturtevant and Dobzhansky

(1936) reported that in meiosis in *sr* males, the autosomes paired but the X and Y chromosomes remained unpaired. Furthermore, they reported that the X chromosome replicated twice during meiosis I rather than once as is normal, and an X chromosome passed to each pole at anaphase I. The Y chromosome did not line up on the equatorial plate, and, subsequently, formed a micronucleus in the cytoplasm of one of the daughter cells, which later degenerated. In this way, virtually all X-bearing sperm cells were produced. A later study by Novitski, Peacock, and Engel (1965) led to conflicting observations. They reported normal pairing and disjunction of X and Y chromosomes at the first meiotic division in *sr* males, but the Y-bearing secondary spermatocytes degenerated at metaphase II or anaphase II. Further analysis of complete anaphase II cysts revealed that the X-bearing cells were normal, but the Y-bearing cells had only autosomes at the poles, the Y remaining as a chromatin mass at the equator because it showed no centromeric activity. Obviously, the two accounts differ significantly and are difficult to reconcile. In either case, it remains an example of genic meiotic drive associated with inversions on the X chromosome and observed in natural populations of a number of species of *Drosophila*.

Rhoades, however, reported (1942, 1952) chromosomal meiotic drive in maize, involving an abnormal chromosome 10. This chromosome, which has a large terminal knob, is recovered in expected frequencies through the heterozygous pollen parent, but the knobbed chromosome is present in about 70% of the female gametes produced by heterozygotes. Analysis showed that a preferential segregation occurred at anaphase II so that the knobbed chromatids were segregated to the outer nuclei of the linear tetrad of megaspores. Since the basal megaspore always becomes the functional egg nucleus, this preferential segregation accounts for the excess of knobbed chromosomes in the female gametes. This type of chromosome behavior has been found in maize populations in the southwestern United States and in Latin America.

A comparable phenomenon has been observed in *D. melanogaster* females by Novitski (1951, 1967), where, if heteromorphic homologues were present, the two shorter chromatids tended to wind up in the two outermost nuclei of the linearly arranged products of meiosis, and the two longer chromatids tended to lag behind in the two interior nuclei. Since, in *Drosophila* females, one of the two outer nuclei becomes the egg nucleus, this chromosome behavior meant that a disproportionate number of the shorter chromatids were incorporated in the egg nucleus, and an excess of the longer chromatids wound up in the polar bodies. This, too, is a form of chromosomal meiotic drive, for the chromatid behavior in these asymmetrical dyads was dependent on their length and not on their genetic contents.

Another example of meiotic drive, very widespread in natural populations of the house mouse, *Mus musculus*, has been extensively studied (Dunn, 1955, 1956, 1957a, 1957b; Lewontin and Dunn, 1960; and others). This case involves abnormal segregation ratios at the T locus in the 9th linkage group. The dominant T allele causes the short-tailed or brachyury condition when heterozygous (T/t^+) and is homozygous lethal (T/T). Widely distributed in natural populations are a number of recessive t alleles, which are usually homozygous lethal or male sterile (t/t). While the t alleles are recessive to t^+—that is, t^+/t mice are normal—the T/t heterozygotes are tailless. Moreover, heterozygotes for different lethal t alleles (tI/tJ) are usually normal, suggesting that this locus is in some way compound (Dunn, 1956, 1957b). An abnormal chromosome region rather than a point locus seems to be involved because it generates new alleles fairly often and influences crossing-over in its vicinity (Lyon and Meredith, 1964). Given these properties, it may seem surprising that the recessive t alleles are so common in wild populations. The reason is that the male heterozygotes (but not the females) show aberrant gametic ratios with as high as 95% of the sperm cells bearing the t allele, and only 5% the other allele. Thus, t/t^+ females produce 50% t and 50% t^+ gametes, but t/t^+ males produce up to 95% t gametes and as low as 5% t^+ gametes (Dunn, 1957a). The most efficient test cross is a tailless male (T/t) to a homozygous wild-type female (t^+/t^+) in which the expected ratio is 50% short-tailed (T/t^+) to 50% wild type (t^+/t). The observed proportion of wild type usually exceeds 50%, and for t alleles from natural populations, it often exceeds 90% and indicates a very strong form of drive favoring these otherwise deleterious recessive t alleles. Whether it is meiotic drive remains an open question, for there is some evidence that the effect of the t alleles is exerted on the sperm cells between meiosis and fertilization (Yanagisawa, Dunn, and Bennett, 1961). This phenomenon certainly qualifies as a mechanism for maintaining a balanced polymorphism. The other lesson it teaches is that deleterious and even lethal genes may reach high frequencies in natural populations. Clearly, it does not lead to adaptation, and if widespread, it could be disastrous to populations containing driven deleterious alleles. Because mouse breeding populations are usually small, t alleles are constantly being eliminated in homozygotes; even so, the populations may contain up to 50% heterozygotes (Lewontin and Dunn, 1960). Dunn (1957a) called the "male segregation ratio," such as that generated by the t alleles, a fifth independent evolutionary force in addition to mutation, natural selection, migration, and random genetic-drift. However, like Darwin's sexual selection, it generally seems to be treated under the rubric of natural selection rather than as a separate evolutionary force, and can be regarded as a form of prezygotic selection.

Another case of meiotic drive widespread in natural populations is Segregation-distorter (SD) in *D. melanogaster*, first reported by Sandler, Hiraizumi, and Sandler (1959). SD has been localized in or near the centromeric heterochromatin of chromosome II and is (with apparently one exception) invariably associated with chromosome inversions on II, but not with any particular inversions. Segregation distortion occurs in heterozygous (SD/+) males, but segregation in heterozygous females is normal. The proportion of SD-bearing progeny recovered from heterozygous males may be as high as $k = 1.0$, but k is always 0.5 from SD/+ females. The phenomenon is not caused by zygotic mortality of SD^+-bearing flies, however.

It was formerly thought that normal synapsis between the SD-bearing and the SD^+-bearing homologues was necessary for segregation distortion to occur, but more recent work (Trippa et al., 1974) indicates otherwise. Considerable research has been done on segregation distortion and has been reviewed, for example, by Zimmering, Sandler, and Nicoletti (1970), Peacock and Miklos (1973), and Hartl and Hiraizumi (1975). The genetic situation is apparently complex and involves at least two major loci and a number of modifiers. The major loci are SD and a closely linked locus first called AC for activator, and more recently (e.g., Hartl, 1975), Rsp for responder or receptor. In addition to these loci and inversions, the SD chromosomes carry a polygenic system of enhancers and suppressors of SD. Hartl (1975) has shown that SD and Rsp must be in coupling phase (SD Rsp/+ +) for an excess of SD-bearing offspring to be produced. If they are in repulsion (SD +/+ Rsp), there will be a gross deficiency of SD-bearing offspring. Either one acting alone (SD+/++ or + Rsp/++) will give a normal one-to-one segregation ratio.

The segregation-distortion effect has been traced to a breakdown of the SD^+-bearing gametes in the spermatid stage (Hartl, Hiraizumi, and Crow, 1967; Peacock and Miklos, 1973). Within the packets of 64 spermatids located in the testis of SD/+ males, 32 normal and 32 abnormal spermatids have been observed. Thus, SD induces sperm dysfunction of the SD^+ sperm, apparently in the early primary spermatocyte by an unknown mechanism.

Here again is a fairly common and widespread feature found in natural populations all over the world (Zimmering, Sandler, and Nicoletti, 1970). If SD were as fit or fitter than the undriven alleles, it should become fixed; but it is associated not only with inversions but also often with lethals, so that it is not fixed but forms a balanced polymorphic system. Since the overall fitness of the population usually suffers when SD is present (Hiraizumi, 1962), selection could be expected to counteract the effects of SD. This is true, for in Japanese wild populations of *D. melanogaster* lacking SD, the SD+ alleles are very sensitive to the effects of SD on test, giving such k values as 0.99 or 1.00 (Hiraizumi, Sandler, and Crow, 1960). SD+-alleles from Wisconsin populations containing SD were much less sensitive to SD, giving a k value of only 0.83. Therefore, the deleterious effect of SD has been somewhat reduced in the Wisconsin populations by the presence of insensitive SD+ alleles.

As a cautionary note, consider the aberrant sex ratios, also observed in *Drosophila*, which appeared among the progeny of certain females. This phenomenon, which seemed to parallel in females the *sr* trait in males, resulted in nearly 100% female progeny from affected mothers. This condition is due to an infectious microorganism, which caused the zygotic death of male offspring and was transmitted cytoplasmically from a mother to her offspring (Poulson and Sakaguchi, 1961; Poulson, 1963). It was even possible to transmit the condition by injection, as in the case of any infectious agent, even though the agent is not contagious. The organism occurs naturally in several species of *Drosophila* and was artificially transferred to *D. melanogaster*. Its effects were influenced by both the genotype of the host and its own genotype. Before the discovery of the microorganism, the maternal *sex-ratio* condition was variously attributed to a plasmagene, a cytoplasmic particle, or a virus. Obviously, caution and thorough investigation are needed in the interpretation of aberrant sex ratios. One of the more surprising aspects of this relationship is that it is found in so many different species of *Drosophila* despite its seemingly adverse effects.

THE SEX RATIO

The sex ratio at fertilization is known as the primary sex ratio. With chromosomal sex determination and normal segregation of the X and Y chromosomes, the expected primary sex ratio is one-to-one. However, because the males and females may subsequently suffer differential mortality, the sex ratio is not necessarily constant, but may vary. Secondary sex ratios may be determined at the time of birth (hatching), or weaning (fledging), or when parental care ceases, or at sexual maturity. In *The Descent of Man and Selection in Relation to Sex*, Charles Darwin (1871) stated his belief that natural selection must somehow influence the sex ratio, but found the subject too intricate to handle. Subsequently, R. A. Fisher (1930, 1958) dealt with it in more detail.

Fisher based his argument on the relation between all sorts of "parental expenditure" on the offspring before they can lead an independent existence and the "reproductive value" of these offspring when parental expenditure ceases. He stated that when parental care ended, the "total reproductive value of the males in this group is exactly equal to the total value of all the females, because each sex must supply half the ancestry of all future generations of the species. From this it follows that the sex ratio will so adjust itself, under the influence of Natural Selection, that the total parental expenditure incurred in respect of children of each sex, shall be equal." Thus he argued that natural selection operated to equalize the parental expenditure on their male and female offspring and that this, in turn, would produce a one-to-one sex ratio at the time parental expenditure ceased. He further stated that differential mortality between the "period of dependence" and the "attainment of maturity" could affect the "relative numbers of the sexes without compensation." Having reached this conclusion, however, he seemed to back away from it, for he added, "Any great differential mortality in this period (between dependence and maturity) will, however, tend to be checked by Natural Selection, owing to the fact that the total reproductive value of either sex, being, during this period equal to that of the other, whichever is the scarcer, will be the more valuable, and consequently a more intense selection will be exerted in favour of all modifications tending towards its preservation." Thus, having first concluded that selection will act to produce a one-to-one sex ratio at the time parental expenditure ceased but not beyond that point, he later seemed to conclude that selection would continue to maintain a one-to-one sex ratio to the time of sexual maturity. Note that he set forth two rather different interpretations of how natural selection might influence the sex ratio. One, based on parental expenditures on male and female offspring, could be considered ecological; the other, based on genetic contributions made by each sex to the next generation, could be considered a genetic hypothesis. Subsequent discussions have centered primarily on one or the other of these hypotheses.

Shaw and Mohler (1953), however, dealt only with the primary sex ratio. They argued that if the primary sex ratio is not 0.5, selection acting on autosomal genes affecting the sex ratio will cause the primary sex ratio to approach 0.5. In their treatment, they purposely ignored parental care, intergroup selection, and sex-linked genes.

Following Fisher's lead, several authors have carried the idea of parental expenditure even further. Bodmer and Edwards (1960) put Fisher's theory on an "analytic basis" and showed that parental expenditure on each sex should be equal and that the sex ratio should be about one-half when parental care ceases. Kolman (1960), Verner (1965), and Leigh (1970) reached the same conclusion, and Leigh added that differential mortality between the sexes "cannot possibly affect the neonate sex ratio." Trivers (1972) wrote in terms of "parental investment" in the young, whereas Emlen (1973) referred to the "energy" expended. Emlen (1968a, 1968b), however, felt that differential mortality of the sexes at ages beyond the end of parental care could affect the evolution of the primary sex ratio. Both Mac-Arthur (1965) and Emlen (1973) concluded that the best strategy for a breeding pair to follow is to maximize the value of the product $m_x f_x$, where m_x and f_x are the values for males and females, respectively, of "reproducing age." However, Emlen (1973) then attempted to reconcile this statement with the idea of parental energy expenditure and later stated that "a breeding pair will contribute maximally to the ancestry of future generations by equalizing the total amount of energy spent on raising males and on raising females, including those individuals which die before reaching independence." According to this ecological approach, natural selection equalizes parental investment or energy expenditure on their male and female progeny and does not equalize the sex ratio itself (Trivers and Willard, 1973). If the parental investment per male and per female is widely different, the sex ratio could deviate greatly from equality.

Genetic theories suggest that natural selection will act directly on the sex ratio. Crew (1937) argued that selection will produce equality in the sex ratio among those at the "threshold of their reproductive prime" because this would provide the greatest opportunity for genetic recombination. Kalmus and Smith (1960) wrote that the function of sexual reproduction is to reduce inbreeding as much as possible, and that the greatest chance of encounters between males and females and the least amount of inbreeding will occur when the numbers of males and females are equal. Again, this reasoning applied not to the primary sex ratio, but to the sex ratio at sexual maturity. In general, genetic theories, including Fisher's afterthought, conclude that natural selection will make the sex ratio equal at the time of reproduction, rather than at fertilization, birth, weaning, or when parental care ceases.

The reasoning behind this conclusion is relatively simple. First, of the gametes making up the next generation, half must come from the male parents and half from the female. For example, if there are 500 offspring, they carry 1000 gametes, 500 from their fathers and 500 from their mothers. If the parental generation consisted of 50 males and 100 females, with random mating and an equal genetic contribution from each individual for its sex, then each male would contribute 500/50 or 10 gametes/male to the next generation, but each female would contribute 500/100 or only 5 gametes/female to the next generation. Therefore, males are twice as successful as females in making a genetic contribution to the next generation, and any individuals carrying genes that shift the sex ratio in their progeny toward an increased proportion of males would be favored by natural selection. This, too, is a form of frequency-dependent selection, for, as the sex ratio approaches equality, these genes will lose their selective advantage. Conversely, if males were in excess, selection would favor the production of a greater proportion of females. Since the rare sex would always be favored by selection under this theory, a balanced 1:1 sex ratio should exist at the age of reproduction, but not necessarily at other stages of the life cycle.

If natural selection does, in fact, control the sex ratio, a pertinent question is whether selection is more apt to act directly on the sex ratio or indirectly on the sex ratio through parental energy expenditure. Since genetic mechanisms are known that directly modify the sex ratio, it seems reasonable to assume that natural selection directly controls the sex ratio. If the genetic theory of the control of the sex ratio is correct, selection might produce widely different parental energy expenditures on the two sexes to equalize the sex ratio and their genetic contributions to subsequent generations. On the other hand, if selection produces equal energy expenditures on the two sexes, this may result in highly skewed sex ratios. It would be very helpful to have reliable data on both sex ratios and on parental expenditures to resolve this question; but, at present, reliable data of either type are rare. (Note that even though the sex ratio is a group characteristic, both of these theories involve individual selection rather than group selection.)

In humans, the sex ratio at birth is about 105 males to 100 females and is thought to be even higher in favor of males at earlier stages. By sexual maturity, the human sex ratio is very close to equality because of higher male mortality. In the United States, this trend continues so that among older people, there is an excess of females. This excess is probably a recent phenomenon related to medical progress and changes in the birthrate. In New England churchyards, it is not uncommon to find a husband buried with two or three wives, who often had died of conditions related to childbearing.

Reliable sex-ratio data are not often available for other species (Trivers, 1972; Selander, 1972). One major reason is the sampling problem associated with the different bahavior of males and females of the same species. For example, leopard frogs (*Rana pipiens*) in Minnesota have approximately a one-to-one sex ratio during most seasons, but at breeding ponds during the breeding season, about 25 males are captured for every female. The males are very conspicuous with their calls and aggressive approach to other individuals, whereas the females are very circumspect and secretive in their approach to the breeding area. Similar problems exist in many other species (Trivers, 1972). In general, however, sex ratios appear to approximate equality even though some rather aberrant ones have been reported (Hamilton, 1967). Moreover, there is not necessarily a direct relationship between the sex ratio and the mating system of a species (Willson and Pianka, 1963), for not all sexually mature individuals will necessarily breed.

HETEROGENEITY IN TIME OR SPACE

Another mechanism by which a balanced polymorphism might be sustained is seasonal selection. One of the first cases to be reported was color polymorphism in the lady-bird beetles (*Adalia bipunctata*) near Berlin (Timoféeff-Ressovsky, 1940a, 1940b). Timoféeff-Ressovsky sampled the population in April and October over several years, and found that the frequency of the black morph ranged from 55 to 70% in the fall but declined to 30-45% by spring. The black morph apparently had a higher mortality rate than the red morph during winter, but they survived summer conditions better than the red form so that by fall, their frequency again rose to 55 to 70%. Timoféeff-Ressovsky interpreted these results to indicate that seasonal selection was maintaining this balanced polymorphism; but Mayr (1963), quoting Kimura (1955), stated that such a fluctuating polymorphism could not be maintained by this balance of selective advantages because such a system would, sooner or later, inevitably lead to fixation of one or the other type owing to random fluctuations unless it was also maintained by heterozygote superiority (Dobzhansky, 1943, 1956; Dubinin and Tiniakov, 1945). However, Kimura treated only two special cases of cyclic changes in selection intensity, one with many generations per cycle and the other with a cycle of just two generations. For the former, he concluded that the most efficient mode of adaptation would be the "direct change of the phenotype" rather than "genetic 'coadaptation'." For the latter, Kimura showed that if the selection coefficient, s, is small, and the gene frequency is close to zero or to one, the gene will ultimately be lost or fixed. However, Timoféeff-Fessovsky's data showed that the gene frequencies were intermediate, and the seasonal changes in color-phase frequency were so large that the selection coefficients must have been large rather than small. Hence, Kimura's conditions do not apply, and it seems reasonable to suppose that regular, seasonal, or cyclical reversals in selection pressure can maintain a balanced polymorphism even without an assist from heterozygous advantage. Seasonal selection, for example, seems to be involved in the polymorphism for the unspotted Burnsi form of leopard frog (*R. pipiens*) in Minnesota and adjacent areas (Merrell and Rodell, 1968; Dapkus, 1976). However, even though Burnsi is controlled by a single autosomal dominant gene, there was no evidence for heterosis in the heterozygotes among the progeny from crosses (Merrell, 1972). Dempster (1955), Mather (1955), and Haldane and Jayakar (1963) have postulated theoretical situations in which seasonal, temporal, or cyclical changes in selection pressure could maintain polymorphism without heterosis. More recently, Hedrick (1974), Gillespie (1975), Hoekstra (1975), and Hedrick, Ginevan, and Ewing (1976) have also considered temporal changes in selection pressure in relation to balanced polymorphism. In general, they conclude that random changes in the selection coefficients are eventually apt to lead to fixation, but that regular cyclical reversals in selection pressure are more favorable for the maintenance of balanced polymorphism.

If the environment is heterogeneous, different genotypes may be favored by selection in different subniches. This is sometimes called *disruptive* or *diversifying* selection. Ludwig (1950) was one of the first to point out that environmental heterogeneity could lead to balanced polymorphism, even in the absence of heterosis. This concept is consequently sometimes called the Ludwig effect or Ludwig's Theorem. Subsequent treatments, offered by Levene (1953), Mather (1955), Dempster (1955), Maynard Smith (1962, 1970), Levins and MacArthur (1966), Levins (1968), Prout (1968), and Wallace (1968a), are primarily theoretical, and the assumptions differ in the different papers. In general, they all concluded that if, in a genetically variable population, one genotype was best adapted to one subniche and another to a different subniche, and so on, and that mating was random (or else a significant amount of gene flow continued between subniches), then a balanced polymorphism could exist. This situation might develop if an insect species lays its eggs on several different host plants, for example. However, clearcut cases of this sort of balanced polymorphism have rarely been reported, the reason being the difficulty in determining the mating patterns in natural populations or in measuring the amount of gene flow. More recent theoretical treatments are to be found in Christiansen (1974, 1975), Gillespie (1974a, 1974b), and Hedrick, Ginevan, Ewing (1976). Dobzhansky (summarized in 1951) suggested that each of the different inversion types in *D. pseudoobscura* was adapted to a different subniche within the general environment of that species. However, other interpretations have been offered (Epling, Mitchell, and Mattoni, 1953; Carson, 1958b) for the greater amount of inversion polymorphism found at the center of a species' range compared to that at the periphery.

If gene flow is sufficiently restricted among the populations occupying different subniches, then disruptive or diversifying selection has been postulated to cause sympatric speciation. However, this controversial topic is better deferred until the discussion of the origin of species.

CHAPTER 7

Polymorphism and Population Dynamics

POPULATION NUMBER

Before continuing the discussion of polymorphism, we must first consider the factors influencing the numbers of individuals in a given species. To deal with this, some definitions are required. One word widely used and often undefined is *population*. For the purposes of this discussion, a *population* is a group of organisms of the same species occupying a particular space at a particular time. Some of the attributes of a population are derived from the nature of its individual members. Other attributes are unique to the group as a whole and are not characteristics of individuals. Some characteristics of populations are the following:

a. Population size
b. Population density
c. Birth rate
d. Death rate
e. Age distribution
f. Dispersion
g. Population growth pattern
h. Biotic potential
i. Reproductive fitness
j. Genetic variability

Because these are group characteristics, they are usually best expressed in statistical terms.

Population size can be defined as the total number of individuals in a population at a particular time. *Population density* refers to the number of individuals per unit of area or volume that they occupy. Population size is of special interest in population genetics as well as in population ecology, whereas population density is more often of interest in population ecology.

How does one proceed to determine the number of individuals in a population? Suppose, as an assignment, you were asked to determine the total number of people living in the eight-county Metropolitan Mosquito Control District, which coincides, more or less, with the greater metropolitan area of the Twin Cities of Minneapolis and St. Paul; or, suppose you were asked to determine the total number of mosquitoes in the Metropolitan Mosquito Control District. How would you proceed? Both of these questions represent real problems. The human census is carried out by the U.S. Bureau of the Census, the mosquito census by personnel from the Mosquito Control District.

For the human census, an effort is made to count all individuals of all ages. The labor involved is so great that the census is taken only once every ten years, and special teams of workers are assembled and every block is canvassed. Hospitals and funeral homes have to be monitored for births and deaths. In addition, special efforts must be made to deal with the transients who occupy motels, hotels, hospitals, colleges, and various other institutions. A complete census of this sort is possible only because individual members of the population can be identified. For the mosquitoes, a complete census of all individuals is clearly out of the question. For one thing, one would then have to get data on all the eggs, larvae, and adults—an impossible task. Where the data are fairly complete, as in humans, it is possible to get not only an estimate of population size, but also an estimate of population density by dividing N, the population size, by the area occupied by the population (A)—that is, the population density equals N/A. For human populations, the density is usually expressed as the number of people per square mile.

In some cases it may be impractical to count all of the individuals in all stages of development. For example, the leopard frog, *Rana pipiens,* exists as three groups in June in Minnesota: sexually mature adults that bred a month or two earlier; sexually immature yearling frogs, the young of the previous year; and tadpoles, the young of the present year. The tadpoles in a single breeding pond are clearly separated from those in other breeding ponds, and, thus, could be considered as a well-defined population. It is conceivable that all the tadpoles in a pond could be counted but, given the delicacy of the tadpoles, probably not without harm. It is also conceivable that a count of all the adult frogs in a breeding pond could be attempted during the mating season. However, such counts are apt to show a heavily biased sex ratio because males behave so differently from females during the breeding season and are much easier to catch. By June, the sexually mature frogs will have scattered widely from the breeding ponds over their summer feeding-habitat, and not only will the frogs be difficult to capture, but unless they have been previously marked at the breeding ponds, there is no way of relating the adults to the population of tadpoles. For the sexually immature yearling frogs, the difficulty is compounded. Because these frogs dispersed the previous summer, after metamorphosis, from their breeding pond to their summer feeding-range, migrated in the fall to an overwintering site in a river or lake, and then dispersed again in the spring to summer habitat, it is difficult if not impossible to relate them to either the adult population or the tadpole population. The adults, of course, have gone through these migrations at least twice. Another factor in the problem is that, apparently because of the difference in size, the dispersal patterns of adults and yearlings are apt to be different. The purpose of this discussion is to bring out that even though a count of all *R. pipiens* adults, yearlings, and tadpoles in a given area might satisfy the definition of a population, the genetic relationships among these different subpopulations would be unknown, and they would not necessarily constitute a single breeding population. Similar problems are apt to emerge in attempts to census populations of almost any species. Plant populations are somewhat easier to census than animals because they are sessile, but wind and insect pollination and seed dispersal can complicate population estimates in plants.

By now it should be obvious that total population counts are usually very difficult if not impossible to make, and the effort involved will far outweigh the benefits gained. Instead, to obtain estimates of population numbers, population biologists resort to sampling techniques by which they make counts of organisms in sample plots or along linear transects of appropriate size and number. These methods are particularly useful with plants, soil organisms, and other relatively immobile species. This approach also has its hazards. If sample plots are used, they must be the appropriate size, number, and distribution so that the areas sampled are truly representative of the entire area. If the individuals are neither uniformly nor randomly distributed in an area, but show some form of aggregation, the sampling problem may become very sticky. For these samples, it is essential to have an adequate statistical test of the magnitude of the sampling error. Here, just as with the total population count, it is necessary to count every individual in the plot or quadrat being sampled. Since this method is often used for estimating populations of small species like insects or ticks, and the reliability of the estimates depends on complete counts in each quadrat, considerable care may be necessary to ensure that all individuals are detected and counted. Here, too, the labor involved must be balanced against the information gained.

For more active species, a different sampling method is used to estimate population size. This is the capture-mark-release-recapture technique. In the study area, a sample of individuals is captured, marked so that they can be identified if recaptured, and released. Then a second sample is taken, and from the number of marked individuals recaptured in the second sample, it is possible to estimate the population size in the area as follows:

Let M = total number of marked individuals from the first sample

R = number of marked individuals recaptured in the second sample

S = total size of the second sample

N = the estimated size of the population in the area.

The population then can be estimated from the following relationship, sometimes referred to as the Lincoln-Petersen index (Scattergood, 1954).

$$\frac{R}{S} = \frac{M}{N} \text{ or } N = S \cdot \frac{M}{R}$$

If, for example, a sample of 100 frogs was captured, marked, and released in an area, and, subsequently, a second sample of 80 frogs was captured, among which 10 were marked recaptures, the estimated size of the population in the area would be

$$N = 80 \cdot \frac{100}{10} = 800.$$

The unmarked frogs in the second sample could also be marked and the entire sample released so that there would then be a total of 170 marked frogs in the area. Another sample could then be taken and the process repeated indefinitely so that additional estimates could be obtained.

In principle, this method seems very good, and in practice, it is widely used to get estimates of population size in species that otherwise might be very difficult to deal with, such as fish. However, the method makes certain assumptions that may not always be realized and thus must always be kept in mind. One assumption is that the marked individuals have distributed themselves homogeneously among the unmarked, uncaught individuals before the second sample is taken. Since the marked individuals are usually released at a single point, it may take a while before homogeneous mixing has occurred; therefore, some time must elapse before the second sample is taken. Another assumption is that marked individuals are as likely to be caught as unmarked individuals. However, if live traps are used, for instance, previously captured small mammals may become trap shy, or, conversely, if the trap is baited with peanut butter, they may become peanut butter addicts and get caught regularly, distorting the estimates. Still another assumption is that the marking itself has no effect on the behavior or survival of the marked individuals. Marking may involve such treatment as toe clipping in frogs, clipping of fins or ears, or tagging of fish and mammals. The marking may well have an effect on the behavior or survival of the marked individuals. The equation also assumes that the recaptures occur immediately after the releases before deaths, births, or migration have a chance to influence the results. As noted above, time must be allowed for homogeneous mixing to occur, and the question becomes How much time should be allowed for mixing, and during this time, what are the effects of births, deaths, immigration, and emigration on the estimate?

A final matter of concern relates to sampling near the boundaries of the population, which may not be ecologically well defined, and where migration may bias the samples.

A variation of the capture-mark-release-recapture method is the release of marked imported or lab-reared individuals into a natural population. This approach is subject to all of the above conditions. The release of lab-reared *Drosophila* with mutant eye colors is an example. The assumption is that the released flies behave like the wild flies, but the rearing conditions and the eye mutant itself may affect the viability and behavior of the flies in nature.

A somewhat different approach is removal sampling, where the population is sampled repeatedly, but the samples are not returned to the population. A constant probability of capture is assumed, and the decreasing size of successive samples can be used to obtain an estimate of the zero point and of the population size. This method, of course, can be rather hard on the population.

In addition to these direct methods for estimation of population number, more indirect methods are also used. For very small organisms, the total biomass of the population may be determined. Then a small sample of the total is weighed and counted, and from these figures, the population size for the total biomass can be estimated. Population estimates for ruffed grouse are obtained from counts of the number of male drumming-logs used during the breeding season. Pheasant populations are often estimated from road counts of the numbers of birds seen during a drive over a prescribed course. For mammals, which are often secretive or nocturnal, scat counts may be used; the shed antlers of deer or the number of pelts purchased by fur companies have also been used to estimate population numbers.

The usual objective of a census is to estimate the absolute number of organisms. However, as we have seen, this objective is usually not easily realized. In some cases, the changes in numbers are of more interest or concern than the absolute number of individuals. Game managers, for example, who deal with populations of ruffed grouse or ring-necked pheasants, populations that fluctuate greatly from year to year, are generally more interested to know whether the populations are greater or smaller than last year's than in the absolute figures. Thus, the relative numbers in samples taken in the same fashion from year to year provide adequate information for management purposes. The indirect methods cited above are often used to estimate relative numbers because they are often simpler to use. Even where direct counts are made, their main value may be to monitor changes in relative numbers because their meaning in absolute terms is obscure. Hawk Ridge, in Duluth, Minnesota, has for years been the site of a continuing study of the fall migration of hawks. When the migrating hawks reach the north shore of Lake Superior, instead of flying directly across the lake, they turn southwest and fly along the escarpment that stretches for miles along the shore. At Hawk Ridge, high above and overlooking the Duluth harbor, the nature of the terrain and the wind currents funnel the migrating hawks past an observation point, which is manned for many hours each fall. More hawks migrate past this point than at any other place in the country. Census figures are recorded each year for thousands of birds ranging from sparrow hawks to eagles. The primary value of this information is to permit a comparison of the relative numbers of each species from year to year. The meaning of the data in absolute terms is difficult to determine because it is not known with any certainty from whence the hawks have come or where they are going.

From the foregoing discussion about the difficulties in estimating population size, it should come as no surprise that so often in population biology, the population size is assumed rather than measured. Instead of estimating what might be considered the primary statistic about a population—its size—the population biologist often assumes a size of 10, or 100, or 1,000, or 10^8 and then attempts to evaluate the consequences of his assumption.

POPULATION GROWTH

The ideal estimate of population size should be instantaneous, because populations are not static, but dynamic, and the numbers are subject to change through time. Since population size is in a constant state of flux, the factors responsible for the changes in number must be considered. There are just four: deaths, births, emigration, and immigration. Let us at first exclude migration from our discussion and consider only death rates and birth rates.

Deaths, of course, reduce the population number, but their effect on the population may be expressed in various ways in terms of the death rate or of mortality and survivorship curves. The *death rate* may be expressed as the number of deaths per unit time, or the number of deaths may be expressed as a percentage of the initial population dying in a given time period. Mortality and survivorship or longevity curves are two different sides of the same coin. If the proportion dying equals M, then the proportion surviving equals 1-M. Longevity in a population is a different phenomenon from physiological longevity, the age reached by an individual living under optimum conditions before it eventually dies of old age. Reliable mortality data in natural populations are not easy to obtain. Even data on physiological longevity, usually based on animals in zoos, are not abundant, and are not always reliable. Princess Alice, for instance, the female consort of the famed circus elephant, Jumbo, died in Australia in 1941 at the age of 157. Unfortunately, she was born an African elephant but died an Indian elephant and Jumbo himself had several reincarnations.

Ecological longevity or realized longevity differs from the physiological maximum possible longevity and will vary according to the ecological conditions and the condition of the population. It is much more meaningful than physiological longevity. In nature, most animals die in extreme youth or in their prime. A very useful way to express longevity is in terms of the *median* life expectancy. Thus, in a population of 1001, the median life expectancy would be the age at death of the 501st individual to die. This measure is also used in bioassay, where the LD_{50} is the dose needed to kill 50% of the population. The median is preferred to the average or mean in both cases because it is less influenced by the exceptional individuals at the extremes of the mortality curve than is the mean. In mackerel, the median life expectancy at the time of fertilization is just twelve hours, which indicates an enormous early mortality. However, a month-old mackerel has a good chance to survive for 3 or 4 more years.

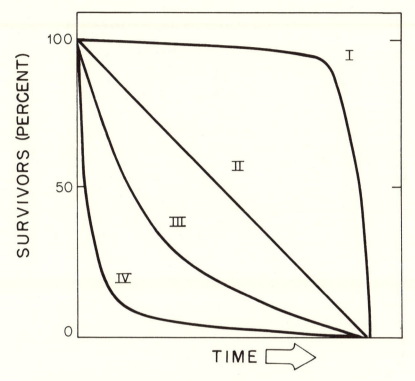

Figure 7-1. Types of survivorship curves. I. Most deaths occur in old animals. II. The number of deaths is constant per unit of time. III. The percentage of deaths is constant at all ages. IV. Most individuals die at an early age. (Source: Slobodkin, 1961).

Several types of survivorship curves are shown in Figure 7-1. In type I, mortality is concentrated in the old animals. In type II, there is a constant number of deaths per unit time. Type III represents a constant risk of death so that a fixed percentage of the animals alive dies at all ages. In type IV, mortality affects primarily the young animals. The type IV curve is probably typical of most wild populations though very few populations will fit any of the curves exactly. In some cases, the population may show a combination of types of curves, for different ages may be subject to different causes of death. Holometabolous insects, for instance, may suffer high mortality as larvae or may be especially vulnerable at the time of metamorphosis. In species with parental care of the young, the offspring may have high mortality at the time of weaning or at the termination of parental care. In many species, the adults show a type III survivorship curve—that is, they are subject to a constant risk of death. In human populations in the developed world, the survivorship curve is a combination of types I and IV, with deaths occurring primarily in infancy and old age and relatively few in between. In the undeveloped world, with relatively poor medical and public health care, the survivorship curves approximate type III.

To determine longevity experimentally, one can isolate a group of individuals of known equal age and maintain them under optimum conditions until the last one dies; however, this is not the most exciting type of experiment ever performed. Even with *D. melanogaster*, a short-lived species, it resembles watching grass grow, but the results

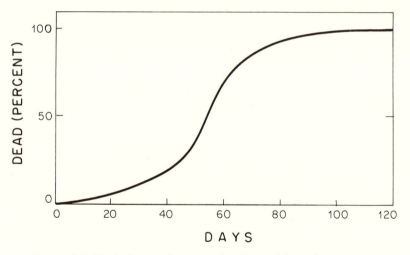

Figure 7-2. Typical mortality curve for *Drosophila melanogaster*.

of such experiments generally look like Figure 7-2. There are usually a few early deaths and also a few exceptionally long-lived survivors that may approach 4 months postmetamorphosis in age. Frequently, it is advisable to make some sort of transformation, such as a log transformation, to convert the curve into a straight line. The most reliable measure of longevity from these data is the median age at death. Such experiments can be used to compare longevity in different strains or stocks of flies.

In wild populations, age determinations are not so easily made as in experimental populations. Teeth are used in seals and black bears, rings on the horns of Dall mountain sheep, and scales in fish. In some species, size classes are indicative of age. Mark and recapture experiments over extended periods of time have also been used and permit the measurement of growth as well as survival.

Although the mortality pattern in humans has changed considerably, especially in the Western world because infant deaths have been reduced, the upper limit on longevity has remained virtually unchanged. The death-rate data on humans are probably the most complete and reliable available because of the detailed records that are kept.

Natality or the *birth rate* refers to the capacity of a population to increase in numbers. Birth rate will be used here to refer to this capacity whether the individuals are born, hatched, germinated, arise by fission, or any other mode of reproduction.

Death in a population is more or less continuous because all members of the population are always at risk. Reproduction, on the other hand, may be continuous or discontinuous. Reproduction in termites and humans is continuous. A termite queen is the closest thing to a continuous reproductive machine, for she will produce more than an egg a minute without stopping for many years. Hospital obstetrical wards stay open year round because babies are born every day of the year. In most species, however, reproduction is discontinuous—there is a definite breeding season, the females producing only one or two broods or litters per year. The 17-year locust reproduces only once every 17 years, the height of discontinuity.

The birth rate, like the death rate, may be expressed in different ways. One of the simplest is litter size, the number of eggs or young produced by a female at a single reproductive event. In humans, one, two, or, rarely, three babies are born at a time, with one by far the most common number. A female leopard frog lays a few thousand eggs at once, and a female oyster produces millions of eggs at a single spawning. Some species always produce the same number of young— a female elephant always has just one offspring, a female armadillo invariably has quadruplets. However, in most species, like humans, frogs, and oysters, litter size is variable. *Daphnia* females, for example, have been observed to carry anywhere from 1 to 150 eggs. A point worth noting now, to which we shall return later, is that despite these great variations in litter size, if population size is constant, each female in the present generation will be replaced, on the average, by just one female in the next generation. Or, put another way, each pair of parents will be replaced by another pair. In other words, if population size is not to fluctuate wildly, the average family size, the number of progeny reaching sexual maturity, must equal two. This statement is true no matter how many hundreds or thousands or millions of progeny may be produced and may be viewed in various ways. It means, for example, that even though a female frog produces 5000 fertilized eggs, two years later when these young reach sexual maturity, one should expect only two of them, on the average, to be alive. If each kernel of wheat that a farmer plants produced only a single kernel to replace itself, we would starve. The difference between the kernels needed for seed each year and the number harvested is the surplus used for food.

Another measure of reproduction is the *reproductive potential*, which provides a measure of the maximum reproductive capacity of a population under optimum conditions. The *realized reproductive capacity* is the reproductive rate actually achieved under a given set of ecological conditions.

A widely used measure for birth rate is the so-called *crude birth rate*, which is the number of births per individual per unit of time — that is,

$$\text{Crude birth rate} = \frac{\text{Births per year}}{\text{Population size}}.$$

In demography, the study of human populations, the crude birth rate is often given as the number of births per 1000 individuals per year. For humans, the physiological maximum crude birth rate is estimated to be about 50 per 1000 per year, whereas a low birth rate would be of the order of 15 per 1000 per year. The crude birth rate may, at times, be misleading because reproduction in a population depends on the number of females of reproductive age. Hence, both the sex ratio and the age structure in the population will influence the crude birth rate.

Because of the difficulties that arise with crude birth rates, *specific birth rates* are sometimes used. The number used in the denominator of the equation is the number of some particular group in the population, say all women from 20 to 25 years of age, rather than the size of the entire population. This type is known as an age-specific birth rate. Studies of such cohorts (groups of individuals of similar age) are often used in demography. Cohorts of women of different age may be studied to detect changing trends in the birth rate, or a single cohort may be studied throughout its reproductive life to study the pattern of its reproductive history.

Fecundity and fertility are often used in discussions of reproductive capacity, sometimes interchangeably. However, we shall use them with different meanings stemming from their use with *Drosophila* populations, where the distinction between them is easier to make than it is in some species. *Fecundity* refers to the number of eggs that a female produces; in other words, it is solely a female trait. *Fertility* will refer to the number of fertilized eggs that develop into living young; thus, fertility is a trait dependent on the characteristics of both male and female parents, and is therefore more complex. In *Drosophila* population studies, fecundity is the number of eggs laid per female, and fertility is the number or proportion of these eggs that hatch. Thus, fecundity and fertility in some ways parallel the difference between reproductive potential and realized reproductive capacity. Not just the words, but the symbols used by different population biologists may lead to confusion. Sometimes different words or symbols are used for the same thing, or sometimes the same word or symbol is used with different meanings by different authors. We cannot hope to reconcile these past differences but will attempt to be internally consistent, at least.

Consider now what Malthus called *geometric population growth*, which, more recently, has been called *exponential* or *logarithmic* growth. Imagine a population reproducing asexually by fission under conditions so favorable that no deaths occur. Let R_O equal the reproductive rate per generation so that in this case R_O equals 2. Furthermore, let N_O equal the population size at the outset, and N equal the size of the population after t generations. The following relation holds:

$$N = R_O^t N_O. \tag{7.1}$$

Then, for example, if N_O were 20, after 3 generations the population size would be

$$N = 2^3 \cdot 20 = 8 \cdot 20 = 160$$

and the following relationship would hold:

Generation (t)	N	(t)	N	(t)	N
0	20	6	1,280	12	81,920
1	40	7	2,560	13	163,840
2	80	8	5,120	14	327,680
3	160	9	10,240	15	655,360
4	320	10	20,480	16	1,310,720
5	640	11	40,960		

When these numbers are plotted graphically, they form an exponential curve (Figure 7-3) that appears to increase without limit. Of

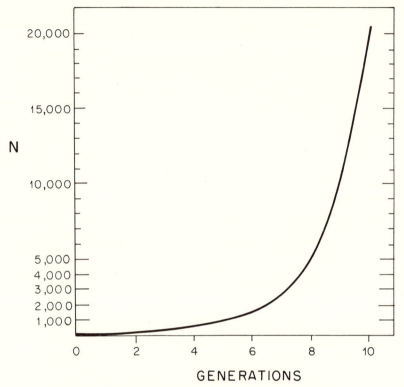

Figure 7-3. Exponential curve of population growth.

course, every population has a death rate as well as a birth rate so that R_O may be treated as the net reproductive rate per generation, and thus may provide a measure of reproductive potential of the population under favorable conditions. The rabbit in Australia and the starling and English sparrow in North America can be considered examples of exponential growth. The striped bass (*Roccus saxatilis*), native to the East Coast of North America, provides still another example. In 1879 and 1881, yearling bass were seined off the New Jersey coast, carried in tanks by train across the United States, and released in San Francisco Bay. Although only 435 fish survived the trip and were released, in 1899 the commercial catch just by nets amounted to 1,234,000 pounds. Thus, in twenty years, there must have been at least a million-fold increase in weight on the Pacific Coast. On the Atlantic Coast, however, during this same period, the population of striped bass remained constant or possibly declined slightly.

In the equation $N = R_o^t N_o$, if R_o has any value greater than 1, the population size will increase exponentially. If R_o equals 1, the population size will remain constant, but if R_o is less than 1, the population size will decrease.

If the exponential population growth were to continue unchecked indefinitely in any species, the mass of organisms would soon be expanding through the universe at virtually the speed of light. Obviously, such population growth does not occur. The population growth levels off at some point, and the population size reaches an upper limit. Rather than being exponential in shape, the usual growth curve is S-shaped or sigmoid, and is often called a logistic curve, as

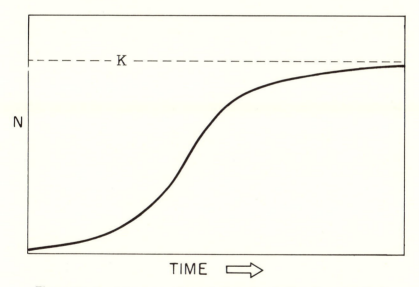

Figure 7-4. Logistic curve of population growth. K = the upper limit to population size or the carrying capacity of the environment.

shown in Figure 7-4. The first part of the curve resembles an exponential curve in shape, but then as environmental conditions begin to place limits on further growth, the increase in numbers slows down and eventually ceases as the population size approaches its upper limit asymptotically.

Consider exponential population growth again in somewhat different terms. In the relation $\Delta N/\Delta t$, ΔN (delta N) is the symbol for the change in the population size N in a stated time interval Δt. The relation dN/dt indicates the instantaneous change in N at time t.

If migration is not occurring, then the change in N can be determined from the following equation:

$$\frac{dN}{dt} = bN - dN = (b-d)N \tag{7.2}$$

where b = average birth rate per individual per unit time

and d = average death rate per individual per unit time.

Multiplying both b and d by N gives the number of births and deaths, and the difference gives the change in N.

If we let $(b-d) = r$, then

$$\frac{dN}{dt} = rN, \tag{7.3}$$

and the value of r obviously is dependent on the values of b and d. If r is positive, N increases; if r is negative, N will decrease; if r is zero — that is, the birth rate and death rate are equal — the population size will not change, for a dynamic equilibrium will exist.

When a population is uncrowded and the environmental conditions are most favorable, r will be at its maximum value. Such r_{max} values can be estimated for populations when they are in their exponential growth phase of the logistic growth curve. Often, r_{max} has been called the *intrinsic rate of natural increase* or the *instantaneous coefficient of population growth*. It has also been regarded as a measure of the biotic potential or the reproductive potential of the population. If estimates of r_{max} are available, say from the exponential portion of a growth curve, and the actual rate of increase can be measured in a laboratory or field population, the difference between the two can be used as a measure of *environmental resistance*, the sum total of the environmental factors that prevent the biotic potential from being realized.

We have already seen that unlimited exponential growth is unrealistic, which means that the birth rate and the death rate are not independent of the value of N, the population size. As N increases in any real population, the rate of growth slows down and stops. This effect can be taken into account by setting

$$b = b_0 - k_b N$$

and

$$d = d_0 + k_d N$$

where

b_o = birth rate when N is small — that is, b_o is an estimate of the maximum birth rate,

d_o = death rate when N is small — that is, d_o is an estimate of the minimum death rate,

k_b = the slope of the decrease in the birth rate as N increases, and

k_d = the slope of the increase in the death rate as N increases.

This refinement in the treatment of the birth and death rates is an attempt to make the mathematical treatment or model more closely approximate the situation in real populations. In this case a linear relation is now assumed between birth and death rates and population size. It takes a form similar to the linear relation between the variables x and y where $y = a + bx$, and a = the intercept of the line on the y axis when $x = 0$, and b = the slope of the line.

If, in fact, b and d change in a nonlinear fashion with changes in N, the model will still not be satisfactory. However, it is clearly an improvement over the previous model, which considered b and d as unchanging, and now the logistic equation for population growth takes the form:

$$\frac{dN}{dt} = [(b_o - k_b N) - (d_o + k_d N)] N.$$

When the number of births is equalled by the number of deaths, the population will cease to grow and will reach an equilibrium, the upper limit on the logistic growth curve. At this equilibrium point,

$$\frac{dN}{dt} = 0$$

and

$$b_o - k_b N = d_o + k_d N.$$

Rearranging gives

$$b_o - d_o = k_b N + k_d N,$$

and solving for N gives the population size at equilibrium

$$\hat{N} = \frac{b_o - d_o}{k_b + k_d} = K.$$

The caret over a letter is often used as an indication of an equilibrium value, in this case of the population size, N. In this case, $\hat{N}$ is usually called K, and K is so widely used for this particular equilibrium value of N that it seems wise to retain this usage.

The logistic equation can now be rewritten in a different but equivalent form, which is derived as follows:

$$\frac{dN}{dt} = [(b_O - k_b N) - (d_O + k_d N)]N;$$

Rearranging,

$$\frac{dN}{dt} = N[(b_O - d_O) - N(k_b + k_d)].$$

Recalling that $b_O - d_O = r$, then let

$$\frac{dN}{dt} = N\left[r - N(k_b + k_d)\left(\frac{b_O - d_O}{b_O - d_O}\right)\right]$$

and

$$\frac{dN}{dt} = N\left[r - N(b_O - d_O)\left(\frac{k_b + k_d}{b_O - d_O}\right)\right].$$

Since $\frac{k_b + k_d}{b_O - d_O} = \frac{1}{K}$, the equation becomes

$$\frac{dN}{dt} = N\left(r - rN \cdot \frac{1}{K}\right)$$

$$\frac{dN}{dt} = rN\left(1 - \frac{N}{K}\right)$$

$$\frac{dN}{dt} = rN\left(\frac{K - N}{K}\right) \tag{7.4}$$

where

$r = b_O - d_O$
$N =$ the actual population size, and
$K =$ a constant, the upper limit to the population size, often known as the *carrying capacity of the environment*.

From the equation it is clear that when $N = K$, $dN/dt = 0$, and no further increase in population size will occur.

The two values, r and K, have assumed considerable importance in population studies — r because it is a measure of reproductive potential and K because it measures the upper limit in size that a population can reach under a given set of environmental conditions.

Both r and K can be estimated. Recall the earlier equation (7.3) for population growth

$$\frac{dN}{dt} = rN$$

from which

$$r = \frac{1}{N}\frac{dN}{dt}.$$
(7.5)

In integrated form this becomes

$$N_t = N_o \, e^{rt}$$
(7.6)

where

N_o = population size at time o,

N_t = population size at time t, and

e = the base of the natural logarithms (ln);

and

$$ln \, N_t = ln \, N_o + rt$$

Hence

$$r = \frac{ln N_t - ln N_o}{t}.$$

This equation shows that if the population size is known at two different times during a period of unlimited growth, it is possible to estimate r, and that r will be given per day, or per week, or per year depending on the time units used to measure t. Conversely, if r and N_o are known, it is possible to project the population size, N_t, at some future time.

Since r is the instantaneous rate of increase, it is possible to compare r values from species with different generation lengths. Table 7.1 shows that three insect species have similar high values, and two small mammals have lower r values. For the human population of the world, the r value is presently estimated to be only about 0.02 per year (or for comparison with the other species, only about 0.0004 per week).

The coefficient of population growth, r, should not be confused with the net reproductive rate, $R_o - r$ is the instantaneous rate of increase whereas R_o is related to generation length. Thus, R_o values for different populations are not comparable unless their generation lengths are similar. However, R_o and r obviously must be related to some extent. For a stable population, as noted above, $R_o = 1$ and $r = 0$. Moreover, in an increasing population, $R_o = e^{rT}$, where T is the mean generation length, or $r = \frac{ln R_o}{T}$. What this means is that if a life table is available that gives the age-specific survival rate and the age-specific birth rate or natality rate for a population, it is possible

Table 7 - 1. Estimates of Population Growth

Organism	r (per Week)	T (Mean Generation Length — Weeks)	R_O (Net Reproductive Rate)	Doubling Time (Weeks)	Source
1. Rice weevil (Calandra oryzae)	0.76	6.2	113.56	0.91	Birch (1948)
2. Flour beetle (Tribolium castaneum)	0.71	7.9	275.0	0.96	Leslie and Park (1949)
3. Human louse (Pediculus humanus)	0.78	4.4	30.93	0.88	Evans and Smith (1952)
4. Vole (Microtus agrestis)	0.088	20.2	5.9	7.90	Leslie and Ranson (1940)
5. Brown rat (Rattus norvegicus)	0.104	31.1	25.66	6.76	Leslie (1945)

to estimate the net reproductive rate (R_O) directly and from the above equation to obtain an estimate of the instantaneous rate of increase (r) as well.

The age specific survival rate (l_x) is usually given as the proportion of individuals surviving to a given age, x, and the age-specific birth rate (m_x) as the average number of female offspring per female aged x. The net reproductive rate (R_O) is obtained by multiplying the fraction of females surviving to each age (l_x) by the average number of female progeny produced per female at that age (m_x) and summing them over all ages, $R_O = \Sigma l_x m_x$. R_O thus is an estimate of the average number of offspring produced by a female over her entire life span. This method of calculating the net reproductive rate with data from only females should give male chauvinists pause. Since data on survivorship and fecundity or fertility can be obtained in populations, estimates of both R_O and r have been possible, but most such estimates have been made for laboratory populations (Table 7-1).

Another estimate, often of interest about a population, is the doubling time, the period required for the population to double its numbers. This value can be estimated from equation 7-6 as follows:

$$N_t = N_0 e^{rt}$$

$$\frac{N_t}{N_o} = 2 = e^{rt}$$

$$ln\,2 = rt$$

$$t = \frac{0.6931}{r}.$$

Such calculations gave rise to the doubling times shown in Table 7-1. It may be added that the doubling time for the human population with an r value of 0.02 per year is only 35 years. A further comment is that the highest reproductive potentials (r) characterize species with high ecological death rates. This relationship seems almost inevitable because the average population size is relatively stable, which means that, on the average, one female leaves one female offspring to survive and reproduce, or one pair leaves two surviving offspring. If a pair produces 2000 zygotes, 1998 of them, on the average, will die before reaching maturity.

Just as estimates of r and R_o have been made, estimates of K, the so-called carrying capacity of the environment, have also been made. Again, these estimates have usually been made in laboratory populations where it is possible to control environmental conditions much more carefully than in natural populations. Since the K values will change with changes in environmental conditions, probably not too much emphasis should be given to any estimate of K. A given K means that that particular population, with its array of genotypes and phenotypes, reached that particular upper limit in numbers under the environmental conditions indicated (Figure 7-4). The same population under different conditions would reach a different asymptote, and a different population of the same species under the same conditions might also reach a different asymptote or K value.

Figure 7-5 compares a curve for population growth, in which population size (N) is plotted against time (t), with a curve for the population growth rate for the same population, in which the increment added to the population per time unit ($\Delta N/\Delta t$) is plotted against time (t). In the graphs, it can be seen that as the population reaches its upper size limit, the growth rate approaches zero. It is worth noting that, if the population started to decline in numbers, there would be a negative growth rate.

POPULATION FLUCTUATIONS

The treatment of population growth thus far has been unrealistic because we have dealt only with populations that are increasing in

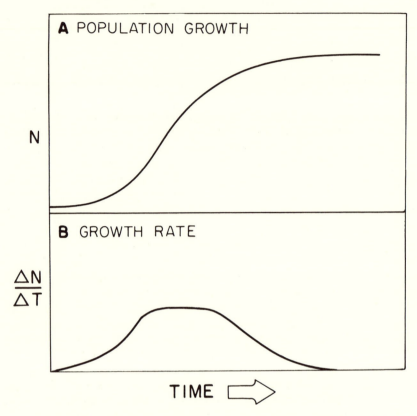

Figure 7-5. Comparison of the curve for population growth (A) with
the curve for the growth rate (B) in the same population.

numbers or have reached an equilibrium in size. Both the exponential
curve and the logistic curve of population growth are oversimplified.
Populations not only increase in numbers, sometimes they decrease;
and they will do so whenever the combined effects of mortality and
emigration exceed the combined effects of natality and immigration.
Whenever a population approaches its upper limit in numbers, it does
not hold at some constant upper limit; rather, it tends to fluctuate
about the asymptotic level or the carrying capacity of the environment (K). The causes of these fluctuations are diverse: a change in
environmental conditions may raise or lower the asymptote, interactions occurring within the population or with other populations
(predators, prey, parasites, plants, etc.) may affect the numbers, or
discontinuous reproduction alone would lead to fluctuations in

population size so that even in the more or less constant environment of laboratory populations, fluctuations may occur.

In natural populations, seasonal changes in size may be observed. They may be related to life history as in the leopard frog (Merrell, 1977), where breeding occurs in early spring and a new age-class is added to the population of frogs each July when the tadpoles metamorphose. In other cases, the changes may be related to seasonal environmental changes in temperature, moisture, and food as they are in insects.

Annual fluctuations in numbers may also be observed, which may be due to extrinsic factors such as variations in rainfall, severity of the winter, etc. from year to year. Intrinsic factors may also be involved, such as competition for food, alteration of the environment by metabolites, or social interactions. The extrinsically caused changes are apt to be irregular and can be correlated with one or more of the physical limiting factors in the environment. The intrinsically caused changes are more apt to show regularity and to appear *cyclical* in nature. There can be little doubt that many species show regular annual population cycles with the passage of the seasons. However, on the question of cyclical oscillations extending over several years, opinion is more divided. Many efforts have been made to demonstrate 3- or 4-year cycles in lemmings and their predators, snowy owls and Arctic foxes, or 9- or 10-year cycles in snowshoe hares and lynxes, and in numerous other species. Although there are clearly large fluctuations over the years, it is not clear that they are always cyclical. Perhaps it is sufficient to realize that the numbers of any population are always subject to both extrinsic and intrinsic factors, and that the problem is not to determine whether intrinsic or extrinsic factors alone are responsible for the observed fluctuations, but rather to estimate the relative importance of each in a given situation.

A study of herons (*Ardea cinerea*) in Great Britain by Lack (1966) over a thirty-year period showed that in two areas the populations were quite stable except that cold winters were associated with a decline in abundance in both areas, with subsequent recovery (Figure 7-6). In this case, the fluctuations clearly seem to result from the severe weather rather than some intrinsic cycle.

The snowshoe hare and the lynx are often cited as representing a classic example of regular cycles of abundance. Every 9 or 10 years the populations grow to a peak and then decline rapidly, or *crash*. This pattern was apparently repeated every 9 or 10 years from the 1840s to the 1930s, and could not be related to any obvious environmental changes (MacLulich, 1937). The peak in abundance of the

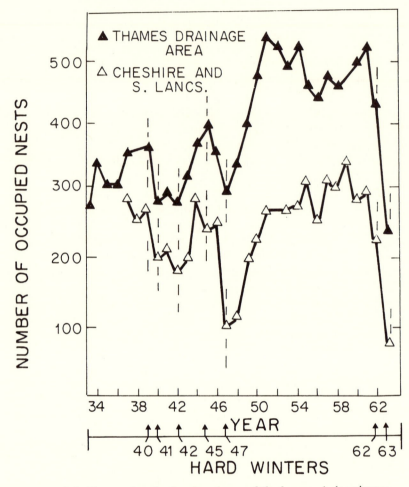

Figure 7-6. Fluctuations in the abundance of the heron, *Ardea cinerea*, related to the severity of the winters. (Source: Lack, 1966).

snowshoe hare preceded the peak in the lynx by about a year (Figure 7-7). Since the snowshoe hare is a primary food source for the lynx, the cycle of the predator appears to be related to that of the prey. The reason for cycling in the hare population is less obvious, for the hares show cycles even where there are no lynxes (Keith, 1963), and thus the two cycles cannot be caused by a simple predator-prey interaction. Another factor to consider about this study is that it was based not on census data in the field but on the records of the numbers of pelts of these species received by the Hudson Bay Company

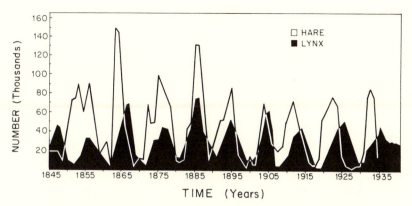

Figure 7-7. Widely cited example of cyclic oscillations in numbers
of predators and prey, the lynx and the snowshoe hare, with
abundance indicated by the numbers of pelts received by
the Hudson Bay Company. (Source: MacLulich, 1937).

over that long period. Thus, the possibility exists that economic and
social factors related to fur trapping could be a factor in these apparent cycles.

More direct evidence has been obtained with respect to the 3- to 4-
year cycles in small Arctic mammals and their predators, the Arctic
fox and the snowy owl (Elton, 1942; Pitelka, 1957). The explosions
in the lemming populations that lead to their famous mass migrations
from overcrowded areas are an indication of the magnitude of the
fluctuations that may occur. The periodic invasions of snowy owls
into the United States every 3 or 4 years have been shown to follow
crashes in the abundance of their prey in the far north. However,
even though fluctuations in population size in many species show
some regularity, they are not completely regular, for the cycles differ
in both amplitude and duration from one oscillation to the next.
Therefore, the idea of cycling in populations quite independent of
environmental conditions seems oversimplified. However, Cole's
concept (1954) that the observed cycles are spurious statistical artifacts may be too extreme in the other direction. A considerable
amount of research has been done on population cycles, especially
in small mammals, and a variety of theories has been proposed to explain the observed fluctuations. These theories range from Cole's idea
of pure chance, to predation, nutrient depletion, sunspots, emigration, reproductive failure resulting from the stress of overcrowding,
changing selection pressures related to aggression, and so on. Perhaps
too much attention has been paid to population cycles and to the

search for a single explanation. For our purposes, it is sufficient to realize that the size of a population does not reach some upper limit and stay there, but instead may fluctuate, sometimes quite wildly. Moreover, if there is any regularity about the factors influencing the population—food, moisture, temperature, predators, or whatever they may be—there is also likely to be regularity or periodicity in the fluctuations in numbers.

The most violent fluctuations in numbers have been observed in relatively simple ecosystems with few species such as are found in the far North, or in pine plantations, or single-crop agriculture. Apparently, in more complex ecosystems with greater numbers of species, the oscillations are damped. The spruce budworm (*Choristoneura fumiferana*), for example, causes more serious damage in pure stands of adult balsam fir than it does in mixed stands of trees. The caterpillars are so destructive in defoliating conifers that, owing to periodic outbreaks in Canada, severe timber losses have occurred every 35 to 40 years. As a consequence, considerable attention has been given to the factors influencing population outbreaks in this species (Morris, 1963). Another group of pest species that cause widespread economic losses are the migratory locusts of the Old World, whose history goes back to Biblical times. The early Mormons in Utah faced a similar plague, and gulls occupy a special place among the symbols of the Mormon church because they came in great numbers to feed on the insects.

It is usually difficult to census such outbreaks, but the moth (*Bupalus piniarius*), whose larvae defoliate conifers, especially the Scotch pine (*Pinus sylvestris*), has provided some interesting data because the larvae drop to the forest floor to pupate and can then be counted rather easily (Varley, 1949). A variation of five orders of magnitude was found, from less than one per 1000 square meters to more than 10,000 per 1000 square meters. Since several generations occur in a season, at high densities these moths can defoliate and kill the trees; but the outbreaks do not show the regularity observed in the snowshoe hare or in some of the small mammal populations.

PREDATION

For our purposes, we shall define predation in the broad sense to include the eating of live organisms by other organisms. Included are the eating of plants by animals called herbivores, the eating of animals by other animals called carnivores, and also insect parasitoids. These parasitoids, for a time, live in or on the body of a living host like a true parasite, but eventually kill the host as they consume its tissues so that the final outcome is much the same as predation, but slower.

Lotka (1925) and Volterra (1926) were the first to attempt to treat predator-prey relationships quantitatively, in a manner rather similar to their studies of competition. In the absence of predators, the prey population is assumed to increase geometrically in accordance with the equation

$$\frac{dN_1}{dt} = r_1 N_1$$

where

N_1 = population size (or density) of prey and

r_1 = intrinsic rate of increase for the prey species.

If prey and predator live together in a limited space, the population growth of the prey will be reduced by a factor dependent on the density of the predators, and the equation for the prey population becomes

$$\frac{dN_1}{dt} = r_1 N_1 - K_1 N_1 N_2$$

where

N_2 = population size (density) of the predators and

K_1 = a constant indicating the proportion of lethal encounters between predator and prey.

The product $N_1 N_2$ is proportional to the probability of a chance encounter between predator and prey. The coefficient K_1 is a constant that measures the fraction of those encounters that are fatal to the prey—in a sense a measure of the skill of the prey in escaping predators.

In the absence of prey, the predators are expected to die at a constant rate independent of density as follows:

$$\frac{dN_2}{dt} = -d_2 N_2$$

where

N_2 = population size (or density) of the predator and

d_2 = instantaneous death rate of predators in the absence of prey.

Thus, in absence of prey, the predator population is expected to decrease geometrically.

When prey are present, the predator population is expected to increase at a rate dependent on the density of the prey population in accordance with the equation

$$\frac{dN_2}{dt} = K_2 N_1 N_2 - d_2 N_2$$

where

K_2 = a constant indicating the efficiency with which predators convert prey into offspring.

K_2 can also be regarded as a measure of the skill of the predator in catching prey. In this simple case, it is assumed that the predator is completely dependent on one prey species, and the number of prey is limited solely by predation by this predator and the prey has no interaction, for example, with its own food supply. In addition, it is assumed that there are no time lags, that there is no age-specific variation, and that the amount of interaction between predator and prey is simply proportional to the product of prey and predator numbers. Recall also that the death rate of the predators is assumed to be independent of density. Obviously, these assumptions oversimplify the situation in natural populations. However, the Lotka-Volterra equations were useful in that they predicted that predator and prey populations would oscillate. The reason is that the number of predators at any given moment is dependent on the materials they have acquired from the prey in the past. Thus a time dimension is introduced into the system, which permits oscillations.

To understand this situation, let us consider what happens at equilibrium when $dN_1/dt = 0$ and $dN_2/dt = 0$. Then

$$r_1 N_1 = K_1 N_1 N_2$$

$$\hat{N}_2 = \frac{r_1}{K_1}$$

and

$$K_2 N_1 N_2 = d_2 N_2$$

$$\hat{N}_1 = \frac{d_2}{K_2}$$

Since r_1, d_2, K_1, and K_2 were all assumed to be constants, at equilibrium $\hat{N}_1$ and $\hat{N}_2$ will also be constants, as shown graphically in Figure 7-8. In the graph, the intersection of the equilibrium isoclines, which are straight lines, defines the number of prey needed to sustain a constant number of predators.

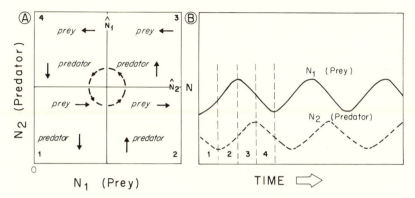

Figure 7-8. The relationship between predator and prey populations based on the Lotka-Volterra equations. A. The circle about the intersection of the population isoclines, N_1 and N_2, shows the trajectory of the joint abundance of the two species. B. The same data plotted against time. The numbers along the abscissa refer to the quadrants in A. The predator and prey populations will continually oscillate out of phase with one another, with the predator cycle lagging behind the prey cycle. See text for further details.

If the predator and prey populations are displaced from the equilibrium point, the arrows in the four quadrants indicate the direction of change in numbers of prey, N_1, and predators, N_2. It can be seen that above the line $\hat{N}_2$, there is an excess of predators so that the number of prey will decrease. Below $\hat{N}_2$, the number of prey can increase because the density of predators is low. To the right of line $\hat{N}_1$, the number of predators will increase because there is an excess of prey; to the left of $\hat{N}_1$, there is a deficiency of prey, and hence the number of predators will decrease.

The joint changes in the size of predator and prey populations have a circular trajectory in a counter-clockwise direction, as shown in Figure 7-8, and the populations will continue to oscillate indefinitely in this manner because the equilibrium is neutral and there are no forces driving the populations back toward the equilibrium point at the intersection of the predator and prey isoclines. Any further perturbation of the system will shift the oscillations about the equilibrium to a cyclical path of different amplitude and duration. Thus the path followed by any particular population of predator and prey will depend on the starting point, and will persist unless some perturbation takes place, after which the populations will move along a new trajectory. It will be noted that the predator and prey populations oscillate out of phase with one another, with the changes in the predator population following but lagging behind those in the prey population. Eventually, these random perturbations may increase the amplitude of the oscillations to the point that the trajectory hits one of the axes and one or both of the populations become extinct.

The Lotka-Volterra equations gave rise to a rather interesting prediction. If individuals of both species are destroyed uniformly and in proportion to their numbers by some outside agency such as a pesticide, the prey species will subsequently increase proportionately but the predator species will decrease proportionately. In other words, if we apply a broad spectrum pesticide that kills 50% of both the predator and its prey, subsequently the number of prey will increase faster than the number of predators and, if the prey is a pest species we are trying to control, we may be worse off than before the pesticide was applied.

If both N_1 and N_2 are reduced proportionately in the predation equations, the effect on the product $N_1 N_2$ will be much greater than on either N_1 or N_2 alone. However, the product $N_1 N_2$ determines the death rate of the prey and the birth rate of the predator. Therefore, the birth rate of the predator will be considerably reduced at the same time that the death rate of the prey is reduced, with the result that the prey population is able to increase relatively faster than the predator population. For example, if $N_1 = 1 \times 10^5$ and $N_2 = 1 \times 10^3$, then $N_1 N_2 = 100 \times 10^6$. If 50% of both N_1 and N_2 are killed by a pesticide, N_1 becomes 50,000, N_2 becomes 500, and $N_1 N_2 = 25 \times 10^6$, a reduction to 25% of its former value. No matter what the proportion killed or the original numbers of N_1 and N_2, under the Lotka-Volterra assumptions, the death rate of the prey will be affected proportionately more than its birth rate, and the birth rate of the predator proportionately more than its death rate, with the consequence that the prey subsequently will be expected to increase more rapidly than the predator.

This effect can also be visualized by examination of Figure 7-8. If it is assumed that before the application of the pesticide N_1 and N_2 are near their equilibrium values, $\hat{N}_1$ and $\hat{N}_2$, then a marked decline in both N_1 and N_2 owing to the pesticide would place the new point (N_1, N_2) in the lower-left quadrant where it can be seen that the prey will increase faster than the predator. This prediction should make us wary of using broad-spectrum pesticides, for they may actually make matters worse.

The Lotka-Volterra equations are of considerable interest because of their prediction of stable cycles in populations of predators and prey, which seem to parallel the observations in natural populations of snowshoe hares and lynxes and of lemmings and snowy owls. However, the assumptions involved are obviously much too simple. Predators ordinarily feed on more than one species of prey, the size of the prey population will normally be influenced by the availability of its own food supply and by the actions of other predators and parasites, there will be time lags and age-specific variation, and so on. Furthermore, r_1, d_2, K_1, and K_2 may not be constant but may show density effects.

In simple laboratory populations of a predator and its prey, the usual outcome is that the predator exterminates all the prey and then dies of starvation rather than that the populations fall into a Lotka-Volterra cycle. Gause (1934), for example, obtained this result with populations of two protozoan species—the prey, *Paramecium caudatum* and the predator, *Didinium nasutum*. Only by such stratagems as the addition of immigrant predators and prey, at intervals, or the provision of a refuge for the prey where they could not be captured by the predator was it possible to obtain population oscillations in these simple systems.

Since the expected Lotka-Volterra oscillations of predator and prey were not usually observed, Rosenzweig and MacArthur (1963) attempted, by graphical analysis and more realistic assumptions, to account for the results actually observed in populations. They assumed first that the prey population is limited by its own food supply, and that at low densities such factors as difficulty in finding mates and reduced productivity would inhibit its growth potential. The result is a convex isocline ($dN_1/dt = 0$) for the prey rather than the horizontal line seen before. The predator population is assumed, as before, to be dependent on the prey for food, but it, too, is expected to reach some upper limit imposed by such factors as territoriality or lack of breeding sites and other density-related phenomena, and its isocline ($dN_2/dt = 0$) also takes a different shape.

With these relatively minor, but somewhat more realistic assumptions, it can be shown that the nature of the predicted predator-prey oscillations depends on the location of the point of intersection of the isoclines. If the intersection is at the peak of the prey isocline, a stable cycle will ensue (Figure 7-9). If the intersection is to the right on the descending part of the prey isocline, a damped cycle will occur (Figure 7-10); but if the point of intersection is to the left on the ascending part of the prey isocline, the oscillations become divergent and an unstable cycle leading to extinction will occur (Figure 7-11). Finally, if it is assumed that at low densities, the prey population is in some way better able to escape predation through the

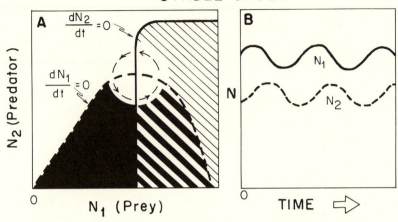

Figure 7-9. Stable predator-prey cycle. A. Predator isocline intersects the peak of the prey isocline. B. Stable oscillations in predator and prey populations persist indefinitely. See text for further details. (Source: MacArthur and Connell, 1966).

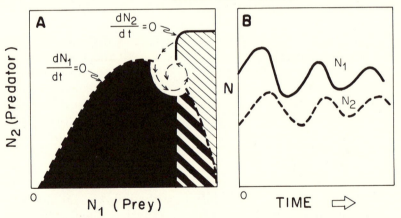

Figure 7-10. Damped predator-prey cycle. A. Predator isocline intersects prey isocline to the right of the peak. The joint abundance curve spirals inward. B. The oscillations are damped, and the populations will approach their equilibrium values at which neither predator nor prey population numbers change. Typical of an inefficient predator. See text for further details. (Source: MacArthur and Connell, 1966).

UNSTABLE CYCLE

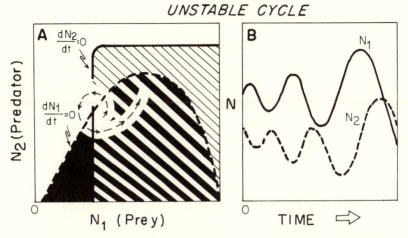

Figure 7-11. Unstable predator-prey cycle. A. Predator isocline intersects prey isocline to the left of the peak. The joint abundance curve spirals outward. B. The amplitude of the oscillations increases to the point of extinction of the prey (and, perhaps, the predator). Typical of an excessively efficient predator. Probably rare in nature. See text for further details. (Source: MacArthur and Connell, 1966).

availability of shelter or favorable habitat—in other words, that the predators feed primarily on the surplus prey—then the prey isocline curves upward rather than downward at low density, and a stable cycle is again generated.

The difficulty with this type of analysis is, again, that it is primarily descriptive. By proper adjustment of the isoclines, it is possible to generate curves that resemble those observed in experimental and natural populations, but predictions are not easy to confirm. In a simple laboratory situation of one predator and its prey, the Lotka-Volterra prediction of regular oscillations of predator and prey numbers was seldom realized; instead, the populations went to extinction. Only if environmental complexity or migration was introduced did the system begin to approach the predicted behavior. In natural populations, there is some evidence that predators do affect the numbers of their prey, but so do many other factors. Since the models are primarily concerned with the interactions between predator and prey, they are generally too simplistic to make useful predictions possible. Although natural populations of predators and prey often appear in equilibrium and sometimes even appear to show cycles, predation is only one of the factors involved in determining abundance in natural populations.

AGE STRUCTURE

The age structure or age distribution of individuals in a population has a direct influence on both the birth rate and the death rate in that population. Hence, knowledge of the age distribution of a population should reveal not only its current reproductive status but also how the population can be expected to develop in the future. In simple terms, an increasing population will tend to have relatively many young individuals, a declining population will have relatively many more old individuals, and a stable population will have a more even distribution of age classes.

There are two sets of conditions that will tend to produce a fixed, constant age structure in a population. First, a population growing geometrically with constant age-specific mortality and fertility rates will maintain a stable age distribution. Second, a population that has reached a constant size, with equal birth and death rates, will also reach a fixed age distribution. Natural populations can seldom be expected to show a fixed age structure for any length of time because no population is able to grow geometrically for very long, and population size is not constant but fluctuates. Nevertheless, it is significant that any population living in a constant environment will tend not only to reach a constant population size (K), but also a fixed age distribution, no matter what the initial age distribution may be. Furthermore, this age distribution can be determined from life-history data if values of l_x (the proportion surviving to age x) and m_x (the age-specific birth rate) are known (Wilson and Bossert, 1971). Once a population approximates this age distribution, perturbations may occur because of fluctuating birth or death rates, but the population will tend to return to the stable age distribution spontaneously. Even though a population may seldom achieve a stable age distribution, the concept is a useful one, for it can serve as a basis for comparison with the age distributions actually observed in natural populations.

One way to consider the age of individuals is in relation to their reproductive status, which means that they can be categorized as pre-reproductive, reproductive, and post-reproductive. The relative length of these three *ecological* ages may vary considerably. In modern human industrial societies, the three periods are about equal, each approximately twenty years. In more primitive or ancient societies where the average age at death is 35 or less, the post-reproductive period is much shorter, on the average. In some species, notably insects like the mayflies and butterflies, the pre-reproductive period may be very lengthy, the reproductive period very ephemeral (whence the name *Ephemeridae* for the mayfly family), and the post-reproductive period nonexistent.

One point worth noting in this regard is the cyclical nature of reproduction in most species. *Homo sapiens* is atypical compared to most species in being sexually and reproductively active the year round. Most birds, for example, concentrate their sexual and reproductive activities within a few weeks of the year. The rest of the time, they are essentially neuter in behavior and sexual activity. Since this pattern is more prevalent than continuous reproduction, it seems probable that discontinuous reproduction has evolved because of the high risk of mortality to both parents and offspring during the breeding season. The difference between cyclical and continuous reproduction may lead to significant differences in the population dynamics of these species.

A useful way to study the age distribution of a population is from an age pyramid (Figure 7-12). The proportions in the different age classes are indicated by the width of the figure, and the youngest class is located at the base of the pyramid. An expanding population, then, will have a very broad-based pyramid, a declining population with few young will have a narrow-based pyramid, and a stable population will show a regular decrease in the width of the figure from one age class to the next. When populations are studied in this way,

AGE PYRAMIDS

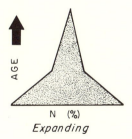
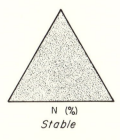
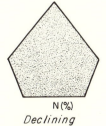

Figure 7-12. Types of age pyramids. A. A growing population is characterized by numerous young individuals. B. A stable population shows a gradual decrease in numbers with age. C. A declining population contains relatively few young individuals.

wide variations in juvenile-adult ratios may be observed. Petrides (1950), for instance, in fourteen studies of muskrats, found as the extremes a ratio of 48% adults to 52% juveniles in one case and 15% adults to 85% juveniles in the other. More unexpected, perhaps, in studies of fish and plant populations, was the discovery of dominant age classes. In some years in long-lived species like trees or fish, reproduction may be unusually successful and a single age class may be far more numerous than the age classes both preceding and following it (Hjort, 1926: Lawler, 1965).

Thus the age structure may be quite informative about the dynamics of a population. However, age determinations in wild populations are not always easy to make. Sometimes, where growth is continuous as in reptiles and amphibia, size can be used as an indication of age. Trees and fish scales may show annual increments of growth, or in some species juveniles can be distinguished from the adults as in muskrats. Mark-recapture data may also be informative; but in general for any species, if information about the age distribution of a population is desired, some knowledge of the life history and development in that species will be needed to make age determinations.

REGULATION OF POPULATION SIZE

Observation of natural populations shows that they vary in abundance from one place to another and that any given population fluctuates in numbers. Moreover, no population continues to increase without limit, and some populations cease to grow at a size well below the apparent carrying capacity of the environment. In some way population size seems to be regulated. The subject of how the natural regulation of population numbers occurs has been a matter of considerable controversy (Nicholson, 1933, 1954, 1957; Andrewartha and Birch, 1954; Lack, 1954; Chitty, 1957; Wynne-Edwards, 1962).

Elements regulating population size may be categorized in two ways. First, the *density-independent factors* may act in either a positive or negative way on population growth quite independently of the size or density of the population. Commonly regarded as typifying density-independent factors are such *extrinsic physical factors* as temperature, rainfall, soil composition, and the like. The *density-dependent factors* are those whose effects are a function of population density or size and are usually considered to be *biotic* factors such as competition, parasites, pathogens, food shortages, predation, alteration of environment by excretory products, and so on.

Ordinarily, for density-dependent factors, one would expect the death rate to increase and the birth rate to decrease as N increases. However, in some cases at very low population densities, the reverse may be true and an increase in N will enhance survival and reproduction. On the other hand, birth and death rates are uninfluenced by density-independent factors as N changes. In an experimental setup, the estimation of birth and death rates at different densities and, for example, different temperatures might provide some clues about density-dependence or independence. However, the classification of factors into physical, extrinsic, supposedly density-independent factors like the weather, and into biotic, density-dependent factors is much too pat. For example, the effects of a hard freeze may be more severe on a small colony of bees than on a large one in the proportion of bees killed and, thus, is clearly density-dependent. If it were density-independent, the same proportion of bees should be killed in both colonies.

In recent years, truly intrinsic physiological, behavioral, or genetic factors have been invoked to explain the natural regulation of animal numbers (Chitty, 1957; Pimentel, 1961, 1968; Wynne-Edwards, 1962; Christian and Davis, 1964). One of the oddities about the controversy is that the position taken by the various proponents seems to be influenced by the type of species population they work with and by its environment. Thus, a biologist working with insects or in a harsh environment is apt to reach very different conclusions about the significant factors controlling numbers in the populations studied compared to a biologist working with birds or mammals or in a mild climate. The former tends to stress density-independent population regulation whereas the latter stresses density-dependence.

One final aspect of density worth attention is that in the logistic growth curve, the growth rate is highest at minimum population density or very low values of N. In fact, such extremely small populations may be in serious difficulty and may be on the verge of extinction for any of several reasons. If they are too rare, males and females may have difficulty finding mates; or if they are colonial nesters, there may not be enough individuals to form a breeding colony. Since the stimulus of other courting pairs is a factor in the endocrine development of members of a breeding pair, courtship, copulation, and ovulation may be impaired in colonial nesters. Although the mass slaughter of the passenger pigeons sped them on their way to extinction, the final step may have come because their numbers were so few that they could no longer breed successfully. For species such as the whooping crane, whose numbers have fallen below 50, inbreeding may lead to a loss of vigor and fertility that may result in extinction despite all the efforts being made to protect and enlarge the population. Therefore, a lower limit may exist below which population growth will become negative and the population will go to extinction. If this lower limit is called M, then the logistic equation can be modified from

$$\frac{dN}{dt} = rN \left(\frac{K - N}{K} \right)$$

to

$$\frac{dN}{dt} = rN \left(\frac{K - N}{K} \right) \left(\frac{N - M}{N} \right).$$

From this equation, it can be seen that if N becomes smaller than M, then dN/dt will become negative and the population size will go to zero. Note that the conception of a lower limit M is certainly plausible, but its determination in a population would be difficult if not impossible, at least in time to do any good. Probably only after the population became extinct would it be apparent that its size had fallen below its M value.

In conclusion, a variety of factors may be involved in the natural regulation of population size. Among them are the weather, predators, pathogens, the quantity and quality of the food supply, the availability of adequate shelter or territory, inter- and intra-specific competition, and the intrinsic physiological, behavioral, and genetic qualities of the individual members of the population. It is hazardous to attempt to categorize these factors as either density-dependent or density-independent because the same factor, for example the weather, may act in a density-independent way on one population and have density-dependent effects on another or may even be density-independent and density-dependent on the same population at different times.

DENSITY-RELATED SELECTION

Apart from the relation between population size and random genetic drift, to be discussed later, population geneticists have not been greatly concerned with the numbers of individuals in a population or with population density. Rather than dealing with absolute numbers or with density, they have been primarily concerned with changes in the relative frequencies of genes, genotypes, and phenotypes. Recently, more attention has been directed to the possible relation between population density and the operation of natural selection. The possibility that selection pressures might change as population density increased stemmed from such work as that of Gershenson (1945), Calhoun (1952, 1962), Chitty (1957, 1960, 1967, 1970), and Christian (1961), Christian and Davis (1964) with small mammals, Cody (1966) with birds, and Dobzhansky and Spassky (1944), Lewontin (1955), Birch (1955), Wellington (1957, 1960, 1964), and Pimentel (1961, 1965, 1968) with insects. Haldane (1956) and Williamson (1958) were among the first to theorize about the relation between density and natural selection.

It is, perhaps, worth noting at this point that even though population size and density are sometimes used almost interchangeably, they are fundamentally distinct. The amount of random genetic drift, for example, is dependent on the effective size of the breeding population; its density, so far as drift is concerned, is immaterial. Density-related selection, on the other hand, is dependent on the density of the population; its total size, so far as density-related selection is concerned, is immaterial. Population density is the number of individuals per unit area or per unit volume. Although population size and density are related if the population is confined to a given area such as an island they are not necessarily related. For example, if a population doubles its range as it doubles in size, population size has doubled, but population density is unchanged. Conversely, if a population is restricted to half its former range with no change in numbers, population density is doubled with no change in population size. Here only the effects of changes in population density are considered.

In the last decade, most of the discussion of density-related selection has been phrased in terms of r- and K-selection (MacArthur, 1962; MacArthur and Wilson, 1967). Recall that the logistic equation for population growth took the form

$$\frac{dN}{dt} = rN\left(\frac{K-N}{K}\right)$$

where r is known as the intrinsic rate of natural increase or the instantaneous coefficient of population growth, and is a measure of the biotic or reproductive potential of the population. K is the upper limit of the population size under the existing environmental conditions, and is known as the carrying capacity of the environment. Thus, r indicates the rate of growth of the population in the log phase of the sigmoid growth curve, and K indicates the upper limit, or the upper asymptote, reached in the size of the population.

MacArthur and Wilson suggested that the selection pressures on a population under favorable conditions at low density in the log phase of growth (*r*-selection) are apt to be very different from the selection pressures on a population crowded together at high density with limited resources (*K*-selection). They argued that *r*-selection would favor individuals able to locate open favorable habitat quickly, who would then rapidly produce many offspring to exploit the available resources before other competing individuals could arrive. Thus, *r*-selection would tend to favor high productivity or a high value for *r*. Furthermore, when the existing resources start to become exhausted and the habitat less favorable, *r*-selection would favor individuals who would tend to disperse from that habitat in search of new, more favorable habitats.

On the other hand, if the population has approached its upper limit in size under the existing conditions, selection will act quite differently. No longer will it be so important to have a high *r* value. Instead, individuals will be selected for their competitive ability, their ability to seize and hold a part of the environment and its resources and to utilize them efficiently. *K*-selection, then, will favor those individuals best able to produce proportionately more offspring under crowded, competitive conditions. Rather than favor high productivity, *K*-selection will favor high efficiency in the utilization of resources to maximize *K*. Relatively less energy may be expended on reproduction, but the fewer offspring produced may receive better care or more sustenance.

The concept of *r*- and *K*-selection has stimulated considerable discussion and has been particularly useful in drawing attention to the possible importance of the density or the numbers of organisms as factors in natural selection. However, a few words of caution may be appropriate. The terms *r*- and *K*-selection stem from the logistic curve of population growth. As seen earlier, the logistic curve is not a very realistic approximation of the changes in number that actually occur in natural populations. Apart from a population of fruit flies placed in a fresh bottle of food, population growth seldom approximates the sigmoid growth curve.

Furthermore, the expressions *r*-selected and *K*-selected populations imply that values of *r* and *K* are known for these populations and that it has been demonstrated that, on the one hand, the *r*-selected population is in the log phase of growth under uncrowded conditions, and, on the other, that the *K*-selected population has reached its upper limit. However, *r* and *K* values are not easily determined in natural populations. Because a population has a low density is not proof that it is undergoing r-selection, for that density may represent its upper limit under existing conditions. Moreover, environmental conditions are seldom constant so that K values are not fixed but are always changing; and even if a population is in the log phase of growth, it soon passes through that phase as it starts to encounter the limits to growth caused by limited resources. Therefore, the population size and density of natural populations fluctuate, and even though density-related selection may occur, it will be along a continuum related to size or density rather than being solely of one type or the other.

Thus, the same population may, at different times, undergo the equivalent of r-selection and K-selection. At all densities, selection will act to maximize fitness—that is, to maximize the number of offspring surviving to reproductive age. To suggest that at low densities selection acts to maximize productivity (or r) and at high densities to maximize efficiency (or K) seems somewhat artificial and substitutes secondary components of fitness for fitness itself.

Carson (1968) and Ford (1971) have taken a somewhat different approach to the relation between population numbers and natural selection. Both argue that during the phase of rapid population growth, which Carson called the "population flush," natural selection is relaxed. Consequently, the amount of genetic variability in the population increases, because under the favorable conditions, the genetically less fit individuals are not eliminated as they would normally be (e.g., Owen, 1964). When a plateau in numbers is reached, selection again becomes more stringent, even though the population may be large, for each pair now leaves but one pair of descendants, on the average. However, selection becomes even more rigorous when the population starts to decline in numbers. There can be little doubt that populations fluctuate in size, but the relation between numbers and the amount of genetic variability is less well established. Whether the postulated changes in variability in relation to population size are caused by relaxed and intensified selection pressures is even more dubious, for other explanations are possible. These ideas contrast with those of MacArthur and Wilson, who argue, not that selection is minimal in an expanding population, but that it is different.

However, perhaps the most important point to note is that MacArthur and Wilson's concept of r- and K-selection treats selection in an expanding population and in a stable population, but not selection in a declining population. Nevertheless, the selective forces in a declining population may be the most significant of all, not necessarily because they are more rigorous than at other phases of the population cycle, but because they determine the characteristics of the small number of survivors that become the progenitors of all subsequent generations when the population again starts to expand. Selection does not cease to operate in small populations, but is effective at all population sizes (Merrell, 1953). Therefore, it can well be argued that the most important type of selection is not r-selection or K-selection, but the selection in a declining population, or d-selection, if you will, which may set the future course of the population for generations to come.

Finally, note that density-related selection can also give rise to balanced polymorphism. For example, if it is postulated that a given allele, A_1, is selectively neutral at density D, has a reduced fitness relative to its alternative allele, A_2, at densities lower than D, and an enhanced fitness relative to A_2 at densities greater than D, then fluctuations in population density about D will tend to perpetuate polymorphism at this locus. In this simple example, the polymorphism will tend to be inherently unstable, for if the population density stays either above or below D for too long, one allele or the other may go to fixation. However, if the fluctuations in density are cyclical and centered around D, both alleles may persist in the population. As many species undergo regular seasonal cycles in population density, it is possible for polymorphism, and seemingly a balanced polymorphism, to be maintained by density-related selection of this kind.

If the fluctuations in density are assumed to stem from environmental effects independent of the genotypes in the population, then the selective forces involved are separate and distinct from frequency-dependent selection. Even though different genotypes may respond differently to different densities, the responses need not be dependent on the relative frequencies of the genotypes. Thus, density-related selection can be separated conceptually from frequency-dependent selection, and is still another mechanism by which balanced polymorphism may be achieved.

The actual changes in population density may be caused by density-dependent factors such as the accumulation of metabolic wastes, or to density-independent factors such as climatic change. Nonetheless, the changes in fitness of the various genotypes may be related to density, no matter whether the changes in density are due to density-dependent or density-independent causes. Although similar conclusions could be drawn if *population size* were substituted for *population density* in the above discussion, it seems more meaningful biologically to think in terms of the effects of population density rather than population size.

It may be quite unrealistic to postulate a simple linear relationship between density and fitness of the sort suggested above, and other relationships can be envisioned. Even so, such polymorphisms are inherently unstable. But if the changes in density are cyclical around D, it may be extremely difficult to distinguish this type of polymorphism from a stable balanced polymorphism maintained by a mechanism such as heterozygous advantage or frequency-dependent selection.

Subsequent to MacArthur and Wilson's development of the concept of r- and K-selection, several other authors have published papers on the theoretical relationship between population density and natural selection (Turner and Williamson, 1968; Hairston, Tinkle, and Wilbur, 1970; Pianka, 1970, 1972; Anderson, 1971; Charlesworth, 1971; King and Anderson, 1971; Roughgarden, 1971; Clarke, 1972; May et al., 1974; Southwood et al., 1974). However, it seems fair to say in the case of density-related selection as in the case of selection in environments heterogeneous in space (the Ludwig effect) or time (seasonal or cyclical selection) that the theory has far outstripped the data, for few well-documented cases of polymorphism maintained by any of these three types of selection have been reported (Hedrick, Ginevan, and Ewing, 1976).

The studies of Wellington (1957, 1960) with the western tent caterpillar, (*Malacosoma pluviale*) were among the first to show that the genetic characteristics of members of dense populations differed from those of members of the population at its low ebb. In essence, he reported that caterpillars from dense populations were sluggish and inactive, formed compact tents, fed less, and were more susceptible to disease than caterpillars from populations at low density in the growth phase of the cycle, which were more active, built elongated tents, and moved out into the foliage to feed.

Studies of the oscillations in population density in both field and laboratory populations of higher vertebrates have revealed that changes in the physiological or genetic characteristics of members of the population may occur. The stress caused by crowding, for example, leads to enlarged adrenal glands, which, in turn, affect behavior, reproductive success, and disease resistance (Christian, 1961; Christian and Davis, 1964; Calhoun, 1962). A few other papers have reported genetic or behavioral changes associated with changes in population density in small mammals (Semeonoff and Robertson, 1968; Tamarin and Krebs, 1969; Krebs, 1970; and Gaines and Krebs,

1971). Chitty (1960, 1967) suggested that cycling in vole populations led to changing selection pressures with respect to aggressiveness, with the genetically more aggressive individuals favored during the growth phase of the cycle but becoming maladaptive at high densities, thus leading to a population decline and selection favoring the less aggressive types. The idea that qualitative genetic changes may play a role in population cycles is intriguing. An effort was made by Krebs et al. (1973) to detect changes in the frequency of alleles at a transferrin locus in voles, and, in fact, such changes were found to be associated with changes in population density. While these results show that the genetic characteristics of populations may change with population density, no direct connection between aggressive behavior and the transferrin alleles has been made so that Chitty's theory of aggression, specifically, remains unproved.

From the foregoing discussion, it should be clear that the polymorphisms in natural populations have many possible explanations. In a practical sense, one must first distinguish between environmentally induced phenodeviants and variation that is primarily genetic. If the variation is, in fact, genetic, a variety of mechanisms may be involved in the maintenance of the variation. The Castle-Hardy-Weinberg law ensures the persistence of genetic variation, but this equilibrium will be influenced by mutation and various forms of selection, and, as shall be seen subsequently, by migration and random genetic drift. If polymorphism in Ford's more limited sense exists, it may be transient or balanced, but to establish the difference will ordinarily require that the population be followed long enough to determine whether gene-frequencies are changing or are remaining more or less constant through time. Moreover, even if the polymorphism appears balanced, a variety of mechanisms may be responsible. Even though we have dealt with each mechanism separately as if they were independent alternatives, the genetic variation in a natural population may, in fact, be the result of a combination of forces acting simultaneously, ranging from mutation and migration to directional selection and various forms of balancing selection. For example, mutation and migration may be feeding genetic variation into a population in which frequency-dependent selection, density-related selection, seasonal selection, the Ludwig effect, and heterozygous advantage all are operative. To seek single causes for the gene-frequency changes observed in such a situation is far too simplistic.

An inkling of the subtlety of natural selection was obtained by Underhill (Underhill and Merrell, 1966) in tests of three DDT-resistant stocks and their corresponding controls. At the time of the tests, the two types of populations had been maintained in population bottles (Reed and Reed, 1948) for twelve years under identical conditions with the single exception of the presence of DDT- impregnated strips of filter paper in the environment of the resistant populations. During this twelve-year period, the exposed populations had not only evolved a high level of DDT-resistance but had diverged from their controls in other ways as well. The median life span of resistant flies in both the fecundity and longevity tests was greater in every case than that of the corresponding control group. The fecundity and fertility data in Merrell (1960) and Underhill and Merrell (1966) seemed to show a tendency for higher fecundity and fertility in the control populations than in the resistant populations. Nevertheless, despite this tendency, the resistant flies, with one exception, actually produced more eggs and, ultimately, more surviving adult offspring than the controls because of their longer life span.

These differences are hardly due to chance, but reflect the different selection pressures in the two types of populations. In the unexposed flies, when a depleted food-bottle was replaced by a fresh food-bottle, there was a selective premium on fecundity and fertility among the first flies to reach the fresh food. On the other hand, when a fresh food-bottle was added to an exposed population, the bottle also contained a fresh DDT-strip, and the maximum exposure of the larvae and adults to DDT came at that time, before the strip became covered with food and debris. Therefore, although fitness in the control populations seemed to depend on an early burst of egg-laying to take advantage of the fresh food (r-selection if you wish), in the populations exposed to DDT, fitness appeared to depend not only on the ability to survive exposure to DDT but also on an extended life span and fecundity. Thus, egg-laying could continue long enough so that at least some of the eggs were laid under conditions favorable to their development. Therefore, it seems possible to account not only for the evolution of DDT-resistance but also for the other differences in fecundity, fertility, and longevity between resistant and control populations.

CHAPTER 8

Genetic Loads

Chetverikov (1926) was the first to report that many "normal" individuals in wild populations carried recessive mutant genes in the heterozygous condition, a finding subsequently confirmed repeatedly. Most of these recessives were found to be detrimental to some degree when homozygous. Therefore, these homozygotes represent a reproductive waste and reduce the overall fitness of the population, and thus constitute a *genetic load* the population has to bear. If the mutants are not completely recessive but have adverse effects in the heterozygotes, the load includes the heterozygotes as well. The concept was first dealt with by Haldane (1937) and the expression *genetic load* first used by Muller (1950). The genetic load has been defined in several ways (Crow, 1970), which usually lead to the same numerical conclusions. The term *load* is used to refer to a population that is more or less in equilibrium to express the change in average fitness of the population associated with maintaining its variability. One definition of genetic load is that it is the fraction by which the population mean differs from the value for some reference genotype. The trait usually used is fitness and the reference genotype is ordinarily the genotype of maximum fitness among the possible genotypes. A similar definition is that the genetic load is the fraction by which the population mean is changed owing to the presence of some factor as compared to the population mean of an otherwise identical population from which the factor is missing. A third, somewhat different definition is the amount of reproductive excess, expressed as a fraction of the

178

reproductive rate, required to maintain the size of the population when it is in equilibrium for the factors under consideration. However, Crow (1970) showed that all three definitions could lead to equivalent loads.

Whereas *load* is used where gene frequencies are in approximate equilibrium, Haldane (1957) introduced the term *cost* for the effect on fitness of the natural selection needed to carry out gene substitutions in evolution. The cost of natural selection, then, refers to a more dynamic situation where gene frequencies are changing. The cost of natural selection has also been called the substitutional load. Genetic load and genetic cost are clearly related concepts, for both refer to the loss in average fitness of a population because of the genetic variability present.

The genetic load of a population in equilibrium is the expressed load, but the total load would include the effect of the recessive or partly recessive factors whose effects are ordinarily not expressed but may be revealed by inbreeding. The load is usually expressed in terms of fitness, but it may be applied to other measurable traits such as viability or sterility as well. When viability was used, Morton, Crow, and Muller (1956) found it helpful to express the load in terms of lethal equivalents. They defined a *lethal equivalent* as a group of mutant genes of such number that, if dispersed in different individuals chosen at random from a population, they would cause, on the average, one death—that is, one lethal mutant, or two mutants each with a 50% probability of causing death, or four mutants each with a 25% probability of causing death, and so on.

There are several kinds of genetic load. Following is a list of some possible components in the genetic load of a population:

A. *Input Load*. This load refers to the inferior alleles introduced into the gene pool of a population either by mutation or immigration.
   1. *Mutational load*. The load owing to recurrent mutations whose average effect is to lower the fitness of the population in the environment where it lives.
   2. *Immigration load*. The load created in a population by the immigration of individuals from other populations. Because the immigrants have evolved elsewhere under other environmental conditions, they are less apt than the local population to be adapted to the local environment. Therefore, the genes they introduce into the population will constitute a genetic load.
B. *Balanced Load*. The load created by selection favoring allelic or genic combinations that by segregation and recombination form inferior genotypes every generation. We have already considered several ways a balanced load might be generated.

1. *Segregational load* (the homozygous-disadvantage load). The load generated when the heterozygote is more fit than either homozygote. Here, the reference genotype is the heterozygote, and the load is the reduction in fitness resulting from the homozygotes produced each generation by segregation.
2. *Recombination load.* The load generated if, for example, two gametic combinations, AB and ab, are favored by selection, but Ab and aB are not favored. In this case, the average effect of recombination increases the frequency of the inferior types, and if all four types exist in a population in some sort of stable equilibrium, recombination will produce a decrease in fitness. As Fisher (1930) pointed out, if this be the case, natural selection would be expected to favor a reduction in recombination between the A and B loci, for example, by a chromosome rearrangement.
3. *Heterogeneous-environment load.* If different genotypes are adapted to different environments or subniches, the frequency of the genotypes should match the prevalence of the environments to which they are adapted. If the population deviates from this ideal matchup between genotypes and environments, it will have reduced fitness and a genetic load.
4. *Frequency-dependent load.* Here, too, at the optimum frequencies there will be no genetic load, for all genotypes are then equally fit, but any deviation from these equilibrium frequencies will generate a genetic load.
5. *Seasonal or cyclical load.* As long as the polymorphism persists, a load will exist because of the presence of a currently less favored type.
6. *Density-related load.* Similarly, as long as the polymorphism persists the genotypes not favored at the existing density constitute a load.
7. *Meiotic-drive load.* If a gene or combination of genes is favored by meiotic drive or gametic selection but is selected against at other stages of development, population fitness will be reduced and a load created.
8. *Incompatibility load.* In mammals, a deleterious maternal-fetal antigen-antibody reaction is sometimes observed. In humans, for instance, an Rh$^+$ child may have a lowered fitness if carried by an Rh$^-$ mother. As the Rh$^+$ infant is heterozygous, one of each type of allele is lost, but a greater proportion of the rarer allele will be eliminated. Therefore, in time the rarer allele will be lost from the population if no type of

compensation occurs. This type of genetic load is called the incompatibility load.

9. *Finite-population load.* In a finite population, gene frequencies tend to drift away from the equilibrium values they would have in an infinite population, in which the fitness would normally be maximized. Thus, the departure from the equilibrium values in finite populations will usually decrease fitness, and thereby create a genetic load.

It has also been suggested that genes for strong dispersal could create a genetic load if they tended to take well-adapted individuals out of the population, leaving the less-fit individuals behind, but thus far, there seems to be little evidence to support this possibility.

C. *Substitutional Load.* This load is generated when selection favors the replacement of an existing gene by a new allele. It was called the cost of natural selection by Haldane (1957), and is the genetic load associated with transient polymorphism.

The concept of genetic load has generated a considerable amount of discussion and criticism, much of which has centered around the relative importance of the mutational load as compared to the segregational load in natural populations, but some has been directed at the concept of genetic load itself. The word *load* is a loaded term, if you will pardon the expression, and Muller, who first used it, intended it that way—to convey the burden of death, illness, sterility, and sorrow imposed on humans by mutation. Others, however, (e.g., Crow, 1970) have suggested that the load is not necessarily burdensome, but rather is a measure of the amount of natural selection associated with a certain amount of genetic variability, and that this variability is essential to the continued existence of the population because it provides the raw material for continued adaptation and evolution.

MUTATIONAL LOAD

First, we shall consider more thoroughly the nature of the mutational load. Haldane (1937) was the first to point out some of the interesting quantitative aspects of the mutational load. It was already seen in Chapter 5 that at equilibrium deleterious genes are eliminated from a population at the same rate they are being produced by mutation. A completely recessive deleterious mutant will reach a frequency of $\hat{q} = \sqrt{u/s}$ at equilibrium. The frequency of the homozygotes, which suffer the loss in fitness and therefore are a genetic load to the population, is $\hat{q}^2 = u/s$. As each homozygote suffers a loss of fitness equal to s, the selection coefficient, the loss of fitness to the population equals the frequency of the deleterious homozygotes (u/s) multiplied by the loss of fitness (s), or the load, $L = s(u/s) = u$. In other words, the mutational load for a deleterious recessive gene equals the mutation rate itself, for the selection coefficients cancel out of the equation. Therefore, it does not matter whether the gene is lethal $(s = 1.0)$, semilethal $(s = 0.5)$, or nearly harmless $(s = 0.001)$, because at equilibrium the loss of fitness to the population depends on the mutation rate, u, and not on the effect of the gene on the fitness of the individuals carrying it, provided selection keeps the gene rare. Moreover, the total mutational load for many such loci equals the sum of all of the individual mutation rates, or $L = \Sigma u_i = U$.

From Chapter 5 it can also be shown that the equilibrium frequency for an incomplete recessive (i.e., a partially dominant) deleterious mutant is approximately $\hat{q} = u/hs$. This value is obtained by simplifying

$$\Delta p = \frac{spq^2 + spq\ [h(2p - 1)]}{1 - 2hspq - sq^2} + vq - up = 0,$$

The spq^2 and vq terms are negligible and can be dropped. Since q is small, $(2p - 1)$ and $(\overline{W} = 1 - 2hspq - sq^2)$ both approximate one. Then $hspq = up$ and $\hat{q} = u/hs$. The frequency of the heterozygotes will be $2pq$ or $2pu/hs$. Again, since q is very small, p is very nearly one so that the frequency of heterozygotes at equilibrium approximates $2u/hs$. The loss of fitness for each heterozygote equals hs, so the loss of fitness by the population equals the frequency of heterozygotes $(2u/hs)$ multiplied by their loss of fitness (hs). In this case, the mutational load owing to an incompletely recessive or partially dominant deleterious mutant equals $2u$ or twice the mutation rate. As before, the selection coefficients drop out, and the mutational load is independent of the effect of the allele upon its carriers. For all such loci, the combined load would equal $2U$, where U is the sum of the individual mutation rates.

Calculations of this sort lead to the conclusion that if the effects of different loci are independent, the mutational load will be between one and two times the total mutation rate to harmful genes per gamete per generation. Because the evidence seems to indicate that most deleterious recessive genes also have some adverse effects when heterozygous, the mutational load is usually thought to be close to twice the total mutation rate rather than the lower figure.

The same conclusions can be reached in a different way. For a complete recessive, the average fitness of a population in equilibrium is $\overline{W} = 1 - sq^2$, and the genetic load can be though of as $L = sq^2$. At equilibrium $\hat{q}^2 = u/s$ and therefore $L = s(u/s) = u$, as before, Similarly, for a partial dominant, the average fitness of a population in equilibrium is $\overline{W} = 1 - 2hspq - sq^2$, and the genetic load is $L = 2hspq + sq^2$. Since q is very small and p is nearly one, the load is approximately $L = 2hsq$. Since at equilibrium $\hat{q} = u/hs$, the load is equal to $L = 2hs(u/hs) = 2u$, as before. This form of the load makes clear that even though the homozygotes may be considerably less fit than the heterozygotes, most of the eliminations are through the heterozygotes rather than the homozygotes.

SEGREGATIONAL LOAD

In Chapter 5 we saw that, when the heterozygote was superior in fitness to the homozygotes, the equilibrium gene frequencies were dependent on the selection coefficients so that $\hat{p} = s_2/(s_1 + s_2)$ and $\hat{q} = s_1/(s_1 + s_2)$. Furthermore, the average fitness of the population was $\overline{W} = 1 - s_1 p^2 - s_2 q^2$. Therefore, the segregational load is $L = s_1 p^2 + s_2 q^2$, which simplifies to $L = s_1 s_2/(s_1 + s_2)$. This can also be shown to equal $L = s_1 \hat{p} = s_2 \hat{q}$. With this last formula, it can be seen that the segregational load can be estimated from the frequency of just one allele and its homozygous effect on fitness—a useful bit of information. Therefore, if the population consisted solely of heterozygotes, it would have an optimal fitness of 1.0. When the homozygotes are present, the average fitness is lowered by the amount $s_1 s_2/(s_1 + s_2)$, which may be of considerable magnitude. It is worth noting that the less drastic allele contributes the greater part of the segregational load. For example, if $s_1 = 0.01$ and $s_2 = 0.99$, $\hat{p} = 0.99$ and $\hat{q} = 0.01$, but $s_1 p^2 = (0.01)(0.99)^2$ while $s_2 q^2 = (0.99)(0.01)^2$. Hence, it can readily be seen that the less drastic allele (p) contributes 99% of the total segregation load while the more drastic allele contributes only 1%.

The worst possible case is sometimes suggested to be one in which both alleles are lethal (e.g., Wallace, 1968a, 1970a; Mettler and Gregg, 1969), supposedly producing a balanced lethal system where half of the offspring die and the genetic load is 0.5 ($s_1 = s_2 = 1.0$, and $s_1 s_2/(s_1 + s_2) = 0.5$). However, heterozygotes for lethal alleles normally die, too, so that unless some extraordinary kind of intracistronic complementation occcurs, such an extreme situation will never arise.

In an attempt to determine the relative importance of the mutational load and the segregational load, the effect of inbreeding on the genetic load has been estimated. The mutational load calculated above was the random mating load, L_R, and for an incompletely recessive harmful mutation $LR = 2u$. If the population were inbred to the point where it consisted only of AA and aa homozygotes and there was no change in gene frequency, then the frequency of aa homozygotes would equal the gene frequency, q. As $q = u/hs$ at equilibrium, and each homozygous aa individual suffers a loss of fitness, s, the average fitness of the population would be lowered by the amount $s(u/hs)$ or by u/h, and the inbred load LI would equal u/h.

The load ratio is taken as the inbred load divided by the random load, or

$$\frac{L_I}{L_R} = \frac{u/h}{2u} = \frac{1}{2h}.$$

Mutations quite detrimental when homozygous have generally been reported to reduce the fitness of heterozygotes only a few percent, at most (e.g., Simmons, 1976). Therefore, the values of h will also be small, of the order of a few percent (depending, of course, on s, too). If h equals 0.02, for example, then the load ratio for genes maintained in the population by recurrent mutation would be 25, and it can be seen that, in general, the L_I/L_R ratio for the mutational load would be expected to be a large number.

On the other hand, the random-mating segregational load is $L_R = s_1 s_2/(s_1 + s_2)$. If such a population were inbred without gene-frequency change so that it consisted only of AA and aa homozygotes, the frequency of these homozygotes would equal the gene frequencies. Thus, the frequency of AA would be $p = s_2/(s_1 + s_2)$, and of aa would be $q = s_1/(s_1 + s_2)$. The segregational load in such an inbred population would equal the loss of fitness by the two types of homozygotes, which is

$$L_I = \left(\frac{s_2}{s_1 + s_2}\right)s_1 + \left(\frac{s_1}{s_1 + s_2}\right)s_2 = \frac{2s_1 s_2}{s_1 + s_2}.$$

In this case, the load ratio is

$$\frac{L_I}{L_R} = \frac{\dfrac{2s_1 s_2}{s_1 + s_2}}{\dfrac{s_1 s_2}{s_1 + s_2}} = 2.$$

Therefore the load ratio for the segregational load is quite small, and should, in fact, equal the number of alleles involved in the polymorphism. This difference in the load ratios expected for the mutational load as compared to the segregational load seemed to offer a means to determine the relative importance of these two types of loads in natural populations. However, these hopes have not been realized, for, too often, the load ratios have fallen into the never-never land between the low values expected if the genetic load were primarily a segregational load and the high values expected if the load were mutational. These results have been reviewed by Dobzhansky (1970) and Lewontin (1974).

Another attempt to distinguish between heterotic and partially dominant genes involved the study of the differing fitnesses of newly arisen mutations and of mutations that have been screened by natural selection. For this purpose, the so-called D/L ratios were determined where L is the severe loss of fitness of homozygotes for lethals and semilethals and D is the loss of fitness of detrimentals and *quasi-normals*. The D/L ratio will have different values for different degrees of dominance of the genes when they come from equilibrium populations where the equilibrium frequencies of different genes with different degrees of detrimental effect will be different. Furthermore, the D/L ratio should be different for homozygotes for new mutations on which selection has not yet acted to change their frequencies, for the ratio will only reflect the total occurrence of the different types of mutations independent of the level of dominance or overdominance. Greenberg and Crow (1960) calculated D/L ratios for *Drosophila*, but Lewontin (1974) criticized the method as suffering difficulties similar to those found in calculations of genetic-load ratios, and the method has not given unequivocal answers to the questions posed.

As pointed out, the mutational load equals the total gametic mutation rate if the mutations are completely recessive, or twice that rate if the mutations have any adverse effect in heterozygotes. The problem, of course, is to determine just what the total mutation rate per gamete might be. The difficulties in estimating mutation rates at individual loci have already been discussed and here the problem is compounded. Morton, Crow, and Muller (1956) estimated that the average person carries 3 to 5 lethal equivalents in the heterozygous state and that the total mutation rate in man is about 0.03 to 0.05 per gamete per generation. They considered this an underestimate, however, and suggested a somewhat higher value of 0.06 to 0.15 per gamete per generation. Crow (1970) used several assumed values for the total mutation rate (0.10, 0.20, and 0.40), and his estimates of L ranged from 0.034 to 0.444, but most of his estimated values for the mutational load were between 0.10 and 0.25, a sizable but not intolerable load.

The segregational load, however, poses a problem. If the two homozygotes are only slightly less fit than the heterozygote so that $s_1 = s_2 =$

0.02, then $L = (s_2 s_2)/(s_1 + s_2) = 0.01$. Thus, there is a 1% loss of fitness to the population because of segregation at a single locus. If there were many such loci segregating, each independently causing a 1% loss of fitness, the population fitness would equal $(0.99)^n$, where n is the number of loci. If n were 500, for example, then the mean fitness of the population would only be 0.007. This argument by Lewontin and Hubby (1966; Lewontin, 1967, 1974) was presented in relation to their discovery of the high level of allozyme heterozygosity in natural populations. King (1967), Milkman (1967), and Sved, Reed, and Bodmer (1967) independently pointed out that some form of threshold model with truncation selection could help to decrease the segregational load. The model assumes that selection favors individuals above a certain level of heterozygosity but acts against those more homozygous individuals falling below this level. The model seems to echo some of Lerner's (1954) concepts of genetic homeostasis. The excessively high fitness of the hypothetical multiple heterozygote was more a product of the model than of reality, for individuals heterozygous at every locus will not, in fact, exist.

Other possibilities have been suggested for decreasing the segregation load. One is that the genes do not act independently, but are linked in inheritance and in gene action as well as under selection. Another is that the selection coefficients are so small—that is, less than the reciprocal of the effective population size—that the alleles behave as if they were neutral, with their frequencies largely determined by random drift and mutation, and thus, there will be no segregational load.

Another possibility is that the heterozygotes at a given locus are rarely superior in fitness to both homozygotes. If only a small proportion of deleterious mutations were overdominant, they would tend to accumulate in natural populations, trapped in a stable equilibrium, while the other deleterious mutations, not favored in the heterozygous condition, would tend to be eliminated. Wallace (1968a), in fact, has argued that such overdominance for deleterious and even lethal genes exists in natural populations, but that it is only expressed in the particular genetic and environmental background in which it arose. Thus, the very crosses necessary to test for over-dominance under controlled conditions are apt to make it disappear. In view of the experimental difficulties in testing for single locus heterosis (Lewontin, 1974) and the rarity of reliable reports of such heterosis, perhaps the problem itself will disappear because, in reality, it does not exist. A long-standing issue in this connection has been the heterozygous fitness of lethals in *D. melanogaster*. Although there have been some reports of heterotic lethals (e.g., Wallace, 1958a), the preponderance of the evidence now seems to indicate a slight deleterious effect of lethals in heterozygotes of the order of a 1 to 5% loss in fitness, usually measured as viability (Cordeiro, 1952; Prout, 1952; Stern et al., 1952; Hiraizumi and Crow, 1960; Crow and Temin, 1964; Yoshikawa and Mukai, 1970; Simmons, 1976).

As pointed out above, many other forms of balancing selection are possible in addition to heterozygous advantage, not all of which generate such large genetic loads. There can be no doubt now of the high levels of genetic polymorphism in natural populations. Perhaps the realization that it is unlikely that heterozygous advantage can be responsible for more than a small proportion of this polymorphism will broaden the search for the mechanisms that maintain it.

SUBSTITUTIONAL LOAD

The cost of evolution was first dealt with quantitatively by Haldane (1957, 1960) and his conclusions seemed so bizarre that they have come to be called Haldane's dilemma. The question he raised was, How many genetic deaths would be necessary to bring about the replacement of one allele in a population by another? In other words, he wanted to know the genetic cost of transient polymorphism. As the substitution of one gene for another imposes a genetic load, this load is also called the *substitutional* or *transient* load as well as the cost of natural selection.

One conclusion he reached was that the number of generations required for a given change in gene frequency is inversely proportional to the intensity of selection. The less intense the selection, the longer the time required to effect a given change. In a sense, the selection intensity multiplied by the time required is a constant so that the total amount of selection required is a constant.

He then attempted to estimate the loss of fitness in a population during the time gene substitution is occurring, and he found that for a favored dominant, the cost (C), measured as the total number of genetic deaths is equal to

$$C = - \ln p_0$$

where p_0 is the initial frequency of the favored dominant and s is not large. For a gene without dominance, $C = - 2\ln p_0$. Thus, the least cost is incurred with a fully dominant mutant, the cost is greater for a gene without dominance, and greatest for a complete recessive. The value $(- \ln p_0)$ then is an estimate of the lower limit for the cost of a gene substitution (Crow, 1970).

The treatment is somewhat simpler for haploids than diploids, but the conclusions are similar, and the cost of a gene substitution in haploids also is $C = - ln\ p_0$ if s is less than about 0.1. The table below, from Crow and Kimura (1970), gives values of C in haploids for different values of s when the initial frequency of the favored allele (p_0) is 0.01.

p_0	s	C
.01	1.00	99
.01	.99	52
.01	.50	6.2
.01	.10	4.8
.01	.01	4.63
.01	Limit	4.61

First, it should be understood that C, the cost, is the number of non-survivors or genetic deaths required per survivor to effect the substitution. Put another way, if the adult population size is roughly constant from generation to generation and $s = 0.1$, then 4.8 times the average population size per generation must die to achieve the gene substitution, starting from a frequency of $p_0 = 0.01$. It can be seen that the cost is extremely high; at best, the number of deaths is several times the total population size in a single generation. Moreover, the total cost is less if s is small. However, the cost does not change greatly if s is below 0.1. Therefore, when s is small or below about 0.1, the cost is essentially independent of s, but is dependent on the initial frequency of p_0. It is for this reason that the formulas given above, for small s, were in the form $C = - ln\ p_0$. Following is the cost for different values of p_0:

p_0	C
10^{-6}	14
10^{-4}	9
10^{-2}	4.6
10^{-1}	2.3

Because genes previously selected against are going to be rare when selection starts to favor them, the cost of substitution is going to be rather large. For example, if p_0 is 10^{-4} and the population size is 10,000, there would have to be 90,000 genetic deaths in that population if the gene substitution were to occur in a single generation. Obviously, this would seem to be an intolerable genetic burden, which would have to be spread over many generations to make it more tolerable for the population.

In a diploid species, the cost of substituting a moderately rare mutant, previously maintained in the population by a mutation rate of 10^{-4} to 10^{-5} is from about 10 to 100. Haldane took 30 to be a representative figure, and further assumed that a species could devote 10% of its reproductive excess per generation to this allelic substitution. From these assumptions, he calculated that a species could carry out only one gene substitution every 300 generations on the average. Therefore, if two species differ at 1000 loci and 10% of the reproductive excess each generation were devoted to gene substitution, it would have required some 300,000 generations for these substitutions to have occurred sequentially.

Haldane's *dilemma* lay in the fact that the cost of evolution appeared to be so extremely high that to bring the cost within reasonable bounds, it appeared that the rate of evolution had to be inordinately low. Evolutionary changes generally involve a number of loci rather than just one. If Haldane's assumptions were correct, the cost of simultaneous substitutions at the rate given above would be so great that for many species, they would lead to extinction rather than evolution.

Various suggestions have been made in an effort to resolve Haldane's dilemma, of which, incidentally, he was quite aware. One is that most gene substitutions are neutral and have no cost. Another is that, as cost depends on gene frequency, genes of relatively high frequency (and thus those with rather minor deleterious effects) would more often be involved. Another is that Haldane's argument assumes that the genes act independently, whereas this may not, in fact, be the case. Here, too, as in the paradox of the segregational load, truncation selection has been suggested as a possibility, which is, in effect, the equivalent of strong epistasis. With truncation selection, the number of loci on which selection could act simultaneously with the same total effect is greatly increased. Last, it should be noted that the lowest cost is associated with the substitution of a fully dominant allele. This finding supports the point made in Chapter 5 about the possible importance of the evolutionary role of dominant genes.

The conclusion from these calculations—that evolutionary rates must be slow—is clearly contradicted by the rapid evolutionary changes that have occurred in the past 10,000 years in numerous species of domesticated plants and animals, by the rapidity with which industrial melanism has been established in many moth species over the past two centuries, and by the numerous examples of pesticide resistance that have developed in the present century. These biological facts seem to require some revision in the assumptions of the model.

Although the concept of genetic load has stimulated much discussion, it has also been subjected to considerable criticism (Li, 1963a, 1963b; Sanghvi, 1963; Crow, 1963; Van Valen, 1963; Brues, 1964, 1969; Feller, 1966, 1967; Maynard Smith, 1968; Wallace, 1968a; Lewontin, 1974). One criticism is that the homozygote (AA) is used as the optimum genotype for the mutational load, whereas a different optimum genotype, the heterozygote (Aa), is used for the segregational load. Sanghvi (1963) argued that the same standard should be applied to both and that this should be the best homozygous genotype. He also suggested use of the ratio of fitness $(1 - L_I)/(1 - L_R)$ as more appropriate than L_I/L_R, but Crow (1963) rejected this suggestion.

Li (1963a, 1963b) pointed out that inbreeding eliminates the fittest (Aa) individuals if the load is segregational, but if the load is mutational, the Aa individuals eliminated by inbreeding are intermediate in fitness. Thus, a population in equilibrium owing to heterozygous advantage would be expected to show a greater decrease in fitness than a population in equilibrium owing to a balance between mutation and selection. However, inbreeding does not permit distinction between mutational and segregational load equilibria because the actual difference between the two for a rare harmful gene is too small. Oddly enough, the more alike two populations are, the greater the difference in their load ratios, and the easier it becomes to distinguish between them by this method, supposedly. For example, two populations with very similar biological characteristics are given in the accompanying tabulation. The fitness and frequency of the homozygous recessives (aa) are the same in both populations and the fitness of AA and Aa are the same in one case and differ by only 0.002 in the other. Therefore, as it would be virtually impossible to detect this slight a difference in fitness, in a practical sense, it would be impossible to distinguish between these populations. However, if it is assumed that the load is segregational, L_R is 100 times greater than if the load is assumed to be mutational. Moreover, the load ratio (L_I/L_R) is 100 if the load is mutational but only 2 if it is segregational. Small wonder that Li felt that the load ratio is operationally backward in that prior knowledge of the type of equilibrium is required before the ratio can be calculated properly, and he suggested that the whole concept of genetic loads is based on statistical artifacts. Of course, it might be argued that that 0.002 difference makes all the difference in the world, just as Mr. Micawber's misery or happiness depended on whether he spent a trifle more or a trifle less than his income of 20 pounds, but it is doubtful that nature keeps books as accurately as Mr. Micawber's creditors.

	Mutational Load W	Segregational Load W
AA	1	$1 - s_1 = 0.998$
Aa	1	1
aa	$1 - s = 0.8$	$1 - s_2 = 0.8$

$$s = 0.2 \qquad\qquad\qquad\qquad s_1 = .002 \quad s_2 = 0.2$$
$$u = 2 \times 10^{-5}$$

$$\hat{q}^2 = u/s \qquad\qquad\qquad\qquad \hat{q} = \frac{s_1}{s_1 + s_2}$$
$$\hat{q} = 0.01 \qquad\qquad\qquad\qquad \hat{q} = 0.01$$

$$L_R = sq^2 = u = 2 \times 10^{-5} \qquad\qquad L_R = \frac{s_1 s_2}{s_1 + s_2} = 2 \times 10^{-3}$$

$$L_I = sq = 2 \times 10^{-3} \qquad\qquad L_I = \frac{2 s_1 s_2}{s_1 + s_2} = 4 \times 10^{-3}$$

$$\frac{L_I}{L_R} = 100 \qquad\qquad\qquad\qquad \frac{L_I}{L_R} = 2$$

To estimate the mutational and segregational loads, the gene and genotype frequencies were assumed to be in equilibrium. Moreover, the size of the population was assumed to be constant from generation to generation, not only in the calculation of mutational and segregational loads but in the calculation of the substitutional load as well. This assumption has been questioned by Van Valen (1963), Brues (1964), and Feller (1966, 1967), and it is clearest, perhaps, in the third definition of genetic load above, which stated that the load is the amount of reproductive excess required to maintain a constant population size. The implication, then, is that the environment is constant and has a certain carrying capacity for members of the species in question. The dubious nature of this assumption has been raised in earlier chapters.

If the population size were constant, one would be dealing with the situation Wallace (1975) has called "soft selection," which is both density- and frequency-dependent. In other words, in a given environment, there are a certain number of individual slots to be filled by a species (the carrying capacity), and no more. How they are filled will depend on the frequency and density of the different types of individuals in the species. It has become almost axiomatic to relate fitness to the environmental conditions. However, "hard selection" represents recognition that some genes, lethals particularly, have adverse effects on their carriers under all known environmental conditions. Therefore, if soft selection prevails, the carrying capacity would be equalled each generation, but if hard selection prevails, this would not necessarily be the case.

Perhaps the easiest way to visualize the implications of the concept of the cost of evolution or the substitutional load is to consider a population of bacteria exposed for the first time to an antibiotic. If, among several million bacteria, there is only one resistant to the antibiotic, it will survive and become the progenitor of subsequent generations of that population. The cost, in this case, is measured in the millions of individuals, and is paid in a single generation. The population size certainly did not remain constant during the period of substitution, which lasted, not 300 generations, but only one. The cost certainly was great, but in light of the alternative, the cost of not evolving was greater.

Another anomaly of the genetic load, as Li pointed out, is that the more beneficial a mutation is, the greater the genetic load it creates. The quickest way to eliminate this genetic load would be to kill off the beneficial mutation. The point made by Brues, Feller, and Van Valen was that the absolute population size had to be considered in discussions of genetic load. A population consisting of only one genotype, by definition, can have no genetic load, yet it may very well decline in numbers to extinction. A population with many different genotypes will always have some sort of genetic load, yet it may be thriving. Therefore, it is not enough to know the nature of the genetic load and the relative frequencies of the different genotypes—the absolute numbers of individuals are needed to determine the reproductive success of the population under existing conditions.

In the example with bacteria above and in Haldane's calculations of the cost of evolution, the assumption was made that an environmental change made the existing wild type unfit and a rare, mutant type beneficial. However, even if the environment does not change, a beneficial mutation might occur of greater fitness than the existing wild type. Suddenly, under load theory, a genetic load would exist where none existed before, even though the old genotype, as the environment has not changed, is just as well-adapted as it ever was. As a whole, the fitness of the population would be enhanced by the appearance of the beneficial mutation. Rather than paying a *cost*, the population would receive a benefit. However, even if the environment does change, the cost of not evolving is greater than the cost of evolving, as Brues (1964) has shown. Morevoer, even the segregation load is only a load if the standard used is universal heterozygosity, which is not attainable. In reality, population fitness is enhanced by the presence of heterozygotes compared to what it would be without them.

One of the more troubling conclusions from genetic load theory was Haldane's (1957) oft-quoted calculation that the substitutional load would be so great that a population could tolerate only one gene substitution every 300 generations. This conclusion has been challenged by Van Valen (1963), Brues (1964, 1969), Feller (1966, 1967), and Maynard Smith (1968). Haldane's calculations were based on certain assumptions: that gene substitution causes a net loss in fitness to the species, that population size remains constant, and that the total cost of a number of gene substitutions equals the sum of their separate costs. However, if gene substitution owing to selection produces a gain in fitness, if population size is not constant, and if selection does not necessarily act independently on genes at different loci (threshold selection, again), then there is no such limit, based on selection, at least, on the number of substitutions or the rate at which they can occur. This conclusion seems more in accord with observed cases of rapid evolution cited previously than with Haldane's estimate.

The genetic load concept has been extremely useful in helping to clarify thought about the nature and effects of genetic polymorphism in natural populations. Although it has been subjected to criticism, there can be little doubt that the theory will continue to develop because of the insights it provides.

CHAPTER 9

Chromosomal Polymorphism

Thus far, we have dealt primarily with diploid, sexually reproducing species in which the chromosome complement has been assumed to be constant among individuals and the variation confined to allelic variation at many different loci within the fixed framework provided by the chromosomes. However, the broad definition of mutation in Chapter 4 included chromosomal changes as well as gene mutations as a possible source of variation in natural populations. The fact is that chromosomal changes are found in natural populations, both within a given species and between closely related species. Thus, chromosomal variation appears to play a significant role in the evolution of natural populations and the nature of this role will now be considered. Three major categories of chromosomal variation can be recognized: changes involving complete haploid sets of chromosomes (polyploidy); changes in the number of intact chromosomes to some number other than a multiple of the haploid number (aneuploidy); and changes involving the gain, loss, or rearrangement of chromosome fragments (duplications, deficiencies or deletions, inversions, translocations, and insertions or transpositions).

POLYPLOIDY

Polyploids typically originate in one of two ways. In the first, *auto-ploidy*, the polyploid contains three or more sets of homologous chromosomes of common origin (triploid AAA, tetraploid AAAA, pentaploid AAAAA, etc., where A represents a haploid set of chromosomes). In the second, interspecific hybridization (species AA x species BB) gives rise to F_1 hybrids (AB) that are sterile because the chromosomes are nonhomologous and fail to pair. Chromosome doubling converts the hybrid into a fertile *allopolyploid* organism because each homologue in each set now has a homologous chromosome like itself with which to pair (AABB). Unfortunately, polyploidy is seldom quite so simple, for depending on the origin of the chromosomes, a complete gradation may be found between homologous and nonhomologous chromosomes. Different subspecies of the same species, for example, may have chromosomes that are almost, but not quite, structurally identical while different but related species may have structurally rather different chromosomes that still retain some homologous segments. For this reason, a third type of polyploid has been identified, the *segmental allopolyploid,* in which some, but not all, chromosome segments from the parents of the hybrid remain homologous.

Because homologous chromosomes tend to pair, autopolyploids will form multivalents rather than bivalents at meiosis, which leads to significant sterility. As noted above, allopolyploids from parents with nonhomologous chromosomes will be fertile because normal bivalents are formed at meiosis. However, segmental allopolyploids will exhibit different degrees of multivalent formation and fertility, depending on the degree of partial homology between the chromosomes of different origin. One further note is that at the hexaploid level or higher, an organism could be both an autopolyploid and an allopolyploid. Such an autoallopolyploid would be formed, for example, if an autotetraploid hybridized with a diploid with a different genome and the resulting triploid underwent chromosome doubling to form an AAAABB hexaploid.

Under changes involving whole haploid sets of chromosomes, we should also mention *haplodiploidy* (White, 1973). In haplodiploidy (or *male haploidy* or *arrhenotoky*), the males develop from unfertilized eggs and are haploid, but the females, developing from fertilized eggs, are diploid. Haplodiploidy is characteristic of the highly successful insect order Hymenoptera (bees, ants, wasps), consisting of more than 150,000 species, but it is otherwise rare, having arisen only eight known times in the evolutionary history of the Metazoa, six times in insects, once in arachnids, and once in rotifers. This rarity is in contrast to *thelytoky*, a form of parthenogenesis in which the progeny are ordinarily all diploid females, for thelytoky has evolved repeatedly in nearly all of the animal phyla. The population genetics of the Hymenoptera might be expected to differ from that of cross-breeding diploid species because selection against deleterious recessives should be extremely rigorous in the haploid males, and the genetic variability present in the population should be smaller than in a comparable population in which both sexes are diploid. Although, theoretically, this might be expected, there is virtually no supporting evidence, and the evolutionary success of the Hymenoptera, which include many species with complex social organizations, hardly suggests a lack of genetic variability on which selection can act to bring about the remarkable adaptations seen in this group. One obvious possible explanation is that the expression of many genes is sex-limited to the females, and, therefore, much of the genetic variation is not exposed to haploid selection in males because it is not expressed in the males. In this connection, it is worth noting that the different castes of the social Hymenoptera are all diploid females while the sole function of the haploid males appears to be to inseminate the queen. In the other large group of social insects, the termites (order Isoptera), both males and females are diploid, and the workers and soldiers may be of either sex.

Even though in lower forms of plants and animals the haploid condition may be predominant, or in species with an alternation of haploid and diploid generations the haploid generation may occupy a significant portion of the life cycle, generally in the higher forms apart from the Hymenoptera, haploidy does not appear to be of great evolutionary significance. Similarly, polyploidy appears to be of limited evolutionary significance in higher animals and is found primarily in those species that reproduce parthenogenetically. Polyploidy seems theoretically possible in hermaphroditic animal species as well, but the available evidence (White, 1973) does not indicate that it is at all common among them. In higher plants, however, polyploidy has been of marked evolutionary importance. Typical of recent estimates of polyploid frequency among higher plants are those of Grant (1971), who estimated an overall frequency of polyploidy in angiosperms of 47% (43% in dicotyledons; 58% in monocotyledons), 38% in Gnetales, 95% in pteridophytes, but only 1.5% among the Coniferales. The evidence is now quite clear that nearly all of the naturally occurring polyploids in higher plants are of hybrid origin— that is, they are allopolyploids, or amphiploids, as they are sometimes called. For this reason we shall defer further discussion of polyploidy in plants until Chapter 11, which contains a discussion of hybridization.

In contrast to plants, most of the rare polyploids in animals appear to be autopolyploids, although, recently, Maxson, Pepper, and Maxson

(1977) have suggested that the North American tree frog, *Hyla versicolor,* is an allotetraploid derived from hybridization between eastern and western diploid races of *H. chrysoscelis.* The rarity of allopolyploids in animals may be related to the fact that natural hybridization may be less common among animals than plants (see Chapter 11), but it is probably caused primarily by the disruption of the chromosomal sex-determining mechanism in animal species with separate sexes. Furthermore, if an occasional tetraploid arises in an obligatory cross-fertilizing species, its potential mates will be diploid and its offspring will be sterile triploids so that its chances of becoming established are minimal unless some form of parthenogenesis can become established.

ANEUPLOIDY

Aneuploids contain chromosomes in some number other than an exact multiple of the basic haploid number. In an otherwise diploid organism, for example, if one of the chromosomes is missing, a *simple monosomic* $(2n - 1)$ is formed. If one of the homologues is represented three times rather than twice, a *simple trisomic* is formed. Monosomy is apt to be detrimental or even lethal, for any deleterious recessives on the single remaining homologue are apt to be expressed and depress the fitness of the monosomic individual. Trisomy might be expected to be less harmful, but the discovery that Down's syndrome or mongolism was caused by trisomy 21 in humans has been paralleled by the discovery of detrimental effects for other trisomies in humans and in other species. Therefore, an increase in chromosome number also seems to have adverse effects on development, presumably because the extra chromosome, by introducing a complete extra set of genes, disturbs the developmental balance conferred by the normal genome. Therefore, aneuploidy seems not to have played a major role in the evolution of plants or animals (Grant, 1971; White, 1973). Polysomy is a mechanism whereby additional genetic material could be incorporated into the genome, but the evidence suggests that such increases are not common. This is not to say that changes in the basic haploid number have not occurred, but rather that the increases or decreases in the haploid number have usually not been caused by the simple gain or loss of intact chromosomes.

Accessory chromosomes, also known as *supernumeraries* or *B chromosomes,* are found in natural populations of a variety of species of both plants and animals (Stebbins, 1971a; White, 1973; Müntzing, 1974). In a broad sense, supernumeraries fall under aneuploidy, but they differ from polysomy in several respects. First and foremost, acessory chromosomes are additional to—but not homologous with—the chromosomes of the normal karyotype. They are usually smaller than the normal chromosomes, heterochromatic, and irregular in number. Most supernumeraries are acrocentric or telocentric; if metacentric, they are usually isochromosomes with the arms homologous to one another. At meiosis, pairing between normal chromosomes and supernumeraries rarely occurs, but supernumeraries may pair with one another if they are present in even numbers; however, chiasmata do not form. Their presence or absence has not been shown to have detectable external morphological effects, but subtle biometric effects on viability and fertility have been shown to be associated with different numbers of supernumeraries. Thus, it is not possible to tell from the external phenotype whether or not an individual carries accessory chromosomes.

The functional role of supernumeraries is not clear. The number of accessory chromosomes may differ in different tissues of the same organism or in different individuals from the same population. Furthermore, there are geographic differences in the numbers of supernumeraries. Within the same species, some populations may lack them entirely while others differ markedly in the mean number of accessories present. There appears to be a genetic equilibrium in natural populations with respect to the number of supernumeraries. One possible explanation for this is that low numbers confer an adaptive advantage to individuals carrying them over individuals lacking them, but that high numbers of accessory chromosomes are deleterious. It has also been suggested that supernumeraries are deleterious, but selection against them is opposed by some sort of accumulation mechanism that tends to increase their numbers. Another possibility is that they are inert or adaptively neutral, but the observation that high numbers of acessories reduce viability casts doubt on this.

The difference in frequency and distribution of supernumeraries in natural populations suggests that they may have ecological significance, but its nature remains unknown. Similarly, the evolutionary origin of these chromosomes is a matter for conjecture. Presumably, if they originated from the basic chromosomes in the genome, it occurred some time ago, for they lack homology with the normal chromosomes and differ in size and structure. Although they raise some fascinating cytological, ecological, and evolutionary questions, supernumeraries can hardly have played a fundamental role in speciation and evolution, for they are not universally a part of the genome of all species.

CHROMOSOMAL REARRANGEMENTS

Chromosomal rearrangements involve the gain, loss, or rearrangement of fragments of chromosomes following chromosome breakage. Loss of a fragment leads to a *deletion* or a *deficiency* for a block of genetic material. Deficiencies are often lethal when homozygous, and, if large enough, may act as dominant lethals in heterozygotes. Hence deletions are not apt to play a significant role in evolution.

Sometimes, rearrangements occur in such a way that a chromosome segment is duplicated or repeated. The effects of *duplication* on fitness are usually not as drastic as the effects of deletion, and duplication has long been regarded as a mechanism by which additional genetic material can be added to the genome (Metz, 1947; Stephens, 1951; Ingram, 1963; Ohno, 1970). Divergence from the ancestral genes in function has been assumed to occur through mutation, for the presence of one gene to carry on the original function frees the duplicated gene to mutate and assume a new function. As the long-term trend in evolution has been toward increased amounts of DNA and greater functional complexity (Britten and Davidson, 1969; Hinegardner, 1976), the role of duplications has been thought to be of considerable evolutionary importance. Dupications and deficiencies may result from a variety of types of chromosomal rearrangement and also from unequal crossing over. If the duplicated segments are adjacent and in the same order, i.e., ABCABC, it is called a *tandem duplication* or *repeat. Reverse repeats,* i.e., ABCCBA, are also known and are somewhat more stable because they are less apt to be broken up by crossing over.

A chromosome *inversion* occurs when a chromosome breaks in two places and the segment with two broken ends rotates 180° before reuniting with the other two fragments. Thus, the chromosome *ABCDEFG* may become *ABCFEDG.* Inversions may be *paracentric* (beside the centromere), involving only one chromosome arm, or *pericentric* (including the centromere), and involving both chromosome arms. Diploid individuals may be homozygous for one or the other of the inversion types or may be inversion heterozygotes, carrying one chromosome of each type. Crossing over at meiosis in a heterozygote for a paracentric inversion leads to the formation of a *dicentric* chromosome and an *acentric* fragment, which are characterized by duplications and deficiencies. Crossing over in a heterozygote for a pericentric inversion also results in duplications and deficiencies in the resulting chromosomes, but each has just one centromere. Because of the duplications and deficiencies, inversion heterozygosity frequently leads to reduced viability and fertility. An inversion leads to changed linkage relations among the genes on the chromosome, but they all remain in the same linkage group. Although position effects might be expected to be associated with inversions and other types of rearrangements, this apparently is seldom the case.

Translocations involve the exchange of chromosome fragments between two different chromosomes. The exchange is ordinarily reciprocal between nonhomologous chromosomes, the genes in the translocated segments being found in different linkage groups than before. Because homologous genes continue to pair, synapsis in translocation heterozygotes leads to the formation of rings of chromosomes at meiosis, and crossing over may lead to duplications and deficiencies and to dicentrics and acentrics, all of which may reduce viability and fertility. Translocation homozygotes will have normal viability and fertility if they carry complete genomes on monocentric chromosomes—if not, they will suffer from sterility and reduced viability. If there are three chromosome breaks, *insertions* or *transpositions* of chromosome segments to a different location in the genome may occur. Thus duplications may be found separated from one another within the genome and not just adjacent to one another.

INVERSION POLYMORPHISM

Despite its potential for causing sterility and inviability, inversion polymorphism has been found in a number of species. Although it has been observed in plant genera such as *Paris* and *Paeonia* (Stebbins, 1950), it is probably more common and has certainly been more extensively studied in animals, especially insects of the genera *Drosophila, Sciara, Chironomus, Simulium,* and *Anopheles,* all of the order Diptera. It may be no coincidence that these two plant genera are characterized by some of the largest chromosomes known in the plant kingdom and the Diptera are characterized by extremely large salivary gland chromosomes. Large chromosomes permit the ready detection of chromosomal rearrangements that might escape notice in other species with smaller chromosomes. As the inversion heterozygotes do not ordinarily show any visible external differences from the inversion homozygotes, detection depends, to a large extent, on the ability to make favorable cytological preparations for study. Therefore, inversions and inversion polymorphism may be more widespread in natural populations than is presently known.

The most extensive studies of inversion polymorphism have been conducted in *Drosophila* (Dobzhansky, 1971; White, 1973). However, even within this genus there are wide differences in chromosomal poly-morphism. In some species such as *D. willistoni, D. paulistorum, D. robusta,* and *D. subobscura,* paracentric inversions occur in every chromosome pair in the genome, and more than 40 different inversion types have been identified in *D. willistoni* and *D. subobscura.* In other species such as *D. pseudoobscura* and *D. persimilis,* the inversions are found primarily on a single — the third — chromosome pair, where over 20 inversions have been recognized in the former and more than 10 in the latter. In chromosomally monomorphic species such as *D. simulans, D. virilis,* and *D. repleta,* inversion polymorphism is absent.

In some polymorphic species, which have spread widely in associa-tion with humans, the same chromosome inversion types are geograph-ically widespread. In the cosmopolitan *D. melanogaster, D. busckii, D. ananassae,* and *D. immigrans,* the same inversions appear to be almost universally distributed, some with a high frequency and others with a low frequency.

In other species not closely associated with humans, the type and frequency of the inversions varies geographically. Thus, in species such as *D. funebris, D. robusta,* and *D. pseudoobscura,* the incidence of inversion polymorphism will be quite high in some areas, quite low in others, and the most frequent type of arrangement will differ in different regions. The reasons for these differences within and be-tween species in frequency and pattern of distribution of inversion polymorphisms, for the most part, remain unknown.

The most thoroughly studied case of inversion polymorphism is in *D. pseudoobscura* and its sibling species, *D. persimilis,* which primarily involves paracentric inversions on the third chromosome, where more than 30 different inversion types have been identified.

The so-called Standard (ST) arrangement was named thus because it is common to both species, which suggests considerable age for this chromosome type. If, after one inversion has occurred, a second in-version occurs in the same chromosome, the second may lie outside the limits of the first. This is called an independent inversion, and heterozygotes for Standard and a chromosome with two independent inversions will show chromosome pairing with two separate inversion loops in the large polytene chromosomes of salivary gland nuclei. If the second inversion lies inside the first, it is known as an included inversion. If one end of the second inversion lies inside and the other end lies outside the limits of the first, it is known as an overlapping inversion. Both included and overlapping pairs of inversions, when heterozygous, give rise to characteristic, complex, double-loop con-figurations.

Overlapping pairs of inversions are of particular interest because they have been used to determine the phylogenetic history of the chromosome types that are found in natural populations. For example, if three sequences are known in a chromosome — (1) ABCDEFGHIJ, (2) ABC<u>GFEDH</u>IJ, (3) ABCGF<u>IHDE</u>J — heterozygotes for (1) and (2) or for (2) and (3) will form single inversion loops, but heterozygotes for (1) and (3) will form double loops. It should be clear from the sequences that (1) can give rise to (2) and (2) can give rise to (3) through a single inversion, or, conversely, (3) might have given rise to (2) and (2) to (1) through single inversions, but (3) cannot arise directly from (1), nor (1) from (3), without going through the intermediate stage (2). Therefore, the phylogenetic relationship of these three chromosome types may be (1) → (2) → (3), or (3) → (2) → (1), or (3) ← (2) → (1).

The phylogenetic chart for the inversion types found in *D. pseudoobscura* and *D. persimilis* (Figure 9.1) was drawn so that each sequence differs by a single inversion from those to which it is connected by a single arrow. The phylogeny is remarkably complete, for only one sequence (Hypothetical) has not actually been found in nature. Standard (ST) differs from Santa Cruz (SC) by two overlapping inversions, and, presumably, the Hypothetical sequence that once connected them is now extinct. Of course, it may be found yet, but the already extensive collecting makes this improbable.

The proper direction for the arrows has been inferred from the location of the sequences on the chart and also from their geographic distribution. The oldest sequences are thought to be the ones on the central axis of the chart (Standard, Hypothetical, Santa Cruz, and Tree Line). Because it is found in both *D. pseudoobscura* and *D. persimilis,* Standard may be the ancestral chromosome type, but Hypothetical may be even older. The inversion types one step removed from the central axis are presumably older than those two steps removed from the axis. In general, the sequences on the periphery of the chart also have a more restricted geographic distribution than the more central sequences, which further supports their relatively recent origin. Although it is possible to infer the relative antiquity of the various sequences, it is quite another matter, at present, to estimate their absolute age. It should be realized that underlying these discussions is the implicit assumption that all chromosomes of a given type are descended from a single ancestral rearrangement and that each rearrangement represents a unique event in the history of the species.

D. pseudoobscura is a western species, ranging from British Columbia to Guatemala and from the Pacific coast to Texas. There is also

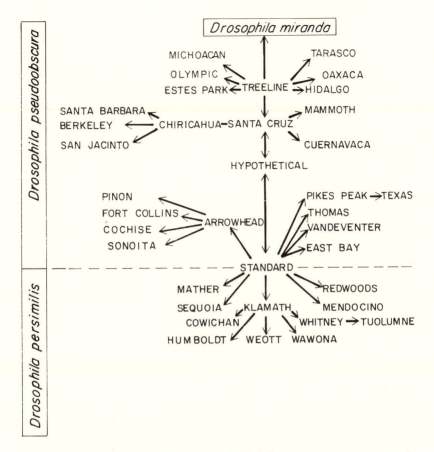

Figure 9-1. Phylogenetic relationships of third chromosome arrangements in *Drosophila pseudoobscura* and *D. persimilis*. Arrows connect arrangements differing by a single inversion event, with the direction of the arrow pointing from the putative ancestral type to its derivative. See text for further details. (Source: Anderson et al., 1975).

an isolated population in Columbia. The most frequent chromosome type varies in different parts of the range: Standard (ST) is most common in California, Arrowhead (AR) is most frequent in Arizona and New Mexico, Pike's Peak (PP) is predominant in Texas and on the eastern slope of the Rockies, and Chiricahua (CH) is most common in northern Mexico. Although the most frequent type of third-chromosome sequence varies from one part of the range to another, all the populations show a fair degree of third-chromosome polymorphism, ranging from 6.8% heterozygotes for third-chromosome inversions at the Grand Canyon to 76.2% heterozygotes at Mather, California. The lack of external visible differences between inversion homozygotes and heterozygotes, and the absence of any obvious signs of heterosis such as larger size, plus the fact that the frequencies of inversion homozygotes and heterozygotes in natural populations seemed to be in agreement with expectations based on the Castle-Hardy-Weinberg law led, originally, to the conclusion that the inversions were adaptively neutral (Dobzhansky, 1941).

However, subsequent findings made clear that the different inversion sequences were not selectively neutral. For example, when a population was sampled at regular intervals in a given locality, the relative frequencies of the different chromosome types were found to undergo cyclical seasonal variations. At Mount San Jacinto, California, for instance, the three most frequent chromosome types were Standard (ST), Chiricahua (CH), and Arrowhead (AR). AR stayed at about the same frequency from March to October, the period when the flies were most active, but ST decreased during the cool season from March to June and increased in frequency during the warm summer months. CH was favored during the cool months and declined in frequency in the summer (Dobzhansky, 1943). Presumably the ST/ST homozygotes were relatively more fit during the warm months and the CH/CH homozygotes were more fit during the cool months. Comparable results were obtained by Strickberger and Wills (1966) for *D. pseudoobscura* and by Dubinin and Tiniakov (1945) for *D. funebris*.

In another study (Dobzhansky, 1948) in the Yosemite National Park area of the Sierra Nevada mountains, some 300 miles north of Mount San Jacinto, altitudinal as well as seasonal changes in inversion frequency were observed. In this case, ST was most frequent at low elevations and gradually decreased in frequency with increasing elevation, AR increased in frequency over the same range of elevations, but the frequency of CH did not change significantly with elevation. In addition, during the summer, the frequency of ST increased while that of AR decreased so that during the warm season, the populations at the higher elevations came to resemble those at lower elevations during the cool season. These seasonal and altitudinal changes in inversion frequency suggest that these chromosome types are not adaptively neutral, but that the gene contents of the different chromosomes confer different fitnesses in relation to temperature. Furthermore, even though ST seemed to thrive best under warm temperatures at both Mount San Jacinto and in the Sierra Nevadas, the chromosome increasing in frequency during the cool season differed in the two localities. At Mount San Jacinto, CH was favored during the cool months and AR remained unchanged; but in the Sierra Nevada, the reverse was true. This last fact indicates that the gene contents of the same inversion type may be different in different geographic areas.

The cyclical seasonal changes in inversion frequency could maintain a balanced polymorphism for these chromosome types if these cyclical changes are regular, as was indicated in Chapter 6. However, such a polymorphism is inherently unstable, and the retention of the alternative types in the populations can be ensured by a different mechanism. This mechanism is heterosis, for it was found that in many cases, the heterokaryotypes appeared to have greater fitness than either of the homokaryotypes (Dobzhansky, 1970).

However, it must be added that seasonal cycles in inversion frequencies were not always observed nor could heterokaryotype advantage always be detected in wild populations. Thus, neither of these mechanisms can be regarded as a universal mechanism by which balanced inversion polymorphism is maintained in natural populations of *Drosophila*. Carson (1958b), for instance, did not find significant seasonal changes in the frequency of inversions in *D. robusta* populations in Missouri, but Levitan (1951a, 1957) found marked seasonal variations in some populations of this species on the East Coast, but not in others (1951b).

Furthermore, an excess of structural heterozygotes in wild populations over the number expected in a Castle-Hardy-Weinberg equilibrium is not adequate proof of a selective advantage for the heterozygotes over the homozygotes, although Dobzhansky and Levene (1948) originally interpreted their data that way. As Novitski and Dempster (1958) and Wallace (1958c) subsequently showed, if genotype frequencies are known only for a single generation, it is impossible to determine whether or not selection is acting simply from the distribution of the frequencies. Even if the adult frequencies appear to fit a binomial distribution, selection may still be acting in the population. Conversely, if the frequencies do not fit the binomial distribution, it is still hazardous to attempt to infer the nature of the forces involved from the distribution of genotypes (Spiess, 1977). What is needed is information from more than one generation to determine whether or not an equilibrium exists. If the population is in equilibrium so that the values of p and q remain the same from generation to generation, then these values can be used to calculate the expected frequencies of genotypes. However, natural populations are seldom in equilibrium so that such calculations are seldom valid.

Nevertheless, if the frequency of inversion heterozygotes significantly exceeds 50%, as it does in some cases in *Drosophila* (Dobzhansky and Pavlovsky, 1955, 1958), this provides clear evidence for a selective advantage for the heterozygotes because, if the three possible geno-types were adaptively neutral, no values of p and q would give an expected frequency for the heterozygotes ($2pq$) greater than 50%.

Although the results in natural populations with respect to heterosis were somewhat ambiguous, the results in experimental populations of *D. pseudoobscura* were more clearcut. When populations, with known initial frequencies of the inversion types ST and CH from the same locality in California, were established in cages, the frequencies changed initially in a systematic way, but eventually they reached an equilibrium, with neither chromosome type being eliminated. In eleven different experimental populations of this type run at different times, at different initial frequencies, and, otherwise, under different condi-tions, they all reached an equilibrium with about 70% ST and 30% CH (Dobzhansky and Pavlovsky, 1953). Thus, even though they were not exact replicates, the outcome was essentially the same in all cases. In some populations, the initial frequency of ST was as low as 10%. The simplest interpretation for this balanced polymorphism was that the ST/CH heterozygotes were superior in fitness to both the CH/CH and the ST/ST homozygotes, with the ST/ST individuals, in turn, being about twice as fit as CH/CH. For four of these populations the relative fitnesses were estimated as ST/CH = 1, ST/ST = 0.89, and CH/CH = 0.41. However, it should be noted that these differences in fitness existed only in populations maintained at 25°C. At 16°C no significant differences in fitness among the three genotypes could be detected; they behaved as if they were adaptively neutral. Further-more, if the two chromosome types had different geographic origins, e.g., ST from California ancestors and CH from Mexican ancestors, the results were quite different, for in that case the ST/CH heterozy-gotes showed no heterozygous advantage but were intermediate (in one case, inferior) in fitness to the two homozygotes. However, even though heterosis was initially absent, it subsequently developed in two out of six such populations (Dobzhansky and Pavlovsky, 1953).

COADAPTATION

These contrasting results with chromosomes of the same or different geographic origin merit further discussion. In the first place, they again suggest that the same chromosome type from different sources may carry different complements of genes. Furthermore, different inversions from the same locality appear to interact favorably, but if they come from different areas, no such favorable interactions occur in heterozygotes. The favorable interactions have been explained in the following way (Dobzhansky, 1970). A major effect of inversion heterozygosity is that genetic recombination is effectively suppressed in the chromosomes heterozygous for the inversions. Thus, genic recombination can occur freely in inversion homozygotes, but is blocked in inversion heterozygotes. Therefore, natural selection can act to preserve favorable gene complexes within each inversion type (leading to the development of what Mather in 1943 called "internal balance.") However, selection will operate not only in inversion homozygotes but also in inversion heterozygotes, and, thus, will favor those linked gene-complexes that are favorable not just in homozygotes, but that interact favorably (i.e., produce heterosis or optimum fitness) in inversion heterozygotes as well (leading to what Mather referred to as "relational balance"). Since these favorable linked gene complexes in the different chromosome types are inherited as blocks and are rarely broken up by crossing over between the different inversions, they have sometimes been called *supergenes*. Where the gene complexes of different inversion types interact favorably, they are said to be *coadapted,* and natural selection is said to have produced an *integration of the genotype.* Note that the phenomenon described above, which we have characterized as coadaptation or relational balance, is known in other contexts by other names, e.g., *heteroselection* and *rigid polymorphism,* to be considered later, and *specific combining ability* by plant and animal breeders.

As mentioned above, heterozygotes for inversions of different geographic origin failed to show heterosis initially, but the fitness relations changed in two of six such populations so that they eventually showed heterozygote superiority. This result was interpreted to illustrate the evolution of coadaptation through natural selection in these populations. The failure to observe the evolution of coadaptation in the other four populations, in which the ST chromosome seemed destined to go to fixation and CH to extinction (Dobzhansky and Pavlovsky, 1953), was attributed to the failure to generate the appropriate coadapted genotypes on each chromosome type in the limited time and with the limited numbers of genotypes available. Thus, the results in these populations were indeterminate in contrast to the consistent results in experimental populations with chromosomes of a common origin.

Although the evolution of coadaptation is widely accepted as the explanation for the development of heterosis in some populations with inversions of different geographic origin, it should be realized that this is not the only possible explanation. The fitnesses measured are relative fitnesses; therefore, the observed changes in these populations could occur if the heterozygotes improved in fitness and that of the homozygotes remained unchanged. (This increase in the fitness of heterozygotes is what is stressed in the evolution of coadaptation.) However, the same changes in relative fitness could occur if the heterozygotes' fitness remained unchanged, but the fitness of the inversion homozygotes decreased. The third possibility, of course, is that the fitness of both heterozygotes and homozygotes changed in such a way that the fitness of the heterozygotes exceeded that of the better homozygote. Given the relatively few strains of each inversion type used to start the populations, the limited size of the populations, the number of generations, and the magnitude of whole chromosome mutation rates, it is conceivable that each of the inversion types could accumulate its own assemblage of deleterious genes. In that case, the development of heterosis could be caused by the decline in fitness of the homozygotes rather than the evolution of coadaptation in the heterozygotes. Apparently this possibility was never considered in these populations although it could have been easily checked.

After "coadaptation" appeared, if a sample of ST chromosomes and another of CH chromosomes from one of these experimental populations were isolated and individually made homozygous to check for the presence of deleterious recessive genes, tests among the ST chromosomes and among the CH chromosomes would reveal the degree of allelism for deleterious genes within each inversion type. Tests between ST and CH chromosomes would reveal the degree of allelism for deleterious genes between the two inversion types. If the degree of allelism for deleterious recessives is higher within chromosome types than it is between chromosome types, this finding would support an alternative explanation for the appearance of heterozygous advantage in some of the populations with inversion types of different geographic origin. It is, simply, that the apparent fitness of the homozygotes relative to the inversion heterozygotes has declined owing to the accumulation of different assemblages of deleterious genes on the ST and CH chromosomes.

If selection for coadaptation were occurring in such populations, the outcome might be expected to be uniform as it was in the populations with inversions from the same locality, rather than indeterminate as was actually the case. The alternative explanation, dependent upon chance events, mutation, and random drift, seems more compatible with the indeterminacy observed.

The concept of coadaptation requires that the coadapted gene complexes evolve together in the same population as the result of natural selection. A particularly awkward finding for the theory of coadaptation was made in several different studies of interlocality and intralocality hybrids. The maximum values for the various components of fitness studied were generally found in the interlocality F_1 hybrids (Vetukhiv, 1953, 1954, 1956, 1957; Brncic, 1954, Wallace, 1955). This result was not an isolated finding but was reported for a number of different crosses in four different species of *Drosophila* (*D. pseudoobscura, D. willistoni, D. paulistorum,* and *D. melanogaster*). The contrast between this result and that of Dobzhansky and Pavlovsky (1953) should be noted, for they reported that, initially, interlocality structural heterozygotes had fitnesses intermediate to the homozygotes. However, these flies of intermediate fitness were apparently not the F_1 interlocality hybrids. Vetukhiv's comments (1957, p. 359) in the last of this series of papers are worth repeating:

> "In a widespread, common, and ecologically versatile species like *D. pseudoobscura*, processes of coadaptation are certainly taking place within the populations of most localities where this species lives. However, the heterosis observed in the F_1 hybrids between the populations would have to mean that the coadaptation of the genes found in the gene pools of populations of different localities is superior to that within the gene pools of these localities themselves. It seems rather difficult to imagine a mechanism of natural selection which would bring about such a state of affairs."

Dobzhansky (1952a) and Wallace (1955) drew a distinction between *euheterosis* (for the heterosis observed within Mendelian populations due to selection for coadaptation) and *luxuriance* (for the heterosis observed in interpopulation hybrids, which could not very well be due to natural selection of this sort). Wallace also stated, "The higher average viability of interpopulation F_1 hybrids is a measure of the price paid by local populations for an integrated gene pool capable of being transmitted successfully from generation to generation," a statement whose meaning remains obscure.

Both Vetukhiv (1954) and Dobzhansky (1970) suggested that the parental populations could not have been inbred because each population from a given locality was started from eight or ten different strains, each derived from a single impregnated female caught in nature. The fact that the F_1 interlocality hybrids showed maximum fitness compared to all other types in their experiments thereby became even more difficult to explain. As Dobzhansky said (p. 306), "the origin of this luxuriance is unclear." Vetukhiv (1957) wrote, "The weight of the evidence seems to favor the heterozygosis 'per se' hypothesis. Heterozygosis does, under some conditions which at present cannot be specified, bolster the adaptive value of a genotype, even without a previous history of coadaptation by natural selection."

Further studies of "coadaptation" and the "integration of the gene pool" by Brncic (1954), Wallace (1955), and Kitagawa (1967) were made in such a way that the relative viabilities of several classes of flies could be compared. In particular, the effect of recombination on fitness was tested. The classes of flies were the following: (1) interlocality hybrids, flies with intact wild chromosomes from two different localities; (2) intralocality hybrids, flies with intact wild chromosomes from the same locality; (3) flies with one recombinant chromosome derived from crossing over between chromosomes from two different localities and the other chromosome intact and from a single locality; (4) flies with both chromosomes of a pair the products of recombination between chromosomes from different localities. Although the results varied occasionally, the usual sequence for viabilities was (1) > (2) > (3) > (4). These results were interpreted as supporting the concept of coadaptation, and the relationship of (2), (3), and (4) certainly seems to do so, for both (3) and (4) appear to produce the hybrid breakdown sometimes observed in the F_2 from species crosses, which is thought to act as an isolating mechanism (Chapter 14). These results were summarized and discussed by Wallace (1968a) and Dobzhansky (1970).

However, Vetukhiv (summarized in Wallace 1968a, p. 316) showed that the interlocality F_1 hybrids were about 20% better than the intralocality F_1 hybrids in larval survival, eggs laid per female, and longevity. These differences are too great to be ignored. The consistent finding in these and a number of other experiments on coadaptation in *Drosophila* that the F_1 interlocality hybrids exceeded all the other genotypes in fitness would seem fatal to the concept of coadaptation as set forth by Dobzhansky, Wallace, and others. Calling the phenomenon "luxuriance" and postulating "heterozygosity *per se*" as the explanation, does not explain away the basic inconsistency of this result with the whole concept of coadaptation among chromosome types being built up by natural selection within populations.

At this late date, it is hazardous to attempt to find alternative explanations for these data. Apparently no other possibilities were considered at the time that would explain all the F_1 and F_2 data within a single theoretical framework. Nonetheless, one aspect of these results warrants some comment.

The fitnesses measured are relative fitnesses and not absolute fitnesses. The relative fitnesses must be referred to some standard of fitness as a basis of comparison. Dobzhansky (1970, p. 117) wrote, "For many purposes, it is more convenient to express the results of experiments in percentages, not of wild-type flies in the cultures, but of the average (or 'normal') viability of flies heterozygous for pairs of chromosomes taken at random from the gene pool of the population." As pointed out in Chapter 3, the use of wild-type flies heterozygous for two chromosomes drawn from the same breeding population carries a greater risk that these chromosomes carry detrimental genes in common than if the chromosomes are drawn from two different breeding populations. Wallace (1968a, p. 40) made the point that the standard of comparison used in studies of relative fitness was unimportant since it was being used merely as a standard against which to measure the fitness of other genotypes. If this were the case, then the Cy L/+ (and similar heterozygotes) could just as well have continued to serve as the standard of comparison. The switch to flies heterozygous for two different wild-type chromosomes from the same locality as the standard of comparison was more than just a switch to a new and better standard, but was, as indicated by Dobzhansky above, an attempt to define a "normal" array of genotypes.

There is abundant evidence that dominant mutations in *Drosophila* have an adverse effect on viability and fitness (e.g., Pavan et al., 1951). Even though the standard of fitness used was the heterozygotes for two wild-type chromosomes from a single locality, these heterozygotes still were being compared to heterozygotes for the dominant markers, whose fitness must be somewhat reduced because of the presence of the dominants. If the fitness of the heterozygotes for the dominant markers and for two different wild-type chromosomes from the same locality were equal, the expected proportion of wild-type heterozygotes would be 33.3%. In several reports (e.g., Wallace and Madden, 1953, Table 5; Dobzhansky and Spassky, 1953, Table 1; Dobzhansky and Spassky, 1954, Table 1) the viability of these two types of heterozygotes was very similar, with the wild-type heterozygotes either equal to or, at best, only slightly greater than the heterozygotes for the dominants. Dobzhansky and Spassky (1954) called the wild-type heterozygotes "normal by definition" and then concluded that "the viability of the flies which carried the marking chromosome was approximately normal." As these are relative fitnesses, the alternative conclusion is that the fitness of the heterozygous wild-type "normal" flies is surprisingly low because it does not markedly exceed that of the flies heterozygous for the deleterious dominant marker genes. In other words, even though the wild-type heterozygotes could serve as *standards*, it seems inappropriate to define them as *normal.*

A further point to note, made by Wallace and Madden (1953), is, "The problem is complicated, however, by the lack of a reliable standard of 'normal' viability; heterozygous individuals are not uniform in viability . . . and, consequently, there is a range of viabilities that must be regarded as normal." Therefore, not only was the viability of intralocality heterozygotes surprisingly low, but the heterogeneity in the viabilities of individuals for two different chromosomes from the same locality was surprisingly high. This heterogeneity was also documented in Wallace and King (1952) and in Greenberg and Crow (1960). Wallace and King excluded lethal and semilethal heterozygous combinations from their analysis in Table 2, and in Table 3 they illustrated some of the extremes in viability observed in the remaining crosses, which ranged from 16.6% to 46.1%, where 33.3% was expected. Greenberg and Crow presented their data in a different form and showed that the viability ratios of the intralocality wild-type heterozygotes ranged from 0.3 to 1.7, where the expected ratio was 1.0.

Therefore, the use of the average fitness of heterozygotes for two different wild-type chromosomes from the same locality suffered from the following drawbacks:

1. Some of these heterozygotes were excluded from the average because the heterozygous combinations were lethal or semilethal.

2. Among the remaining intralocality wild-type heterozygotes there was a high degree of heterogeneity in fitness relative to the heterozygotes for the dominant marker genes.

3. Even with the lethal and semilethal heterozygotes excluded, the average fitness of the remaining heterozygotes equaled or only slightly exceeded the fitness of the heterozygotes for the dominant markers.

These results suggest that the use of intralocality heterozygotes as a standard of fitness was a poor choice and to call them "normal by definition" was misleading. The simplest interpretation of these data is that the lethal and semilethal combinations were caused by allelism for a common lethal or semilethal gene in chromosomes drawn from the same population. Similarly, the reduced fitness and heterogeneity in fitness of the remaining heterozygotes could be due to the allelism in some of these hybrids for genes of less drastic effect.

If this interpretation is correct, it also accounts for the otherwise surprising finding by Vetukhiv that F_1 interlocality hybrids exceeded F_1 intralocality hybrids in fitness by some 20%. Chromosomes drawn from populations in two different localities have a much lower probability of carrying the same deleterious alleles than chromosomes drawn from the same population.

Therefore, the interlocality heterozygotes would be a more appropriate standard of fitness than the intralocality heterozygotes. The heterogeneity cited above would be minimized, and a number of studies confirm that the interlocality hybrids have optimum fitness as compared to a variety of other chromosome combinations.

When the average of the intralocality heterozygotes is used as the standard, some homozygotes turned out to be *supervital,* significantly more fit than the *norm.* This anomalous situation seems attributable more to the low average fitness of the intralocality heterozygotes than to the inherent supervitality of some homozygotes. If the interlocality hybrids were the standard, the *supervitals* as a group would essentially disappear.

The choice of intralocality heterozygotes for wild chromosomes as "normal by definition," despite their obvious drawbacks, apparently stemmed from adherence to the concept of coadaptation. If these chromosomes had been selected for favorable internal and relational balance within a given breeding population, then it is logical to use such heterozygotes as the standard or *norm* of fitness. However, the concept of coadaptation implies a rigid population structure, with each population isolated from the others so that within each population, integrated gene complexes can be built up. Relatively little is known about vagility, dispersal, and gene flow in *Drosophila,* and existing studies have led to widely divergent results (Chapter 11). Given the potential mobility of *Drosophila,* it is questionable whether many *Drosophila* populations retain their identities sufficiently long for intralocality integration of the gene pool to develop.

At the species level, the argument for an integrated gene complex is persuasive. Hybrids from species crosses are generally less fit than the parental species for a variety of reasons. Similarly, at the subspecific level, ecotypes, geographic races, and the like seem to consist of integrated gene complexes well-adapted to the particular circumstances of life for these groups. Whether natural selection will be able to bring about integration of the genotype at the level of the individual deme will depend on the strength of the selection pressure and the amount of gene flow involved.

Wallace (1968a, p. 309-310) presented what he regarded as decisive evidence for coadaptation at the level of the deme in *D. pseudoobscura.* However, he did not report the numbers of flies involved or the tests of significance. Examination of the original data (Dobzhansky, 1950) reveals that the numbers were small and, rather than a clearcut case for coadaptation, the tests of significance reveal that the data are ambiguous. Wallace (p. 709) states that in Table 18-3, "In every experimental population without exception, one homozygote (in one case, both homozygotes) exceeds the heterozygote in fitness." However, only in four of seven such comparisons were the homozygotes significantly more viable than the heterozygotes. In the other three cases the viability of one of the homozygous chromosome types and the interlocality chromosome heterozygote did not differ significantly. Therefore, the data do not provide unequivocal support for the idea that the interlocality chromosome heterozygotes should be intermediate in fitness between the homozygotes for chromosomes from two different areas, the expectation under the theory of coadaptation.

Wallace cites an experiment of Dobzhansky (1950, Table 3) (see Table 9-1) as providing decisive proof for the coadaptation hypothesis. In this experiment Dobzhansky compared the viability of inversion homozygotes with the same inversion drawn from two different localities with inversion heterozygotes in which the different inversions both came from the same locality. Wallace interpreted the data to show that "the intrapopulational structural heterozygotes exceeded the two structural homozygotes in fitness despite the hybridity of the latter arising from their remote geographic origins." In Dobzhansky's (1950) Table 3, reproduced here in its entirety (Table 9-1), each of the chi-squares has one degree of freedom. Therefore, only in one comparison out of five, the males from cage 51, did the intrapopulational structural heterozygotes significantly (at the .05 level) exceed the interpopulational structural homozygotes in fitness. Thus, rather than supporting Wallace's argument for coadaptation, these data seem to refute it.

INVERSION POLYMORPHISM AND THE ENVIRONMENT

As mentioned previously, there is evidence in *D. pseudoobscura* that different inversions carry different complements of genes and that the same type of inversion from different localities may carry different gene complements. Moreover, in certain other *Drosophila* species, populations in the center of the geographic range have been shown to have a high level of inversion polymorphism, but populations on the periphery have little or none (da Cunha, et al., 1959, and earlier papers cited therein concerning *D. willistoni*). *D. willistoni* ranges from Florida to Argentina, with the center of its geographic range in central Brazil. Inversions were found in all of the chromosomes, and in the center of the range there was an average of nine inversions per individual. The inversion heterozygosity decreased from the center toward the periphery so that at the margins of the range, the populations were almost monomorphic, with a mean of two, one, or fewer inversions per individual. Carson (1959, 1965) and Carson and Heed

Table 9 - 1. Observed Numbers and Viability Quotients of Homozygotes and Heterozygotes[*] for Chromosomes with AR and ST Gene Arrangements of Pinon Flats (P) and Mather (M) Origin

		ST^M/ST^P	ST^P/AR^P and ST^M/AR^M	AR^M/AR^P
Cage No. 50	Obs. ♀♀	32	64	17
	Viability	0.94	1	0.48
	x^2	0.13	-	8.55
Cage No. 50	Obs. ♂♂	30	57	18
	Viability	0.99	1	0.64
	x^2	0.01	-	5.03
Cage No. 51	Obs. ♀♀	29	58	8
	Viability	0.94	1	0.24
	x^2	0.11	-	16.94
Cage No. 51	Obs. ♂♂	23	67	20
	Viability	0.63	1	0.54
	x^2	4.50	-	21.46
Total	Obs.	114	246	63
	Viability	0.87	1	0.46
	x^2	2.19	-	35.25

Source: Dobzhansky, 1950, Table 3.

[*]The viability of the heterozygotes is taken to be unity.

(1964) reported a comparable situation in *D. robusta,* an eastern North American species. Dobzhansky (1957), in a study of island populations of *D. willistoni,* found that they resembled peripheral populations in having very low levels of inversion heterozygosity and that the smaller islands tended to have lower levels of inversion polymorphism than the larger islands.

To interpret these findings, Dobzhansky and his associates postulated that a variety of ecological niches is available to a species at the center of its geographic range, and that chromosome polymorphism enables the populations at the center of the range to exploit them. At the periphery of its distribution, a species is running up against the factors limiting its distribution, and the number of available niches is apt to be quite limited. Thus, chromosome polymorphism is considered to be adaptive and the number of inversions present is related to the number of ecological niches open to exploitation. The data, in general, appear to be consistent with this explanation. Small islands are usually less diversified ecologically than large islands, and ordinarily the higher the environmental rating, the higher the degree of inversion heterozygosity (da Cunha et al., 1959). However, there were exceptions, for some populations in extremely favorable habitat for *D. willistoni* in Columbia, Ecuador, and the coastal rain forest of Brazil were nearly monomorphic.

This interpretation is traceable to the ideas of Ludwig (1950) and to the mathematical treatments of Levene (1953), Li (1955), and Dempster (1955), which showed that environmental heterogeneity could support genetic polymorphism with or without heterosis. (See Hedrick et al., 1976 for a recent review.) However, Carson (1959) has proposed a somewhat different interpretation of these same data. He suggested that the marginal populations of a species would also be ecologially marginal and that these populations would tend to be small and relatively inbred. Natural selection would act to adapt these populations to the relatively limited habitats open to them. Thus, inbreeding, random genetic drift, and natural selection would all tend to promote homozygosity, both genic and chromosomal, within these small, marginal populations. He postulated that homozygosity for chromosome arrangements would permit an *open* recombination system so that whatever genic variation was present in these populations could recombine freely. In this way novel gene combinations could be generated in these small, isolated populations that could become the basis for adaptive experiments in the exploitation of the marginal habitat. Thus, under Carson's concept, the peripheral populations would be the primary focus for the origin of new races and even new species. The concept was considered applicable only to geographically differentiated, polytypic species in which the peripheral populations were relatively small and partially isolated. This type of natural selection Carson called *homoselection.*

In contrast to the idea of Dobzhansky (da Cunha et al., 1959) that the diversity of chromosome types in the center of the range of a species represented an adaptive response by the species to the diversity of habitats open to it, Carson (1959) suggested an alternative explanation. Selection favors heterozygosity, not because of the match between specific variants and particular niches, but rather because the heterosis and general vigor of the heterozygotes enable them to perform well in a variety of niches. They are "buffered" against environmental variations and show high fitness in a wide range of environmental conditions. The heterozygotes have a high level of adaptability or "plasticity." Such a population can be expected to generate a number of relatively less fit homozygotes as the price it pays for "heterotic buffering." However, the large, successful populations at the center of the range can afford the luxury, so to speak, of producing the less fit homozygotes in order to exploit heterosis. The price can be minimized if the heterotic system can be fixed in such a way that it is not easily broken up by recombination. Inversion and translocation heterozygosity provides such a mechanism. Natural selection then in these central populations was thought to act to capture and stabilize heterosis. Carson called this *heteroselection*. The high level of heterozygosity, both genic and chromosomal, and selection for favorable heterotic interactions and general adaptability would retard rather than promote the formation of new races or species at the center of the range.

Heteroselection then was conceived as favoring a high degree of generalized fitness through heterosis, and produced, not adaptation to any particular environment, but adaptability, the ability to function well in a variety of environments. In this case, the dogma that fitness is dependent on the environment in which the genotype resides is no longer of primary importance, for the heterozygotes were thought to be relatively more fit than the corresponding homozygotes in almost any environment. Heteroselection can be thought of as producing coadaptation (or selecting for special combining ability). Under this assumption, testing for relative fitness is simplified, for the superiority of the heterozygotes is assumed to hold under almost any circumstances, and the tests can be run without too much concern over the exact environmental conditions of the test. This fact may help account for the attractiveness of heterosis as an explanation for balanced polymorphism.

Heteroselection then would be expected to give rise to *rigid* polymorphism (Dobzhansky, 1962b), examples of which have been reported (Carson, 1958b, *D. robusta;* Brncic, 1961, *D. pavani;* Dobzhansky, 1962b, *D. willistoni*). In these cases, the inversion frequencies held remarkably constant despite environmental heterogeneity or seasonal changes in environmental conditions. This is in sharp contrast to the cyclic seasonal changes and marked geographic differences in inversion frequencies discussed earlier, which have been called *flexible* chromosomal polymorphisms. Rigid and flexible chromosome polymorphisms have been described, often within the same species (for a review, see Dobzhansky, 1971). The most remarkable case was reported by Crumpacker and Williams (1973), who found a seasonally rigid third-chromosome polymorphism in *D. pseudoobscura* populations in the foothills of the Rockies north of Denver, while the populations in the foothills south of Denver showed seasonal cycling and flexible polymorphism. The same chromosome arrangements were present in both types of polymorphism and the environments also were similar. Although Crumpacker and Williams sought an explanation for this difference, no satisfactory interpretation was found.

Carson's concept of homoselection and heteroselection implied that speciation, the differentiation of a species into new subspecies and species, is most likely to occur in marginal or peripheral populations. Dobzhansky's concept of *annidation* (Ludwig, 1950), the adaptation of specific genotypes to specific habitats within the same region, on the other hand, implied that speciation occurred most actively in the center of the range. This idea is probably traceable to Vavilov (1926), who argued that centers of diversity in plant species were the probable centers of origin of domesticated plants, and more recently to Brown (1957), who argued that marginal populations were relatively depauperate genetically and that *centrifugal speciation* occurred.

When electrophoretic techniques made it possible to compare genic variation with chromosomal variation in marginal populations, a number of such studies were made. What emerged was called by Soulé (1973) the "*Drosophila* paradox," for even though the inversion polymorphism was, as noted above, greatly reduced in the marginal populations, genic polymorphism was not (Prakash et al., 1969; Ayala et al., 1971, 1972; Prakash, 1973; Zouros et al., 1974; Zouros, 1976). Moreover, there was considerable geographic variation in the kinds and frequencies of inversion types, but the level of electrophoretic or genic variation not only remained relatively high throughout the species' range, but the configuration of allelic frequencies from one locality to another throughout the distribution of the species remained remarkably similar (e.g., Ayala et al., 1972). In addition, not just the mean but the distribution of electrophoretic variation was found to be relatively constant over homologous chromosomes even in different species of *Drosophila* (Zouros, 1976). This finding is in sharp contrast to inversion polymorphism, for the same chromosome element can be completely monomorphic in one species and highly polymorphic in another.

This situation is not only paradoxical, it presents quite a dilemma. It seems generally agreed that the variation in level of inversion polymorphism is in some way adaptive, but that the adaptive value of the various inversion types rests not with the inversion per se, but with its gene contents. However, the gene contents, at least as measured by electrophoretic techniques, seem to bear little or no relation to the inversion polymorphism. One of the arguments in favor of the study of protein polymorphisms is that it provides a random sample of the genes in the genotype (at any rate, of the structural genes). However, this lack of correlation between the cytological and electrophoretic variation has forced the conclusion (Zouros, 1976) that the "allelozymes are not members of the coadaptive complexes of genes characterizing inversions, and they do not by themselves differentiate the fitness of the gene arrangements with which they become associated." It has been further assumed, as pointed out earlier, that the gene contents of different inversions would differ. If this were the case, there should be a nonrandom association between genes (or electromorphs) and inversions. The fact is that linkage disequilibrium (or non-random association) has rarely been observed in natural populations except where recombination is virtually absent (Zouros et al., 1974). For the most part, whether the comparisons were between alleles at a locus and inversions, or between alleles at two different loci, they were found to be in linkage equilibrium; the association was random even where recombination was as low as 1%. These results, at the very least, suggest that if selection is working on these two kinds of variation, electrophoretic and karyotypic, it must be working in very different ways. It is, of course, possible that these differences are not due solely to selection but that other factors are involved as well. Several possible explanations have been advanced for this paradox, to which we shall return after dealing in the next few chapters with random genetic drift, gene flow, and non-Darwinian evolution.

In contrast to *Drosophila* where genic polymorphism seems to remain high even in marginal populations, genic variability may be greatly reduced in vertebrates inhabiting islands or caves or otherwise living under conditions where the populations are apt to be small, isolated, or subject to periodic *bottlenecking* or reductions in size (Selander and Johnson, 1973; Soulé, 1973; Selander, 1976). Therefore, *Drosophila* in marginal or peripheral populations tends to become chromosomally monomorphic but remains genically polymorphic, but vertebrates tend toward allelic monomorphism in marginal populations. Studies similar to the *Drosophila* work comparing genic and chromosomal polymorphism have apparently not been done with vertebrates.

Note that the persistence of high levels of genic polymorphism in marginal populations of *Drosophila* that are virtually free of chromosomal polymorphism is a situation highly favorable to Carson's concept of homoselection in marginal populations. With plentiful genic variability and free recombination, the marginal populations should be able to adapt to their immediate environmental conditions. Here, the puzzling fact is that genic variability seems to be uniform through-out the species, contrary to what might be expected under Carson's hypothesis. Lewontin (1974) argued that the temporal instability of the environment of marginal populations is responsible for their high level of genic heterozygosity. In the unstable and unpredictable environment of marginal populations, the favored genotypes are con-stantly changing through time so that genic heterozygosity remains high. However, the selection pressures in marginal populations would seem likely to vary in space as well as in time. If so, then different populations might be expected to diverge and differentiate from one another genetically. Therefore, Lewontin's explanation also seems to founder on the uniformity of the genic variability observed through-out a species.

Epling, Mitchell, and Mattoni (1953, 1955, 1957) developed still another explanation for inversion polymorphism in *D. pseudoobscura*. They failed to find consistent excesses of inversion heterozygotes in either natural or laboratory populations, which led them to question heterosis as the mechanism maintaining this inversion polymorphism. Instead, they pointed out that the presence of inversion heterozy-gosity in *D. pseudoobscura* is confined chiefly to the third chromo-some and, to a lesser extent, to the X chromosome. This inversion heterozygosity effectively suppresses crossing over and recombina-tion in these chromosome regions, but produces the so-called Schultz-Redfield effect (1951) in the other bivalents of the nucleus and in other regions of the same bivalent — that is, the chiasma frequency actually increases in those regions of the genome free of inversions. Like Dobzhansky, they postulated that "the different gene arrange-ments of a given chromosome are genetically and adaptively different." They assumed that the total amount of crossing over in a nucleus remains more or less constant, but that the different kinds and levels of inversion heterozygosity would alter the pattern of crossing over and recombination in different populations. Thus, the inversions would be selected not just for their gene contents but for their second-order effect in increasing recombination in the rest of the genome. In this way new gene combinations could be generated on the second and fourth chromosomes that are adaptive when combined with the particular arrangements on the third and X chromosomes. Thus, they envisioned inversion heterozygosity as an adaptive mechanism operat-ing through its effect on crossing over and recombination — more or less a fine-tuning mechanism for the recombination process. In some ways their ideas parallel Carson's, but whether selection can effectively act on such a second-order process is an open question. In other words, the gene contents of the inversions may turn out to be of greater significance than their effect on recombination. Furthermore, this is an *ad hoc* theory for *D. pseudoobscura*. As pointed out earlier, in some geographically widespread species of *Drosophila (D. repleta, D. simulans, D. virilis)* chromosomal polymorphisms are apparently absent. In others *(D. paulistorum, D. robusta, D. subobscura, D. will-istoni)*, inversion polymorphism is found in every chromosome of the set. The reasons for these differences remain a puzzle. However, lest we tend to dismiss the effect of inversions on recombination in *Drosophila* as unimportant, it is worth recalling that crossing over is usually absent in the males of this genus—a most unusual situation— suggesting that at some point in the evolution of the genus strong selection against recombination occurred. The low haploid chromosome numbers in this genus have also been suggested as a means to minimize genetic recombination.

A further dimension to the study of inversion polymorphism in *D. pseudoobscura* was added by the discovery of *long-term* frequency changes (over a period of three decades) for some of the third-chromosome inversion types. The most striking trends were a decline in frequency of Pike's Peak (PP) and an increase in frequency of Tree Line (TL) all along the Pacific Coast (Anderson et al., 1975). Although various explanations have been offered for these long-term trends, none of them has held up, and "the problem remains unsolved." Among these explanations were a correlation with rainfall (Dobzhansky, 1952b, 1956), the widespread introduction of pesticides, especially DDT, after WW II (Dobzhansky et al., 1964, 1966; Anderson et al., 1968; Cory et al., 1971), or various other environmental factors; however, evidence in support of any of these environmental factors could not be obtained.

The other type of explanation proposed was the formation and spread of new, adaptive gene complexes. One great difficulty here is the apparent rapidity of the genetic changes in frequency as compared to estimated rates of dispersal and gene diffusion in *D. pseudoobscura* (Dobzhansky and Wright, 1943, 1947; Crumpacker and Williams, 1973; Dobzhansky and Powell, 1974), which are far too low to account for the observed changes. Another difficulty is that if an adaptively superior gene complex arises in association with TL, for example, it must confer a general form of fitness in a wide variety of environments, for the increase in TL was observed over a wide geographic area containing a diversity of habitats. Thus, presumably TL is involved in a *rigid* rather than a *flexible* polymorphism. This presumption was not borne out in experimental populations, however, for TL fared no better in experimental populations than before (Anderson et al., 1967, 1975). Alternatively, it was proposed that the inversion type developed different adaptive gene complexes as it spread, related to the habitats it occupied so that the same gene arrangement would be genetically different in different localities (Dobzhansky, 1971). If this were the case, one may ask why the adaptive gene complexes could not have developed just as well in chromosome types other than TL. Note that, in all of these explanations, the implicit assumption is made that the observed changes in inversion frequency are due to some form of selection. Because these changes occur over a significant portion of the species range, this assumption seems plausible. Nonetheless, the nature of the selective forces remains elusive, and it would be helpful to know more about dispersal in *Drosophila*. Dobzhansky recognized this in his suggestion (1973) that "passive transport" might be responsible for the rapid spread of a new inversion type.

A footnote to the work on inversion polymorphism in *Drosophila* is an intriguing report by Dubinin and Tiniakov (1946a) on inversion polymorphism in *D. funebris*, which apparently has never been followed up. They found that the percentage of inversion heterozygotes in urban areas in Russia was very high, an average of 54%, with one population consisting of nearly 90% heterozygotes, whereas the percentage of inversion heterozygotes in rural populations was very low, of the order of 1%, and in some populations none was found. Although they postulate that natural selection is responsible for this difference, there is no evidence of the nature of the selection pressures that might be involved.

Although most of the discussion has been devoted to *Drosophila*, it should not be forgotten that inversion polymorphism has been found in other groups of animals and plants as well. Whereas paracentric inversions are characteristic of *Drosophila*, pericentric inversions have been found regularly in a number of species of grasshoppers and crickets (review in White, 1973). Because polytene chromosomes are lacking in these groups, the pericentric inversions are detected when they convert acrocentric chromosomes (with one short and one long chromosome arm) to metacentrics (chromosomes with arms approximately equal in length) or vice versa. Paracentric inversions and symmetrical pericentric inversions are very difficult to detect in grasshoppers where the rearrangements are usually studied in meiotic chromosomes. Therefore, it is possible that paracentric inverions also occur in grasshoppers but have simply gone undetected.

An interesting finding, both in the grasshoppers and in the Diptera, *Drosophila* and *Chironomus,* is the non-random association of inversions. In the grasshopper, *Keyacris scurra,* a non-random association of inversions on different pairs of chromosomes was found, while in the Diptera, marked non-random associations of inversions were found in the same chromosome pair. Although it might be expected that independent inversions would be associated at random in natural populations, this is certainly not always the case. Natural selection seems by far the most likely explanation for these non-random associations of inversions even though the exact mechanisms involved are not yet understood.

TRANSLOCATION HETEROZYGOSITY

Polymorphisms for duplications, deficiencies, and translocations are rare in natural populations. For the most part, when such rearrangements occur in wild populations, they are rapidly eliminated by natural selection because of their detrimental effects on development or through the formation of sterile gametes. However, translocation heterozygotes have been reported frequently in the plant genera *Oenothera, Clarkia,* and *Gaura* of the evening primrose family, Onagraceae, as well as in *Paeonia, Rhoeo,* and the Jimson weed, *Datura* (Stebbins, 1950, 1971a). Such cases are even rarer in animals, but have been reported in species of cockroaches of the genera *Periplaneta* and *Blaberus* and in species of Brazilian scorpions of the genera *Tityus* and *Isometrus* (White, 1973). Most of the cases of translocation heterozygosity have not yet been studied extensively, an exception being *Oenothera,* first studied by de Vries and more recently by R. E. Cleland and others (Cleland, 1962; Grant, 1971; Stebbins, 1971a).

In the subgenus *Euoenothera,* the group most thoroughly studied, all members have a diploid number of 14 and the chromosomes have median centromeres. The species are self-compatible and capable of selfing even though self-incompatibility occurs in other species of *Oenothera.* Some populations of *Euoenothera* show mostly paired chromosomes at meiosis with an occasional small ring, stemming from heterozygosity for one or two translocations. These plants are usually cross-pollinated. Other populations show heterozygosity for numerous interchanges so that rings of chromosomes of various sizes are formed at meiosis. However, these rings are not permanent, for these plants are also open-pollinated and may sometimes have progeny with seven bivalents. In some groups of *Euoenothera,* however, permanent structural heterozygosity has become established. At meiosis rings of chromosomes are formed; in some species all 14 of the chromosomes are included in a single ring. These extreme translocation heterozygotes are normally self-pollinated, and the situation is perpetuated by the mode of separation of the chromosomes at meiosis. The disjunction of the chromosomes is regular, with alternate chromosomes going to the same pole and adjacent chromsomes to opposite poles. In this way two complete, haploid sets of chromosomes regularly segregate from one another, and no aneuploid gametes are formed. Thus, each of the two chromosome complexes (known as *Renner complexes*) is transmitted as if it were a single chromosome and all the genes in a complex behave as if they belong to a single linkage group. (It was mentioned earlier that the blocks of genes held together within inversions by the absence of crossing over are known as *supergenes*. Each Renner complex is the ultimate in supergenes, for an entire haploid genome is transmitted as a unit.)

Furthermore, the progeny from such a plant resemble one another phenotypically as much as the pure-line offspring of a typical self-fertilized inbred plant despite their highly heterozygous origin. Two mechanisms have been discovered that ensure permanent heterozygosity. In one, a balanced-lethal system is established, with different, non-allelic lethals associated with each chromosome complex. Zygotes homozygous for a given complex are also homozygous lethal and die at an early stage. However, zygotic lethals reduce the plant's reproductive output 50%. The second mechanism avoids even that wastage of gametes. In this case each Renner complex contains a different gametophytic lethal. One complex (alpha) is found in all the egg cells; all the egg cells with the other complex (beta) fail to develop. Conversely, all the functional pollen grains contain the beta complex, for the microspores containing alpha do not mature. In this case there is a full seed set despite the 50% pollen abortion. Moreover, the regular combination of an alpha-bearing ovule with a beta-bearing pollen grain results in a true-breeding structural heterozygote. The mutants that de Vries observed so many years ago were later shown to result from rare viable recombinants or new arrangements involving the Renner complexes. In this way the genetic variation usually locked up in these complexes may occasionally be released.

The history of this group suggests that it has evolved from self-incompatibility and a free-recombination system polymorphic for an occasional translocation toward self-compatibility, increased translocation heterozygosity, and balanced-lethal systems to stabilize and maintain the structural heterozygosity. The selective advantages inherent in the system appear to be permanent heterosis and an ensured seed set. Although the *Oenothera* appear to be very successful, it has been suggested that such heterogamic complexes have limited evolutionary potential because recombination is so severely restricted by the very mechanisms that now make the genus so successful.

The study of natural populations has revealed a surprising amount of chromosomal variation, much of which is sporadic and is presumably rapidly eliminated by natural selection. Some of it, however, is part of the normal genetic system of the species. A puzzling aspect of this chromosomal polymorphism is the differences that exist in the degree of chromosomal polymorphism, not only between closely related species, but even between different populations of the same species. It is generally assumed that these chromosomal polymorphisms have adaptive significance, but in many cases the selective forces involved are not at all well understood. The most useful approach to an understanding of the role of chromosomal polymorphisms in natural populations is the comparative study of populations of the same species or of closely related different species that differ in the nature or degree of their chromosomal polymorphism.

CHAPTER 10

Random Genetic Drift

Natural selection, mutation, and migration or gene flow can be thought of as directional evolutionary forces, producing predictable changes in gene frequency if the mutation rates, the selection coefficients, and the migration coefficients are known. For this reason they are also sometimes known as evolutionary pressures. Although mutation is treated as a directional force, it also has its random aspects. For example, even though mutation rates can be estimated, it is not possible to predict which particular gene will mutate at a given time. Moreover, even though mutagenic environmental agents such as radiation and chemicals are known to increase mutation rates, the mutations that occur are usually as unpredictable as spontaneous mutations and are not an adaptive response to the agent inducing them. We have already dealt with the effects of mutation and selection within a single breeding population. Because migration or gene flow involves study of the genetic effects of the movement of individuals from one breeding population to another, its treatment will be deferred to a later chapter. Here we shall consider the effects of the third evolutionary force to act within a single breeding population, random genetic drift. Genetic drift differs from mutation and selection in that it is not a directional evolutionary force, and thus predictions as to the direction and magnitude of change in gene frequency due to genetic drift are not possible in the way that they are for mutation and selection.

A number of expressions synonymous with random genetic drift are used, among which are *accidents of sampling,* the *scattering* or *decay of the variability, random loss* or *random fixation, non-Darwinian evolution,* and the *Sewall Wright effect* in honor of the man who has done most to draw attention to the evolutionary implications of random genetic drift. Genetic drift signifies the random fluctuations in gene frequency that occur in effectively small populations. Note that Wright originally used the word *drift* in several contexts (Wright, 1970), one of which related drift to the directed process, selection, which he referred to as *steady drift. Random drift* was used to refer not only to the effects of accidents of sampling but also to those stemming from fluctuations in selection pressure and the like. Here, use of the term will be confined to *sampling drift* or the effects of accidents of sampling.

In discussions of random genetic drift, it is customary to consider the alleles involved to be adaptively neutral. It is possible to argue, perhaps, that theoretically no genetic difference should be expected to be adaptively neutral under all environmental conditions, that, given the right set of conditions, differences in fitness would appear. However, some traits do appear to be adaptively neutral in populations; at least, any selective differences between them defy detection.

To grasp the essence of random genetic drift, consider the simplest possible case, that of a single, heterozygous, self-fertilized plant (Aa) in which $AA = Aa = aa$ in fitness, and the population size at sexual maturity remains constant from generation to generation at N = 1. What is the probability that the single surviving F_1 progeny from the Aa heterozygote will also be Aa? There are four possible combinations between the male and female gametes: $AA; Aa; aA; aa.$ Therefore, the probability that the surviving F_1 individual will be heterozygous (Aa) equals 0.5. The probability of fixation of either A or a in the homozygous condition (AA or aa) is also 0.5. The fixation or loss of one or the other of the alleles would be due solely to the chance union of gametes and the chance survival of one of the seedlings produced because they are equal in fitness. Moreover, the initial gene frequencies of A and a were $p = q = 0.5$. If fixation or loss of an allele occurs, p will change from 0.5 to either 1.0 or 0.0 in a single generation owing solely to chance.

In the discussion of the Castle-Hardy-Weinberg equilibrium, it was pointed out that the gene-frequency equilibrium will persist in the absence of mutation, selection, and migration in an infinitely large Mendelian population. Actually, of course, all populations are finite, and some are quite small. The principles involved in the study of genetic drift are the laws of probability used in ways similar to those applied to coin-tossing or to drawing samples of different-colored marbles from a pot.

Although it is not possible to calculate the expected direction and magnitude of gene-frequency change due to genetic drift, it is possible to estimate the probability distribution and range of the possible amounts of genetic drift for a given population size, for the size of the deviations from the C-H-W equilibrium is inversely related to the population size, N. The N individuals in a diploid population are the products of N sperm and N egg cells, or $2N$ gametes. Fertilization is random so that an F_1 generation represents a sample of $2N$ gametes drawn at random from the gametes produced by the parents. Millions of gametes may be produced, especially by the males, but only $2N$ are used to form the F_1 generation.

If in the parents $f(A) = p$ and $f(a) = q$, then among N offspring the gene frequencies can vary with the following probabilities based on the binomial expansion, $(p + q)^{2N}$. Suppose that $N = 5$. Then

$$(p + q)^{10} = p^{10} + 10p^9q + 45p^8q^2 + 120p^7q^3 + 210p^6q^4 + 252p^5q^5 + \ldots$$
$$+ 10pq^9 + q^{10}$$

where p^{10} is the probability that all ten of the alleles carried by the five F_1 individuals will be the dominant A, $10p^9q$ the probability of $9A$ and $1a$, $45p^8q^2$ the probability of $8A$ and $2a$, and so on down to q^{10}, the probability of ten recessive a. In this case where $N = 5$, if $p = q = 0.5$, the chance of fixation of A equals $(1/2)^{10}$ or $1/1024$ and the chance of loss of A (or fixation of a) also equals $1/1024$. However, if $p = 0.1$ and $q = 0.9$, the chance of fixation of A equals $(0.1)^{10}$, a very small probability indeed, and its chance of loss equals $(0.9)^{10}$.

If $p = q = 0.5$ and $N = 5$, what is the probability that the gene frequency will remain exactly at 0.5? In this case, the term of interest is the p^5q^5 term, which gives the probability of 5 dominant A and 5 recessive a genes. In the binomial expansion above, the coefficient of this term is 252 and the probability that the gene frequencies will remain exactly at 0.5 is, therefore, $252/1024$. The chance of some other value of p and q — that is, of drift — is then $772/1024$. Thus, the chances are slightly better than three to one that in the next generation the gene frequencies will be either greater or less than 0.5 owing to chance.

Now consider the relation between population size and the variance in gene frequencies due to drift. The change in gene frequency because of drift may be designated $\delta q = q_1 - q_0$. The variance in gene frequency due to δq is the variance of a ratio, and takes the form

$$\sigma_q{}^2 = \frac{pq}{2N}.$$

The standard deviation equals the square root of the variance, and provides a useful way of examining the relation between population size and drift for different-size populations. The table below shows this relationship for $p = q = 0.5$.

				Population Size N				
	1	2	3	5	10	20	50	1250
Standard Deviation								
$\sigma_q = \sqrt{\frac{pq}{2N}}$	.353	.250	.204	.158	.112	.079	.050	.010

It can readily be seen that the smaller the value of N, the larger the standard deviation of the gene frequencies. Therefore, large chance fluctuations in gene frequency due to genetic drift are to be expected only in small populations.

Another way to look at drift is to consider the expected frequency distribution of gene frequencies for a single value of N. For example, consider $N = 50$ and $p = q = 0.5$ from the table above. The standard deviation was obtained from

$$\sigma_q = \sqrt{\frac{pq}{2N}} = \sqrt{\frac{.5 \times .5}{2 \times 50}} = \sqrt{\frac{.25}{100}} = 0.05.$$

Therefore in the next generation, the gene frequency in the 50 individuals will vary about the mean value of $p = q = 0.5$ with a standard deviation of 0.05. The probabilities for the different values of q are as follows:

q	Probability
$<$ .35	.002
.35-.40	.021
.40-.45	.136
.45-.50	.341
.50-.55	.341
.55-.60	.136
.60-.65	.021
$>$ .65	.002
	1.000

From this set of values, it can be seen that the chances are against the gene frequencies remaining exactly at $p = q = 0.5$. On the other hand, in a population of 50, the probability is vanishingly small that either allele will be fixed or lost when both are equally frequent.

Another way to use the table above is to consider a large number of loci, say 1000, all at an initial frequency of $p = q = 0.5$ in a population of 50. Then the probability column in the table gives the expected frequency with which the various gene frequencies will be found in the next generation (if linkage is ignored). Thus, some loci will drift below 0.5 and an equal number above 0.5, but only 2 in 1000 would be expected to have frequencies less than 0.35 and 2 in 1000 greater than 0.65. Over two-thirds of the loci will have gene frequencies between 0.45 and 0.55.

Still another way to use the table is to compare what would happen in 1000 populations all of $N = 50$ and $p = q = 0.5$ for one locus. Then the probability column gives the relative proportion of different populations in the next generation with the various gene frequencies listed.

Although the laws of probability are similar for coin-tossing and for random genetic drift, there is one very significant difference. In coin-tossing, the probability of heads or tails is always one-half. With drift, when the gene frequencies deviate from one-half, these new frequencies of p and q become the source from which the gametes are drawn for the next generation.

Mayr (1963) considered the use of the word *drift* inappropriate for the phenomenon of random genetic drift. He wrote (p. 213), "To apply the term 'drift' to nondirectional random fluctuations is unfortunate, since in the daily language we generally use the term 'drift' for passive movements in a more or less unidirectional manner." From this passage it is clear that he did not appreciate that, when gene frequencies deviate from equality, the less frequent allele always has a somewhat greater probability of decreasing still further in frequency by chance than it has of increasing in frequency. Thus the rarer allele will always be tending to drift to extinction and the more common allele to fixation. Hence, use of the word *drift* for the phenomenon, as suggested by Wright, is in accordance with Mayr's concept of a more or less unidirectional passive change and seems entirely appropriate.

To illustrate this fact, consider a population in which $N = 3$, $f(A) = p = 1/3$, $f(a) = q = 2/3$. Then

$$(p + q)2N = p^6 + 6p^5q + 15p^4q^2 + 20p^3q^3 + 15p^2q^4 + 6pq^5 + q^6.$$
$$\quad\ \ \text{I}\qquad\ \ \text{II}\qquad\quad\text{III}\qquad\quad\text{IV}\qquad\quad\text{V}\qquad\ \ \text{VI}\quad\ \text{VII}$$

Term V gives the probability that the gene frequencies will remain unchanged at $p = 1/3$, $q = 2/3$, and equals 240/729. Terms VI and VII give the probability that the less frequent A allele will decrease still further in frequency, and together they equal $256/729$. Terms I, II, III, and IV give the probability that the less frequent A allele will increase in frequency, and together they equal $233/729$.

These odds against the rarer allele may not be great (about 0.35 versus 0.32 in this case), but they are always there, and in the long run, the rarer allele tends to lose out, just like the gambler in a casino who tries to beat the odds, which always favor the house. Thus, the effects of drift tend to be cumulative over many generations, leading to eventual loss of one allele and fixation of the other.

When fixation of one allele occurs, random variation in gene frequency ceases. An irreversible change to homozygosity takes place. Since intermediate values of q can go to zero or one, but the reverse does not occur, random genetic drift leads to a decrease in the proportion of heterozygotes in the population—that is, to a *decay of the variability*.

RATE OF DECREASE IN HETEROZYGOSITY

It is possible to estimate the rate of decrease in heterozygosity in a population (Wright, 1931). A useful equation to illustrate this is,

$$k = \frac{1}{2N_e}$$

where

    k = rate of decrease in heterozygosity and

    N_e = effective size of the breeding population.

This equation is exact for complete random union of gametes, including self-fertilization, but even if self-fertilization is excluded, it is a reasonable approximation for moderate values of N_e. Furthermore, the number of males and females should be equal, and the initial dispersal of gene frequencies should have occurred. In other words, all classes of gene frequency should be equally frequent from just greater than zero to just less than one. Then in each generation a certain proportion of the heterozygous loci will become homozygous in the population.

This proportion will be $1/(2N_e)$, with fixation and loss each proceeding at the rate $1/(4N_e)$ so that the total rate of decay equals $1/(2N_e)$ (Figure 10-1) in small populations in the absence of mutation, migration, and selection. The frequencies of the terminal classes ($q = 0$ or 1) increase each generation, but the frequencies of the other classes

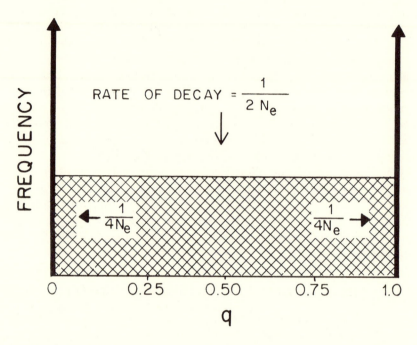

Figure 10-1. The rate of decay in the variability due to random genetic drift. In the absence of mutation, selection, and migration, fixation and loss of alleles each proceeds at the rate $1/(4N_e)$ so that the total rate is $1/(2N_e)$. (Source: Wright, 1931).

remain about equal although their absolute frequency gradually decreases. A useful analogy for this process is an open-ended trough filled with sand and subjected to vibration. The sand in the trough will be level, but as sand grains fall out of both ends (fixation or loss), the level of sand in the trough will gradually decrease.

To give a simple example, if in a population of $N_e = 50$, 1000 loci are heterozygous, then the rate of decrease in heterozygosity is

$$k = \frac{1}{2N_e} = \frac{1}{100}.$$

Among the 1000 loci, therefore, 10 will be expected to become homozygous in the next generation. Other things remaining the same, 1% of the remaining 990 heterozygous loci would be expected to become homozygous in the generation after that, and so on. Obviously, the trend in small populations as the result of genetic drift is to become more homozygous than they would be if they were larger.

It is possible to estimate the rate of fixation over a number of generations from the formula (Wright, 1931)

$$L_T = L_o e^{\frac{-T}{2N_e}}$$

where

T = number of generations,

L_o = frequency of unfixed loci at generation 0, and

L_T = frequency of unfixed loci at generation T.

and L_T will approach zero as T approaches infinity. Note that this increase in homozygosity will occur even though mating within the population is at random.

EFFECTIVE POPULATION SIZE

It may have been noticed in the formulas above that the symbol N_e was used for the effective size of the breeding population. The *effective size* is equivalent to the *actual size* obtained by a census only in a moderate-sized population in which the number of males and females is equal, mating is at random, and all individuals contribute genetically to the next generation. Furthermore, the population size should be constant—that is, the average family size should equal 2, and the distribution of family sizes should fit a Poisson distribution in which the mean equals the variance. Then, in this *ideal* population, the rate of decay will be $1/(2N)$ and the variance of the gene frequencies, as noted earlier, will be $pq/(2N)$.

However, not all populations fit the ideal outlined above, in which case it is useful to convert the actual size of the breeding population to a number equivalent to the number in an ideal population. This abstract number is the effective size of the breeding population and is the size of an ideal population, in the sense described above, that would have the same rate of decrease in heterozygosity as the observed population. From this abstract number, the effective size, an estimate of the expected rate of decrease in heterozygosity in the actual population can be obtained. The effective size of the population may be quite different from the actual size and is generally smaller.

If any of the conditions in the ideal population are not met in the actual population, then the actual size and the effective size will differ. For example, suppose that the number of males and females is unequal. In this case, Wright (1931, 1938) showed that

$$\frac{1}{N_e} = \frac{1}{4N\male} + \frac{1}{4N\female}$$

or

$$N_e = \frac{4N\male\, N\female}{N\male + N\female}.$$

If, for instance, there were 100 breeding females but only 10 breeding males, the size of the breeding population would be 110, but the effective size would be only

$$N_e = \frac{4(10)(100)}{10 + 100} = 36 \; .$$

Therefore, the amount of genetic drift expected in this population of 110 is equivalent to the amount expected in an *ideal* population of only 36. Note that the effective size depends much more on the number of the rarer sex than of the more numerous sex. In general, if one sex is much less numerous than the other, the effective number will be slightly less than four times the number of the less numerous sex.

Another possible departure from the ideal population would occur if the population size varies from generation to generation, a situation often observed in nature. In this case, if there is a regular cycle of sizable changes in number over a few generations, as might be observed in different seasons of the year with insects, for example, an effective number can be approximated from

$$\frac{1}{N_e} = \frac{1}{T}[\frac{1}{N_1} + \frac{1}{N_2} + \frac{1}{N_3} + ... + \frac{1}{N_T}]$$

where T = number of generations.

If the initial population is 10, for instance, and the population increases ten-fold each generation for six generations, then

$$\frac{1}{N_e} = \frac{1}{7}[\frac{1}{10} + \frac{1}{10^2} + \frac{1}{10^3} + \frac{1}{10^4} + \frac{1}{10^5} + \frac{1}{10^6} + \frac{1}{10^7}]$$

$$\frac{1}{N_e} = \frac{1}{7}[\frac{1,111,111}{10^7}]$$

$$N_e = 63.$$

Therefore, in this situation, the effective population size is about equal to the harmonic mean of the different values and is much closer to the minimum size during the cycle than to the maximum. Contrast the N_e of 63 with the arithmetic mean population size, which is 1,587,301. Despite the enormous size of the population in some generations, some drift is possible. If the population census is taken only in the more abundant phases of the cycle, the potential for random drift may never even be suspected.

If the change in population size is not regular and cyclical, but rare and sporadic, it may nevertheless have profound effects on the gene-frequency distribution in the population. The evolutionary role of cyclical and sporadic reductions in population size have been stressed by Wright (1931, 1932, 1938, 1948, 1951, 1969, 1970). If population size passes through a bottleneck, the gene frequencies may be so altered that new evolutionary opportunities or *adaptive peaks* may be open that were not previously available. If these bottlenecks are rare, they may be even more difficult to detect than if they are cyclical.

The actual population may also deviate from the ideal in that the number of surviving offspring left by different parents may differ greatly—far more than the variation expected with a Poisson distribution. Here, too, the effect is to reduce the effective size of the breeding group. Similarly, if there is some degree of inbreeding in the population rather than random mating, the effective size of the population will be less than the actual size. Formulas are available to deal with these cases in Wright (1931, 1938, 1969), Crow and Kimura (1970), and Li (1976).

All the deviations from the ideal population mentioned thus far (aberrant sex ratio, fluctuating population size, reproductive inequalities among individuals, and inbreeding) are similar in causing the effective size of the breeding population to be smaller than the actual size. The only case in which the effective size would be larger than the actual number would be the one in which the variance of progeny number was less than the binomial variance. For example, in the extreme case where each parent contributes exactly the same number of gametes to the next generation, the effective size would be nearly twice the actual size. While this situation might arise with the constraints possible with artificial selection, it is highly improbable in wild populations. Therefore, the factors operative in natural populations will all tend to make the effective size smaller than the actual size of the breeding population, and the potential for random genetic drift greater than might be suspected from the census data.

It might be added that, from a practical standpoint, it is not likely to be easy to measure the degree to which these factors affect natural populations. For example, reliable sex ratio data are not often available for wild populations, but even if they are, they are not sufficient. What is needed is an estimate of the numbers of males and females actually contributing surviving progeny to the next generation. Moreover, the breeding population itself must be identified and delimited — often difficult to do unless the population is in some way isolated from others of the same species on islands or in pockets of favorable habitat. If the population size is fluctuating, it will need to be monitored more or less continuously lest the minimum be missed. Also, it is usually easier to census a population by mark-recapture methods when it is common than when it is rare, so that unless the population is monitored continuously, there may be a bias toward censusing the population when it is large rather than at all stages. The detection of unusually large reproductive inequalities among individuals or of the degree of inbreeding in wild populations may be even more difficult than the detection of cyclical changes in population size or unequal sex ratios in parents.

Such estimates almost necessitate that individuals in the population be identified so that their relationship and degree of reproductive success can be determined. For this reason, it is not surprising that such estimates are far more easily made in humans or in domesticated animals and plants than in wild populations. Estimates of the effective population size of some human populations have been given by Glass et al. (1952), Cavalli-Sforza (1963, 1969), Cavalli-Sforza, Barrai, and Edwards (1964), and Steinberg et al. (1966).

Crow (1954), Crow and Morton (1955), Kimura and Crow (1963), and Crow and Kimura (1970) have drawn a distinction between two types of estimate of the effective population size, the "inbreeding effective number" and the "variance effective number." The inbreeding effective number is related to the number of individuals in the parental generations while the variance effective number is related to the number in the progeny generation. In many cases these estimates are roughly the same, but not always. These estimates stem from two ways of looking at the fluctuations in gene frequencies from generation to generation in finite populations. In one case, one can consider each generation of N diploid offspring as being derived from a sample of 2N gametes from the parental generation, as has already been done. This approach gives rise to an estimate of the variance effective number because it is based on the sampling variance of gene frequencies in the offspring. In the other case, there is a certain probability in a finite population that consanguineous matings will occur even though mating may be at random within the population. Moreover, the smaller the population, the greater the probability of matings between related individuals, or the greater the chances that they will carry alleles identical by descent from a common ancestor. In this case, the inbreeding effective number is determined by the number of individuals in the parental generation rather than in the progeny.

In general, if the population is growing, the estimate of the "inbreeding effective number" will be smaller than the "variance effective number" because it is related to an earlier generation, but if the population is declining, the reverse is true (Kimura and Crow, 1963; Crow and Kimura, 1970). If the population size is more or less constant from generation to generation, the two estimates are about the same.

In a finite population of N diploid individuals mating completely at random, it has already been shown that the rate of decrease in heterozygosity equals $1/(2N)$. The inbreeding coefficient, F, has been defined as the probability that two alleles in an individual are identical by descent from a common ancestral allele. Therefore, F refers to an individual and provides a means for expressing the relationship between its parents. F is the most useful index for describing the properties of a population relative to panmixia, and it can be used in this case as well to express the decrease in heterozygosity (or the increase in homozygosity) resulting from finite population size.

In the initial baseline-population, none of the alleles at a given locus is regarded as being identical by descent so that there are 2N possible kinds of gametes, and the coefficient of inbreeding is zero. In the next generation the probability that two uniting gametes carry alleles derived by replication of a single parental allele is $1/(2N)$. Thus, in a finite population, even with random mating, it can be seen that the rate of decrease in heterozygosity is equivalent to the effect of inbreeding.

The generalized expression for the inbreeding coefficient is

$$F_t = \frac{1}{2N} + \left(1 - \frac{1}{2N}\right)F_{t-1}$$

where F_t = inbreeding coefficient of generation t. F_t consists of two terms, the first measuring the effect of *new* inbreeding, the other measuring the inbreeding in all previous generations. For further information on the inbreeding coefficient, see Falconer (1960), Wright (1969), Crow and Kimura (1970), Merrell (1975a), or Li (1976).

If, rather than random mating, there is some degree of inbreeding as well—that is, matings between individuals more closely related than the average for the population of which they are members,—the effective size of the population will be further reduced, and (Li, 1976) will equal

$$N_e = \frac{N}{1 + F}.$$

The upper limit for F is one, so that the effective size in that case would be one-half the effective size if mating were random.

Crow and Morton (1955) wrote that "In order to determine the effective population number one needs to know (1) the number of reproducing adults, (2) the extent of non-random mating, and (3) the distribution of the number of surviving progeny per parent." With these requirements, it is not surprising that relatively few estimates of effective population number have been made for natural populations. Examples of such studies are Dobzhansky and Wright (1941), Wright, Dobzhansky, and Hovanitz (1942), Wright (1943), Lamotte (1951, 1959), Kerster (1964), Tinkle (1965), Petras (1967), Kerster and Levin (1968), Levin and Kerster (1968), Merrell (1968), Nozawa (1972), and Berry and Peters (1976). It is doubtful that for any of these estimates, reliable data on all three requirements set by Crow and Morton were available. In addition, they assume an isolated, identifiable breeding population, but Dobzhansky and Wright obtained estimates of N_em (where m is the migration coefficient) rather than N_e, and could only estimate N_e by making various assumptions about the amount of migration or gene flow involved.

I shall briefly cite some of the problems involved for such estimates in the leopard frog, *Rana pipiens,* with which I am most familiar (Merrell, 1965c, 1968, 1977). The leopard frog in Minnesota typically breeds in early spring in temporary ponds. They migrate from their overwintering sites in the larger lakes or streams to these ponds as soon as possible after the ice melts, their behavior apparently regulated by temperature (Merrell, 1977).

The calling males assemble on the surface in the warmest sector of the breeding pond where they pursue each other, the females, and just about anything else that moves. In contrast to the conspicuous behavior of the males, the females in the vicinity of the breeding pond are very secretive and circumspect. Because of this difference in behavior, the sex ratio among frogs captured at the breeding ponds was usually about 10 males to 1 female even though the sex ratio among sexually mature individuals sampled at other times of the year was always close to one-to-one. Nevertheless, by the mark-recapture method, an estimate of the number of frogs in a breeding pond could be made. This Lincoln-Petersen estimate was then adjusted to a one-to-one sex ratio. However, if the age or size at sexual maturity differed between males and females, this correction would introduce error.

Once the frogs reached a breeding pond, they seldom if ever moved to other nearby breeding ponds. Because none of the marked frogs was ever found to have moved to nearby breeding ponds, each pond seemed to contain a separate breeding population, and the mark-recapture data gave an estimate of the number of breeding adults in each pond, the first requirement of Crow and Morton.

Moreover, the egg masses were deposited in a small area one or two meters in diameter where it was possible to count the number of egg masses and thus make a different and more reliable determination of the number of reproducing adults. The females are known to extrude all of their eggs in a single mating and to be depleted of eggs for several months following the breeding season so each egg mass is equivalent to a female. Indications are (Merrell, 1977) that each male also mates only once each breeding season. If this is so, the number of breeding adults would be twice the number of egg masses, and the sex ratio would be one-to-one. If it were not so, the number of breeding adults would be somewhat less than twice the number of egg masses.

With these two estimates of size of the breeding population, a direct comparison was possible between the mark-recapture estimates and the egg-mass estimates (Merrell, 1968). It was put in the form of a ratio, N_e/N_A, where N_e was the effective size of the breeding population taken as double the number of egg masses, and N_A was the census number adjusted for an equal sex ratio (Table 10-1). There, a wide discrepancy between the estimates can be seen, for in some cases, the number of adults actually breeding is 10% or less of the number estimated from the mark-recapture data. In all six populations N_e from the egg-mass data is small enough to suggest that random genetic drift may be a significant factor in these populations. Moreover, the comparison suggests that census data alone may be inadequate to estimate the effective size of the breeding population and may be quite misleading about the potential for drift if they are the sole source of information about the population.

Crow and Morton's second point, the extent of non-random mating, could not be estimated in the populations. However, the aggressive mating behavior of the males, which attempt amplexus with almost any seemingly suitable object, including conspecific males as well as females, members of other amphibian species, and even inanimate objects like dead females or beer cans, indicates that they mate at random. Because a female, once seized by a male, seemed unable to dislodge him, mating must be essentially random in both sexes.

Crow and Morton's third requirement was knowledge of the distribution of the number of surviving progeny per parent. Here, too, no estimate was possible in the breeding ponds. However, all of the egg masses in a breeding pond are laid at about the same time in a small area no more than two meters in diameter. The young tadpoles

Table 10 - 1. Comparison of Estimates of Breeding Population Size in the
Leopard Frog from Egg Mass Data (N_e) and Mark-Recapture Data (N_A)

Breeding Pond	N_e	N_A	N_e/N_A
1	80	11,676	0.007
		1,940	0.041
		2,511	0.032
		3,742	0.021
2	48	1,688	0.028
		894	0.054
3	112	213	0.526
		236	0.474
4	46	475	0.097
		419	0.110
5	102	274	0.372
		403	0.253
6	96	144	0.667
		268	0.358

Source: Merrell, 1968.

hatch after several days of development and continue to develop in the pond for three more months before metamorphosis. During this time random mixing of all of the tadpoles in the pond occurs. Mortality is high during this period, but the most reasonable assumption is that deaths are more or less random because of the mixing rather than that a family group survives or perishes as a unit.

Thus, after a fashion, it was possible to estimate the effective population size in these populations, but it still involved a number of assumptions. If it is assumed that mating is at random, that the population size is constant (the average family size is two), that there is a Poisson distribution of family size, and that each egg mass represents two parents, then twice the egg-mass count gives a direct estimate of the effective size of the breeding population. This estimate has greater reliability than a census based on mark-recapture data, primarily because the sexually mature adults without offspring were excluded in the egg-mass count.

However, it must now be added that the parents disperse widely from the breeding pond to the summer feeding habitat, migrate in the fall to an overwintering site in a lake or stream, and again in the spring to a breeding pond—journeys that may take them several kilometers in the course of a year. Similarly, the tadpoles, once metamorphosis is complete, leave the breeding pond for the shores of larger bodies of water, and migrate to overwintering sites in autumn. Thus the well-defined cohesive population of parents and their off-spring at the breeding pond is widely dispersed, and the effective size of the breeding population, complicated in this way by dispersal and migration, becomes a matter of conjecture. Instead of retaining their identities from year to year, the breeding populations must be reconstituted each spring at the breeding ponds. Because the breeding ponds are temporary and dry up periodically, the pattern of ponds used for breeding in an area changes from year to year, and it is un-likely that homing to the pond of origin helps to keep the populations isolated and distinct. Therefore, these estimates of effective popula-tion size involve assumptions and uncertainties that limit their reli-ability. The other estimates for natural populations cited above are all subject to similar limitations.

EVOLUTIONARY SIGNIFICANCE

Sewall Wright (1931, 1938, 1948, 1949a,b, 1955, 1969, 1970) has been the primary advocate of the significance of random genetic drift as a factor in evolution. Because of drift, small isolated popula-tions may diverge from one another in gene frequencies, and local groups may differentiate owing to random non-adaptive changes. Moreover, in effectively small populations, the less favorable alleles may sometimes be fixed owing to drift. The level of homozygosity will also, on the average, be higher in small populations than large. As a consequence, the fate of small populations will usually be extinc-tion, not just because their small numbers make them more subject to the vagaries of nature that may wipe out the entire population, but also because their loss of variability and high level of homozy-gosity may lead to a loss of ability to adapt to changing conditions. Furthermore, if unfavorable alleles or gene combinations are fixed by chance in the population, its fitness may be lowered to the point of extinction even if conditions do not change. Large populations, on the other hand, retain their genetic variability and ability to adapt and are not subject to the loss of fitness owing to random fixation of unfavorable alleles.

Wright argued that the significant role of random genetic drift lay in its relation to his *shifting balance* theory of evolution (1931, 1932, 1955, 1970). In essence, he felt that even though the fate of most small populations was extinction, occasionally new combinations of genes could become established by chance in small populations. These distinctive gene combinations could never arise in large populations, whose genetic composition is more rigorously controlled by natural selection. These small populations could then become evolutionary experiments, permitting the population to drift away from the *adaptive peak* it is confined to in larger populations, across *saddles* of lower fitness to the slopes of different, and perhaps higher *adaptive peaks* toward which selection would now move the population. If, in fact, such a peak were higher, the deme occupying it would be more successful than neighboring demes, would produce relatively more progeny than neighboring demes, and by dispersion of this excess to its neighbors, in time would cause them also to shift to this new and higher adaptive peak.

Fisher (1958), however, dismissed random genetic drift as an evolutionary factor because, he said, even though populations may sometimes be small, they must remain completely isolated for many generations for drift to be significant. Even if perfect isolation were possible, which he doubted, he felt that extinction of the population would occur before anything of evolutionary significance took place, and the area would be repopulated from neighboring, larger populations or else the population would be absorbed into other populations of different genetic constitution.

The difficulties in assessing the importance of random genetic drift as an evolutionary force are shown by the controversy between Fisher and Ford (1947, 1950) and Wright (1948, 1951). Here the same set of data on the moth, *Panaxia dominula*, was interpreted by Fisher and Ford as being due solely to selection but by Wright as being the result of a combination of selection and random drift; neither view prevailed because of the uncertainties involved. A similar case has arisen in studies of snails of the genus *Cepaea*, where Lamotte (1951, 1959) and Goodhart (1962, 1963) have invoked both selection and random genetic drift to interpret their observations. However, Cain et al. (1968), in a summary of their studies of this genus over many years, are strongly selectionist in their interpretation.

In view of the doubts about the evolutionary significance of random genetic drift and the difficulties associated with its study, indicated earlier, certain steps seem desirable. Because the census figures may be so far above the actual number of breeding adults, and the latter figure is usually so difficult to estimate, it would seem wise to work with populations whose size, even in a census by mark-recapture or similar methods, is so small that a significant amount of drift could be expected. Furthermore, to eliminate the problems associated with migration, isolated populations should be studied, or populations in which the members can be individually identified. One other way to minimize the problem introduced by migration is to work with relatively sedentary small populations, of course.

It would seem that random genetic drift is apt to be much more important to the evolution of some species than others. Populations of top carnivores such as bears and wolves or eagles and falcons are never large so that the probability would seem much better of finding significant effects of drift in such species than in houseflies or mosquitoes or some of the species that have been studied.

Another opportunity lies in the study of populations invading areas previously uninhabited by the species. The numbers of the founding population are usually small, which, as seen, may lead to significant divergence from the ancestral population even though the population subsequently becomes large. In the recent past there have been many introductions of species into new areas, both accidentally and intentionally (Elton, 1958; Baker and Stebbins, 1965), which would seem to provide unusual opportunities for evolutionary studies. Not only are the founding populations often small, but they are also exposed to new and somewhat different selection pressures in their new physical and biological surroundings. Mayr (1942, 1954, 1963) has stressed, in addition, that a *genetic revolution* is possible in effectively small founder populations with only a limited sample of the gene pool of the parental species. He has labeled this combination of natural selection and random genetic drift the *founder principle*, but it does not seem to differ in any significant way from Sewall Wright's earlier *shifting balance* concept of evolution leading to new adaptive peaks. The expression has become well entrenched in the literature, but *founder effect* would be more appropriate, for it is not a single well-defined principle like one of Mendel's laws, for example. In this connection, the exchange between Mayr (1959) on the one hand and Wright (1960) and Haldane (1964) on the other is of considerable interest for the insight it provides into the way two of the men most involved viewed the history of the development of population genetics and modern evolutionary theory.

Some attempts to study random genetic drift in laboratory populations of *Drosophila* have been made. One advantage to this approach is that migration, always a possible complicating factor in natural populations, can be excluded. Small population size was ensured by using very small population units (Merrell, 1953), or by limiting the number of parents used in each generation (Kerr and

Wright, 1954a, 1954b; Wright and Kerr, 1954; Buri, 1956), or by using different-sized founder populations (Dobzhansky and Pavlovsky, 1957; Dobzhansky and Spassky, 1962). Random fluctuations were observed, but even in these extremely small populations, in some cases selection continued to be effective against deleterious mutants (Merrell, 1953; Wright and Kerr, 1954). In such cases, the combined effects of strong selection and drift led to more rapid fixation of the favored allele than would occur in larger populations. In other cases fixation was opposed by selection favoring the heterozygotes (Kerr and Wright, 1954b; Dobzhansky and Pavlovsky, 1957). There were even two instances in which the alleles appeared to be neutral (forked and its wild-type allele, Kerr and Wright, 1954a; and the brown alleles, bw^{75} and bw, Buri, 1956). The effective population size (N_e) was calculated by Kerr and Wright and by Buri and could be compared to the known number of parents used. The N_e/N_A ratio was greater than 0.5, in accordance with similar findings by Crow and Morton (1955), who concluded that N_e would generally be the same order of magnitude as N_A, but somewhat smaller.

CHAPTER 11

Migration and Gene Flow

POPULATION DISTRIBUTION

Even if agreement were possible on just what constitutes a species, it would not be possible to say how many living species there are. New species are being described all the time, but there is no way of knowing how many more unknown species remain to be identified and described. Even the number of known species is uncertain, but the estimates seem to be of the order of a million or more, and the speculations about the total number of living species range from about two to four million. It is even more hazardous to try to guess the total number of species that have ever existed on earth, but it is clear that many more species are extinct than are living. It has been estimated that living species constitute only 0.1% or less of all the species that have ever lived. Thus, living species constitute but a very small fraction of the totality of life that has inhabited our planet. Nonetheless, they are the surviving products and the end result of millions of years of evolution during which their germ plasm has been shaped by evolutionary forces. Moreover, their germ plasm carries the potentialities for all future life and all future evolution as well. Therefore, it is essential to learn as much as possible about living species, for they are the key both to the past and to the future.

One of the first questions about a species, once it has been identified, is, Where does it live, what is its distribution? The discussion of adaptation should lead to the realization that species can exist only where the ecological conditions are favorable. Even within a given geographical area, the environment is not uniform and a great variety of different habitats exists. Since any species is not ordinarily found uniformly distributed over a large geographical area, and yet there are no physical barriers to its dispersal in the area, its distribution under these circumstances must be limited by ecological barriers. In the continental United States, red-winged blackbirds live in association with marshes, their natural habitat, rather than uniformly distributed at so many per square mile. The number of species of amphibians and reptiles gradually decreases in North America from south to north. Apparently not all of these cold-blooded vertebrates are adapted to withstand the cold northern climate, and temperature is a limiting factor in their distribution.

However, as Darwin realized many years ago, the distribution of animals and plants cannot be explained solely in ecological terms. The success, one might say the disastrous success, of many species introduced into areas where they were not previously found, is alone sufficient evidence that something other than ecological factors must also be involved in the present-day distribution of animals and plants. The gypsy moth, the Japanese beetle, the starling, the English sparrow, Dutch elm disease, and the American chestnut blight are all familiar examples of species introduced into the United States with dire consequences because they were so well adapted to their new home. A few muskrats introduced into Europe have spread from Spain to Siberia in less than a century; two-dozen European rabbits, released in Australia, increased to an estimated one billion and damaged millions of acres of rangeland. In New Zealand so many alien species have been introduced from Europe and North America that the native fauna have difficulty competing with them. An obvious conclusion is that species do not live everywhere in the world where conditions are favorable for them.

The absence of a species because of an unsuitable environment is easily understood, but its absence when the environment is favorable raises other questions. The answer, of course, is that the distribution of animals and plants has a historical as well as an ecological basis. Ecological factors alone cannot account for the distribution of animals and plants. Long ago, Wallace (1876) recognized that, despite the variety of climates and habitats found within a given geographical region, the species within such an area tend to resemble one another more than they do species from other parts of the world. For this reason biogeographical realms were delimited within which existing groups of animals and plants tend to show similarities. These regions are: the *Nearctic*—North America down into the Mexican Plateau; the *Neotropical*—Central and South America; the *Palearctic*—Europe,

Africa north of the Sahara Desert, and Asia north of the Himalayan Mountains; the *Ethiopian* — Africa south of the Sahara Desert; the *Oriental* — the tropical part of Asia south of the Himalayas; and last, the *Australian* — Australia, New Zealand, and some associated islands. The species in the Nearctic and Palearctic show enough similarites so that these two realms are sometimes grouped together as the *Holarctic*; and, in general, adjacent realms, despite their climatic differences, tend to have faunas and floras more alike than areas more removed from one another. The concepts of biogeography have developed from studies of the distribution of particular groups of organisms. Especially informative have been the mammals and the freshwater fishes. For both groups the oceans have served as barriers to their worldwide distribution. The vagility of birds and bats has made their dispersal patterns more complex and less informative.

The present-day distributions of plants and animals are best understood if certain assumptions are made. First, that each species has spread from a single *center of dispersal* or *center of origin*; second, that all species have a tendency to expand their range because of their high reproductive capacity and their vagility or tendency to disperse. Despite all the efforts to prevent their spread, which included building thousands of miles of fences, rabbits expanded all the way across Australia in a matter of decades. In North America, starlings and English sparrows, introduced in trivial numbers in the East, have increased their numbers and expanded their range to the point where they are now regarded as pests in nearly all parts of the continent. The limitations to dispersal are either ecological barriers such as temperature or moisture, or physical barriers such as oceans, deserts, mountain ranges, or for aquatic species, land itself. (Of course, these so-called physical barriers are also ecological in the broad sense.) Another assumption is that new species have evolved from existing ones and usually appear in areas adjacent to the range of the parental species from which they are isolated by a barrier. Furthermore, existing distributions can be understood only if it is assumed that species have originated in all parts of the earth throughout geological time since the appearance of life. In other words, there has not been some special time or place where the origin of species occurred. Therefore, an understanding of the present-day distribution of a species requires a knowledge of evolution and geological history as well as an understanding of the ecological requirements of that species.

Starting with the work of Darwin and Wallace, knowledge and understanding of evolution and the evolutionary process have greatly increased. In just the past decade, to quote Colbert (1972),

> The science of geology has during recent years been going through a period of revolution as profound as that which swept through the science of biology after the publication of Darwin's "Origin of Species" in 1859. . . . It is parallel in that, just as Darwin and Wallace changed the concept of the species and of life from a static to a dynamic force, so the geological revolution has changed the large scale concept of the earth from that of a rather static and stable planet to that of a very active, and in certain respects, an unstable globe.

The revolution in geology, of course, has led to the universal acceptance of the concept of continental drift, a hypothesis with a long history (Wegener, 1924; Du Toit, 1937) that was generally dismissed by geologists until, in recent years, the evidence in its favor became overwhelming (Wilson, 1972; Sullivan, 1974; Glen, 1975). Until about 1965, the study of present and past distributions of living and extinct species was based on the assumption that major land masses—the continents—were fixed in place or stable and that the major events affecting distribution were changes in climate and sea level, which led to the formation or inundation of land bridges between the continents (Matthew, 1939; Simpson, 1953). The revolution in geology meant that not just time was a variable in the study of biogeography, but space as well. The variation through time in the spatial relations of the major land masses meant that the study of distribution became much more complex so that earlier conclusions about distribution had to be reassessed to bring them into accord with current geological thought. This reevaluation is still taking place, but in general, conclusions about Cenozoic distributions have been less affected than those occurring during the Paleozoic and Mesozoic when the arrangement of the continents was so different.

In summary, the present-day distribution of organisms has been determined by the combined effects of ecological conditions, the vagility of the group, the historical climatic and geological factors that have aided or limited the expansion of populations, and finally, by the theory of evolution. In a practical sense, the distribution of a species is determined from the accumulated records of observations and collections of individual members of that species. Most distribution maps leave a great deal to be desired in terms of the sampling procedures and the amount of data on which they are based.

THE ECOLOGICAL NICHE

The concept of the *ecological niche* is widely used and is fundamental to ecology. To be useful, a fundamental concept requires definition.

Perhaps surprisingly, for such a basic idea, it is often used without definition, or, when defined, the definitions of different authors are quite different. Following are samples of meanings associated with the concept of ecological niche:

Grinnell (1904, 1917, 1928) was apparently one of the first to use niche "to stand for the concept of the ultimate distributional unit, within which each species is held by its structural and instinctual limitations." This type of niche referred essentially to the micro-habitat of the species; it was primarily a spatial niche. It is note-worthy that Grinnell was apparently the first to formulate clear statements of the concepts of ecological niche and competitive exclusion, which have played such a large role in ecological thought.

Elton (1927) used niche to describe the role of a species in the ecosystem. "It is therefore convenient to have some term to describe the status of an animal in its community, to indicate what it is *doing* and not merely what it looks like, and the term used is 'niche'. . . . The 'niche' of an animal means its place in the biotic environment, its relations to food and enemies." Elton stressed the functional status of an organism in the community, especially its energy rela-tions, so that his definition is sometimes called the trophic niche. However, in another version of his definition (also 1927) he wrote, "By a niche, is meant the animal's place in its community, its rela-tion to food and enemies, and to some extent to other factors also." That last phrase seems to qualify the statement somewhat, and a complete reading of Grinnell and Elton leads to the conclusion that their niche concepts were not so far apart as is sometimes sug-gested.

Subsequently, Hutchinson (1957, 1965) visualized the niche as a multi-dimensional space or n-dimensional hypervolume, with each dimension representing one of the biotic or abiotic variables in the environment of the individual, population, or species. The *funda-mental niche* would then be the hypervolume delimited by the limit-ing values for each variable within which the population can survive and reproduce. Because populations do not necessarily live under all the conditions they can tolerate, or utilize all the resources they are capable of using, a distinction can be made between the *fundamental niche* and the *realized niche*. Furthermore, some species may be much more specialized than others—for example, in their food requirements or habitat requirements—so that the idea of *niche breadth* can be used to describe the degree of specialization of a species. Thus, the *hypervolume* or *multi-dimensional niche* of Hutch-inson is all-inclusive. Therefore, the niche concept has been used to refer to the habitat of a population; to its role, especially its trophic role, in its community; or to a broadly inclusive relationship between the population and its environment.

Although at times it seems that each ecologist has his/her own conception of an ecological niche, in recent years the multi-dimensional niche of Hutchinson has drawn the most attention. As Hutchinson (1957) phrased it, "The fundamental niche of any species will completely define its ecological properties." Thus, to determine the ecological niche of a species, it would be necessary to determine all its relationships to its physical environment (light, temperature, moisture, substrate, and so on) necessary for survival, all its requirements in terms of food and habitat, and all its relationships to other organisms of its own and other species.

Since individuals differ genetically, each individual member of a species can then be expected to have a somewhat different fundamental niche, and each individual would also have its own realized niche. Since different populations of a species have somewhat different gene pools, each population has a somewhat different fundamental niche and its own realized niche. Then, the fundamental niche of the species will be equivalent to the most inclusive n-dimensional space that could be occupied by any of these populations, and the realized niche of the species is equivalent to the n-dimensional space actually occupied by all of these populations.

If it sounds as if the determination of an ecological niche in Hutchinsonian terms is a large order, it is. In fact, the concept of a fundamental niche poses some major practical difficulties and has yet to prove particularly useful. For one thing, it has an infinite number of dimensions so that the niche of any organism, let alone of any species, is impossible to determine. In this, it resembles the measurement of the phenotype or of heterosis. All three are determined by the traits that happen to be measured. The best one can hope to do is to make a perceptive choice of traits to measure that are relevant to the study at hand. Next, it is assumed that all environmental variables can be measured in a linear fashion, which may be very difficult, especially for biotic factors (e.g., food quality, or dispersion pattern of individuals). Moreover, the concept is static in that it defines the niche at one point in time, yet in a living, evolving species both the fundamental niche and the realized niche will be constantly subject to change. Rather than attempt to define the niche completely, the customary approach to the study of the ecological niche has been to measure some of the more obvious environmental and biotic variables. This approach has been particularly useful in comparative studies of the ecological requirements of closely related species

(MacArthur, 1968) for which selected ecological requirements can be measured and compared.

In an abstract sense, the all-inclusive concept of an ecological niche is easy to grasp. In a practical sense, the understanding of the ecological niche of any species will only be as complete as the physical and biological components of the environment that are considered important enough to study. To give a simple example: If one is seeking the key to the distribution and abundance of a species by measuring temperature and moisture when, in fact, a food plant and a parasite are the primary factors, one will be very wide of the mark. Thus, in a very real sense, the ecological niche of a species is going to be what we make it by our choice of components to be measured.

POPULATION STRUCTURE

Thus far, we have dealt only with events within a single, local breeding population or deme. In turn, we have considered the Castle-Hardy-Weinberg equilibrium, mutation, natural selection, and random genetic drift and their effects within a single breeding population. However, the population of an entire species seldom, if ever, forms a single, large breeding population. Instead the species population is ordinarily subdivided in some way into a number of separate breeding populations. Even if it is not, the range of the species is usually large enough that individuals in different parts of the range have no opportunity to mate because they are isolated from one another by distance. The robins in Massachusetts, for instance, do not mate with those in Minnesota because of the sheer physical distance separating them. Therefore, the population structure, or the way the species population is distributed in space, may be of significance. Various types of population structure can be envisioned, of which we shall consider only a few.

To discuss population structure, some additional terms must be introduced. In contrast to the *deme*, or local breeding population, we shall use the more general expression *Mendelian population* to refer to a system of individuals united by mating and parentage bonds, or to a reproductive community having a common gene pool. Thus, a Mendelian population may be as small as a deme or as large as an entire species. *Polymorphic variation* refers to the variation within a single deme. The term *sympatric* means living in the same country and refers to members of the same deme or to individuals that are at least potential mates because of their proximity to one another. A somewhat more quantitative definition states that sympatric organisms are those found within the average distance between the points at which an individual and its progeny are born. This definition is a crude measure of the area within which potential mates for the individual must have lived. *Allopatric* means living in another country and refers to individuals living so far apart that they are not potential mates. *Parapatric*, meaning living in adjacent areas, refers to populations that are geographically contiguous so that mating is possible along their zone of contact. Whereas polymorphic variation is intrademic variation, different demes frequently differ genetically from one another, and this interdemic variation is often called *polytypic variation.*

The extremes for types of population structure are continuous and discontinuous. The ultimate discontinuous distribution is the so-called island model in which each population is distinct and well-defined and isolated from the others. It can refer, of course, not only to island populations separated from one another by water, but to populations inhabiting a series of lakes, or ponds, or puddles, or marshes, or patches of woods surrounded by prairie, or patches of prairie surrounded by woods. Any case in which an area of favorable habitat for a species is separated from other similar areas by unfavorable habitat fits the island model. A continuous distribution would be found in a large area of uniform habitat such as the coniferous forest in northern North America, or the eastern deciduous forest, or the western prairie. Although, superficially, such environments may appear to be uniform, they may not be. Even if they are uniform, true random mating may not be attained because of isolation by distance.

A linear distribution such as might be seen along a stream is another type of population distribution. The fish in the stream or the willows on the bank would show a linear, continuous distribution. The entire stream system of tributaries would have both continuous and discontinuous aspects, for populations in the headwaters of different tributaries would be separated from one another. The *stepping stone* model, which is a favorite of theoreticians, involves a linear array of island populations. Actual distributions may fall into many patterns.

The most significant point to note is that any subdivision of the species population leads to an increase in the proportion of homozygotes in the entire population as compared to the proportion present if mating were random. This relation was first worked out by Wahlund (1928) and is known as Wahlund's formula. Following Li (1976), let us examine a simple example. Assume that a large population

Table 11 - 1. The Effects of Subdivision in a Large Population into K groups

Group	p_i	q_i	p_i^2	$2p_iq_i$	q_i^2
I	.9	.1	.81	.18	.01
II	.8	.2	.64	.32	.04
III	.7	.3	.49	.42	.09
IV	.5	.5	.25	.50	.25
V	.1	.9	.01	.18	.81
	$\bar{p}$	$\bar{q}$	$\overline{p^2}$	$\overline{2pq}$	$\overline{q^2}$
Entire Population	.6	.4	.44	.32	.24
Frequency if Not Subdivided			$\bar{p}^2$	$2\bar{p}\bar{q}$	$\bar{q}^2$
			.36	.48	.16
Difference $(\overline{p^2} - \bar{p}^2,$ etc.$)$			+.08	−.16	+.08

Source: Li, 1976.

Variance of group gene frequencies

$\sigma_q^2 = \Sigma(q_i - \bar{q})/K = 0.40/5 = 0.08$

Wahlund's formula

$f(AA) = \Sigma\, p_i^2/K = \bar{p}^2 + \sigma_q^2$

$f(Aa) = 2\Sigma\, p_iq_i/K = 2\bar{p}\bar{q} - 2\sigma_q^2$

$f(aa) = \Sigma\, q_i^2/K = \bar{q}^2 + \sigma_q^2$

is subdivided into five (K) separate groups equal in size and that mating is random within each group. Table 11-1 shows the gene and genotype frequencies for each group and also for the entire population. The calculations are simplified by the assumption of groups of equal size, but if they are not, a weighting factor proportional to group size can be used. Here, the mean gene frequency in the entire population is $\bar{q} = \Sigma q_i/K$. The variance of the group gene-frequencies is $\sigma_p^2 = \sigma_q^2 = \Sigma(q_i - \bar{q})^2/K$. As can be seen in the table, the effects of subdivision are to increase the overall amount of homozygosity. The effect is comparable to inbreeding even though within each population, mating is at random. The increase in homozygosity equals the variance in group gene frequencies for each homozygous type. If all five subpopulations had $\bar{p} = 0.6$ and $\bar{q} = 0.4$, then the variance would equal zero, and the C-H-W frequencies would be found. The greater the deviations from $\bar{p}$ and $\bar{q}$ in the subpopulations, the greater the variance and the greater the increase in homozygosity.

DISPERSION AND DISPERSAL

We shall use *dispersal* to refer to the movement of individuals (or of spores, seeds, larvae, and the like) out of (or occasionally into) a population. *Emigration* refers to the movement of individuals away from a population; *immigration* to the movement of individuals into a population. *Migration* is used in two senses: first, to refer to the joint effects of immigration and emigration; second, to refer to the regular movements of departure and return to an area as in the migratory behavior of birds in the North Temperate Zone. *Vagility* refers to the inherent ability of the individual members of a species to disperse or to the dispersal of their disseminules such as seeds or larvae. In many species, dispersal involves young individuals. In plants, adaptations for seed dispersal by wind, water, or animals are common. In vertebrates, the parents may drive the juveniles away after they reach a certain stage of development, whereas with invertebrates, the larval stages often have greater vagility than the adults (Wolfenbarger, 1946; Carlquist, 1966). As stated earlier, the size, growth rate, and other characteristics of a population are determined by the birth rate, the death rate, and the effects of migration. Migration may be a minor factor in the dynamics of a population, but it may also have a major impact if there is a large-scale influx or dispersal of individuals. The effects of emigration are particularly hard to assess in a population because the individuals are gone, and it is difficult to know whether they have emigrated, fallen prey, or died from other causes. The population biology of a species will be significantly influenced by the barriers to dispersal and the vagility of its members. Two species occupying the same habitats will have a very different population structure if the vagility of one species is low and the other, high.

Dispersal is of particular significance in the colonization or recolonization of available habitats. The first colonists to arrive may have an impact on the subsequent course of development of the colony far out of proportion to their numbers. Similarly, it will be seen later that a relatively small number of immigrants to a population may have a very significant genetic impact on the gene pool of that population.

At least three types of dispersal can be postulated. The first takes an exponential form, e^{-x}, as shown in Figure 11-1. In this case, the number of individuals falls off logarithmically with distance dispersed (Curve A). Plotted on semilog paper, it becomes a straight

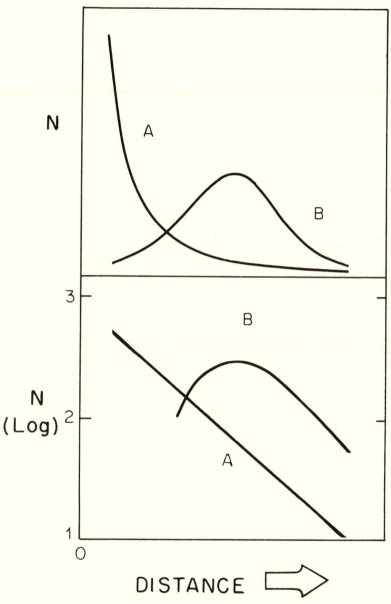

Figure 11-1. Types of dispersal. Curve A shows an exponential pattern of dispersal and Curve B a normal pattern of dispersal. Upper: normal plot. Lower: semilog plot.

line. Another form is a normal distribution pattern of dispersal in which the fraction decreases with distance at the rate e^{-x^2} (Curve B), also shown as a semilog plot. The third type of dispersal pattern would be typical of a migratory species that moves a fixed distance between its summer and winter homes or of species with more or less fixed patterns of diurnal movement.

Dispersion is used in reference to the internal distribution patterns of the members of a population (Figure 11-2). A truly *random*

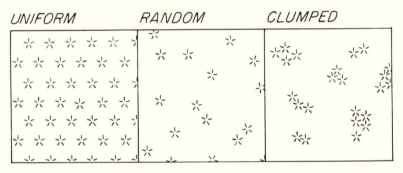

UNIFORM   RANDOM   CLUMPED

Figure 11-2. Patterns of dispersion. See text.

pattern is probably rare, for it would be expected only in a uniform environment in which the individuals show no tendency to aggregate or to interact negatively. A *uniform* pattern of dispersion is observed where competition between individuals leads to even spacing, as in deserts where plants of the same species are often uniformly spaced. In many animal species, members defend a *territory* against other members of the species; and in animals, a *home range*, within which the individual confines its movements, can often be identified. Both of these phenomena will tend to produce a uniform dispersion pattern. The third and most common pattern of dispersion is *clumping*, in which individuals of a species tend to be aggregated in groups. Clumping may result from variation in environmental conditions so that only portions of the environment can be occupied by the species, or it may result from reproductive or other forms of social interaction. To further complicate matters, the clumps themselves may be random, or uniformly dispersed, or clumped. These internal dispersion patterns within a population can pose serious sampling problems, particularly if clumping occurs. It should also be noted that dispersion patterns in animals are not necessarily fixed, but may be subject to change.

THE MIGRATION COEFFICIENT

The amount of genetic divergence between different populations of the same species may be very slight or it may be of considerable magnitude. The question I wish to consider now is What happens if the isolation between two populations of the same species breaks down so that individuals from one population migrate to and become breeding members of another population? This effect is referred to as migration or gene flow. Like mutation and selection, recurrent migration can be treated as a directional evolutionary force, producing systematic changes in gene frequency. Perhaps it should be added that hybridization falls under the rubric of migration.

The expected change in gene frequency due to migration is given by the simple equation,

$$\Delta p = -m(p - p_m) \tag{11-1}$$

where

p = frequency of the A allele in the population receiving the immigrants,

p_m = frequency of the A allele among the immigrants, and

m = the migration coefficient or coefficient of replacement.

It should be noted that m is not simply a measure of the number of individuals that migrate into the population. Instead, it is a measure of the proportion of the gametes contributed by the immigrants to the gametic pool that goes to make up the next generation. In natural populations it is usually difficult to assess migration in terms of individuals, but the relative gametic contribution of immigrants is even more difficult to estimate.

Visualize, if you will, a small, offshore island inhabited by a population of mice. Each generation, a few mice from the mainland manage to reach the island and join its breeding population. Then, if, on the island, $p = 0.3$, but among the immigrants, $p_m = 0.4$, and 10% of the gametes come from the immigrants so that $m = 0.1$, the expected change in gene frequency is

$$\Delta p = (-0.1)(0.3 - 0.4) = +0.01$$

and

p_1 = 0.31 on the island.

With this rate of migration, p would soon equal p_m, and no further change would occur because the mainland and island populations would be in equilibrium with the same gene frequencies. Thus, migration tends to prevent divergence between different populations of the same species or to reduce the differences between them. Furthermore, migration pressure has the potential to produce rapid changes in gene frequency when m is large and p differs greatly from p_m. In addition, many new gene combinations may be formed and exposed to natural selection as the result of gene flow. For these reasons animal and plant breeders commonly introduce widely divergent genotypes into their breeding stocks.

In the discussion of genetic load, mutation and migration were classified together as the input load. Although, biologically, mutation operates at the level of the DNA molecule, and migration involves the movement of individuals from one population to another, in terms of their effects on gene-frequency change, they are very similar, for both are responsible for introducing new genes into populations. One of the niceties of the approach pioneered by Fisher, Haldane, and Wright is that such disparate biological phenomena as mutation and migration (and natural selection and random genetic drift) can be studied in terms of their effects on gene-frequency change.

To comprehend the similar effect of migration and mutation, consider the following. The equation for gene-frequency change due to migration is

$$\Delta p = -m(p - p_m) \tag{11-1}$$

For change due to mutation recall that

$$\Delta p = vq - up \tag{11-2}$$

and therefore

$$\Delta p = v(1 - p) - up$$
$$\Delta p = v - vp - up$$
$$\Delta p = v - p(u + v). \tag{11-3}$$

At equilibrium,

$$\hat{p} = \frac{v}{u + v}$$

and therefore

$$v = \hat{p}(u + v). \tag{11-4}$$

Substituting this value for v in equation 11-3 gives

$$\Delta p = \hat{p}(u + v) - p(u + v)$$

or

$$\Delta p = -(u + v)(p - \hat{p}).$$

When the form of this equation is compared with equation 11-1 for migration, it can be seen that they are similar, with $(u + v)$ equivalent to m and $\hat{p}$ to p_m.

If both mutation and migration are influencing a population, their combined effects can be estimated from

$$\Delta p = -m(p - p_m) + vq - up$$

or

$$\Delta p = -m(p - p_m) - (u + v)(p - \hat{p}).$$

If p_m is considered the equivalent of $\hat{p}$ in a large population, then

$$\Delta p = -(m + u + v)(p - \hat{p}).$$

The equation for migration can also take the form

$$\Delta p = -m(p - \hat{p}) = -mp + m\hat{p} - mp\hat{p} + mp\hat{p}$$
$$= m\hat{p}(1 - p) - mp(1 - \hat{p})$$
$$= m\hat{p}q - m\hat{q}p.$$

Then the equation for the combined effects of mutation and migration can take the simple form

$$\Delta p = m\hat{p}q - m\hat{q}p + vq - up$$
$$\Delta p = (m\hat{p} + v)q - (m\hat{q} + u)p$$

where $m\hat{p}$ and $m\hat{q}$ and u and v are constants.

Even more interesting is the relation between migration and selection. Natural selection will ordinarily cause a population to adapt to local environmental conditions and thus leads to local differentiation of populations, each adapted to its own set of environmental conditions. Migration, on the other hand, by bringing in genes from other populations, tends to oppose local differentiation. Hence migration and natural selection are generally opposed to one another. More exact treatments are available in Wright (1969), Crow and Kimura (1970), and Li (1976); but in general if the selection intensity is much larger than the migration coefficient, that is, $|s| \gg m$, the gene frequencies in a deme will be largely determined by the direction of the selection pressure in the local population, only slightly modified by the effect of the immigrants. Therefore, local differentiation is possible, and gene-frequency differences between different local populations may be quite large. On the other hand, if $m \gg |s|$ so that the extent of migration is much larger than the intensity of selection, the effect of gene flow will overwhelm the effect of selection and local differentiation will not be possible. In such cases, the gene frequencies in the local populations will not differ greatly from the average gene frequency for the entire population. When the percentage of gene flow and the intensity of selection are about the same, that is, $m \cong |s|$, local differentiation is still possible among different groups under differing selection pressures. Thus, local differentiation among populations is to be expected unless $m > s$.

JOINT EFFECTS

The distribution of gene frequencies in populations is influenced not only by mutation, selection, and migration but also by the effective population size (Wright, 1969; Crow and Kimura, 1970; Li, 1976). It has been shown that, in general, if the products $4N_e u$, $4N_e s$, and $4N_e m$ equal less than one, then random fluctuations in gene frequency due to drift will occur, and the various alleles will tend to be fixed or lost from the population. Such populations will be considered *small* and will have relatively high levels of homozygosity. On the other hand, if the products $4N_e u$, $4N_e s$, or $4N_e m$ are greater than 2, then the systematic pressures will control the gene-frequency distribution in the populations, and the populations will be *large*. If the values of the products lie between one and two, the populations will be *intermediate* in size, and there will still be considerable variation in group gene frequencies.

The meaning of *large* and *small* is quite different for the different systematic pressures, as can be seen when values for u, s, and m are substituted in the expressions above (Table 11-2). For example, a

Table 11 - 2. Population Size for Various Assumed Values
of $4N_e u$, $4N_e s$, and $4N_e m$

	"Small" $4N_e x = 0.5$	"Intermediate" $4N_e x = 2$	"Large" $4N_e x = 8$
$u = 10^{-6}$			
N_e	125,000	500,000	2,000,000
$s = 0.001$			
N_e	125.00	500	2,000
$s = 0.01$			
N_e	12.50	50	200
$s = 0.1$			
N_e	1.25	5	20
$m = 0.01$			
N_e	12.50	50	200

reasonable mutation rate based on the discussion in Chapter 4 would be 10^{-6}. In that case, a *small* population that would be largely homozygous owing to random fixation of alleles would consist of more than 100,000 individuals.

The situation is quite different for the selection coefficient where three different values of s were assumed, $s = 0.001$, $s = 0.01$, and $s = 0.1$. Here, the largest of the *small* populations in which considerable random fixation and loss would occur was only 125 (for $s = 0.001$). We have already discussed the practical difficulties in detecting a 1% difference in fitness; a loss of fitness of 0.001 would be even more difficult to detect. However, $s = 0.001$ and $s = 0.01$ are the most commonly assumed values for the selection coefficient, although, as Ford (1971) has stressed, estimates of $s = 0.1$ or more seem common in wild populations. The treatment of selection is more complicated because there may be different forms of selection and also because the selection pressure itself may be variable. This variation in selection pressure is apt to be more prevalent among alleles with very slight effects on fitness rather than those subjected to *hard* selection with large selection coefficients. Only a slight change in conditions might shift the fitness of a type with $s = 0.001$ to neutrality or beyond neutrality to a selective advantage rather than a selective disadvantage. However, Table 11-2 suggests that if selection coefficients are in the range $s = 0.01$ to $s = 0.1$, selection will continue to be effective in all populations except the very smallest.

The migration coefficient may range from $m = 0$ for complete isolation to $m = 1$ if the existing population were completely displaced by immigrants. The value chosen for the table, $m = 0.01$, is not large, but might be detectable. In this case an effectively small population, in which significant drift would occur, would equal about a dozen or so. A more informative way to look at the effect of migration is to set $4N_e m > 2$, say $4N_e m = 4$, so that $m = 1/N_e$. In this case, the population will be *large* — that is, local differentiation due to drift will be prevented — even though only one immigrant per generation joins the population, no matter what the size of the population might be. Thus, what might seem like a low rate of migration is sufficient to prevent local populations from diverging from one another owing to drift.

Still another way to gain insight into the meaning of these relationships is to assume values for N_e as well as for u, s, and m, and to calculate $4N_e u$, $4N_e s$, and $4N_e m$. If $N_e = 100$, $u = 10^{-6}$, $s = 0.1$, and $m = 0.01$, then $4N_e u = 4 \cdot 10^{-4}$ (<1), $4N_e s = 40$ (>2), and $4N_e m = 4(>2)$. Under these circumstances the population is *small* in relation to mutation, but *large* with respect to selection and migration. Different populations might be expected to contain different alleles stemming from different mutations, but their frequencies in the populations, if there is differential selection, will be primarily determined by selection. Migration will tend to reduce the difference between populations to some extent, but, because $s > m$, local differentiation due to different local selection pressures will persist.

In more general terms, in large randomly mating populations, the gene frequencies will tend to remain at a stable equilibrium point determined by the combined systematic pressures of mutation, selection, and migration. If the environmental conditions remain constant, no evolutionary change will occur. If conditions change, the gene frequencies will move toward new equilibrium points, guided primarily by selection.

In effectively small, isolated populations, gene frequencies change at random so that eventually most alleles are fixed or lost by chance. The resulting gene combinations may not be particularly well adapted since they are formed by chance rather than selection. Note, however, that these statements presuppose very small values for the selection coefficients. As shown in Table 11-2, if $s = 0.1$, then $N_e = 20$ is a *large* population in which selection will be a predominant factor. Nonetheless, the small numbers, relatively poor adaptation, and relatively high homozygosity of small populations as compared to large populations of the same species all combine to make extinction the probable fate of most small populations.

Wright argued that the most favorable circumstances for evolution were those in which all the evolutionary factors—mutation, selection, migration, and random drift—could operate simultaneously. These conditions would prevail in a large population subdivided into a number of partially isolated groups. These local populations would be large enough for selection to keep them adapted to local conditions, but small enough for occasional chance events to bring about random fluctuations in gene frequencies. Migration will tend to prevent complete fixation of alleles within a given deme and will also permit the spread of favorable genes or gene combinations throughout the species population. In contrast to the situation in a large breeding population where selection is primarily intragroup, when the population is subdivided in this way, a form of intergroup selection prevails. Each population becomes a separate adaptive experiment. The less successful populations decrease in numbers or even become extinct; the more successful increase in numbers and through dispersal and gene flow, disseminate their more favorable genes throughout the species. Under these circumstances, evolution may continue even though environmental conditions remain constant, with the species becoming progressively better adapted to the existing conditions. However, if the conditions do change, the species is better prepared to cope with the changes because of its subdivision into a number of diversified subgroups.

A few comments may be added at this point. The difficulties in estimating N_e, u, s, and m have already been stressed, yet the basis for the discussion has been the product of two of these elusive values. Because such estimates are so difficult to make, usually *reasonable* assumed values are used, as we have done here, to draw conclusions about the joint effects of the various evolutionary factors. However, the conclusions then become dependent on the nature of the assumptions, and two erroneous assumptions, multiplied together, only compound the error.

Furthermore, we have continued to be concerned with events at a single locus, yet this is obviously too simplistic. Different loci, for example, will have different rates of mutation, and the product $N_e u$ will be different for different loci. Although selection can be treated as if it were genic, it is phenotypic, and an entire genotype, as expressed in an individual's phenotype, is the unit acted on by natural selection. Relative fitness estimates are possible for alleles at a single locus or for the genotypes formed by these alleles although the task is never simple. Fitness estimates for an entire genotype are even more difficult, for fitness must be assessed in relation to the environmental conditions and the other genotypes in the population. Moreover, in a randomly mating, diploid population every genotype would be different, with its own adaptive value. The individual is also the unit of migration. Single genes do not migrate; individuals do. Thus, the genetic input from migration differs from that due to mutation. Mutation does involve a change at a single locus. Migration involves the introduction of entire haploid genomes into the population.

Thus, by concentrating primarily on events at single loci, insight has been gained into the process of evolution, but one should not be misled into thinking that such a limited approach can lead to complete understanding.

Still another aspect of the evolutionary process merits comment. We have dealt with recurrent events: recurrent mutation and recurrent migration from which mutation rates and migration rates can be estimated, and with constant selection pressure, which means that the same selection pressure is exerted repeatedly on a given genotype.

We have dealt with random effects only with respect to random-sampling drift. However, Sewall Wright (1955, 1959, 1969, 1970) has stressed that random drift could also result from random fluctuations in the mutation rate, in selection pressure, in the immigration rate, or in the gene frequencies among the immigrants. Such possibilities would make the course of evolution much less deterministic than would be the case if only recurrent events were involved.

Finally, some events significant to evolution may be unique such as a new favorable mutation, a novel genetic recombination, or a unique hybridization. Other singular events could include a unique selective incident, immigration, or reduction in population size. These unique events could influence the entire subsequent course of evolution in a species, yet by their very nature they are unpredictable: they do not fit into a formula or equation, and they are not useful material for building models. Wright (1949a, 1949b, 1955), more than others, has stressed the possible significance of unique events to evolution. Even though evolutionary theory is not yet equipped to deal with unique events, the possibility of their occurrence cannot safely be ignored.

GENE FLOW

In the discussion of gene flow, we used the term migration in the sense it is used by population geneticists—as a synonym for gene flow. In fact, migration, as used by population geneticists, refers more to what the ecologists call dispersal than it does to true migration, as seen in birds and some fishes and mammals, which involves periodic movements back and forth between different regions. Custom and long usage dictate the use of migration in relation to gene flow, but it is well to realize that most gene flow probably results from dispersal movements by individuals, especially young individuals, or seeds, larvae, and the like, rather than from migration in the restricted sense. The scattering of individuals from their point of origin is the basic process that leads to gene flow between different populations. Dispersal is also a basic phenomenon in biogeography, for the distribution of species depends, in large measure, on their ability to disperse and reach new favorable habitats.

However, much more effort has been spent on the study of migration in birds and other vertebrates than on dispersal. Probably this is because such migrations are regular and predictable and involve mass movements of entire populations. The migration of Canada geese between their breeding grounds and their winter range or the migration of salmon upstream to spawn are striking and recurring phenomena. In contrast, the solitary and bedraggled red squirrel that crawled, exhausted, onto a sandbar after swimming a wide and turbulent river provided a chance observation that has not been repeated. What prompted that venture remains a mystery; nevertheless such movements may be of great significance for gene flow and also for the distribution of the species.

Unlike regular migrations, such events may be so rare that they qualify as unique rather than recurrent events. Still they may play a significant role in the biology of a species, and thus merit attention. Whereas it might be possible, albeit with considerable difficulty, to monitor the number of red squirrels swimming a river, it would be even more difficult to measure their genetic impact. Electrophoretic techniques have permitted analysis of protein polymorphisms in adjacent populations, which in turn have led to inferences about the amount of gene flow, or lack of it, between such populations. However, if adjacent populations have similar gene frequencies, these may merely indicate similar selection pressures, and are not proof of gene flow. On the other hand, if adjacent populations have significantly different gene frequencies, this does not constitute proof of an absence of gene flow, but may merely indicate that s is greater than m in these populations.

Several recent papers (e.g., Ehrlich, 1965; Ehrlich and Raven, 1969; Endler, 1973; Levin and Kerster, 1974; Ehrlich et al., 1975; McKechnie, Ehrlich, and White, 1975; Ehrlich and White, 1980) have tended to discount the importance of gene flow as a significant factor in evolution. Selander (1970) even reported significant gene-frequency differences between subpopulations of the house mouse, *Mus musculus*, inhabiting the same barn. Observations of this sort, or the failure to detect much movement of individuals between populations (Ehrlich, 1965), or very short-range dispersal (Levin and Kerster, 1974) led these authors to downgrade the importance of gene flow. However, the studies are for relatively brief periods, and the picture might be quite different if a broader time perspective were used. For example, despite the sedentary, isolated nature of the three populations of the checkerspot butterfly, *Euphydryas editha*, on Jasper Ridge (Ehrlich, 1965), Ehrlich et al. (1975) reported that one of the three populations became extinct twice and then was reestablished each time. Similarly, the barns inhabited by the mice could not have been built too long ago in terms of geological time, and the mice now living there in seeming isolation, even from one another, must have migrated there. Whenever a new barn is built, it is not long before mice take up residence. The famous recipe of van Helmont for the spontaneous generation of mice—a sweaty shirt plus some wheat germ in a dark corner—is perhaps sufficient testimony to the vagility of even such a sedentary species as the house mouse. Another case in point is the English sparrow, which, unlike most north-temperate birds, is nonmigratory and stays close to its home range. Nonetheless, in less than a century following its introduction, it spread across much of North America. Thus, the problem may lie in the difficulty in detecting rare, singular, or unique events even though they may be of considerable importance. In the absence of selection, as already seen, a single migrant each generation is sufficient to prevent populations from diverging.

However, if selection is strong, populations in close proximity to one another between which there is considerable gene flow may nonetheless diverge from one another and maintain their distinctive attributes. The best studies of this have been done with plants. Pioneering work leading to the ecotype concept was done by Turesson (1922a, b, 1923, 1925). He found that even though a given genotype had a certain amount of phenotypic plasticity when reared under diverse environmental conditions, genotypes drawn from different populations of a single species living in diverse habitats retained much of their distinctiveness when grown together from seed in a common garden. He (1922a) coined the term *ecotype* to describe a subgroup within a species that is genetically adapted to a particular habitat type utilized by the species. He (1923) also introduced the word *genecology* with the meaning usually ascribed to ecological genetics today. This work was subsequently extended by Turrill (1936, 1940, 1946) and by Clausen and his cohorts (Clausen, Keck, and Hiesey, 1940; Clausen, 1951; Clausen and Hiesey, 1958).

More recently, a group in Britain has shown that strong selection can maintain genetically distinct populations within very short distances of one another despite continuous gene flow between them (Jain and Bradshaw, 1966; McNeilly, 1968; McNeilly and Antonovics, 1968; Antonovics, 1968; Antonovics and Bradshaw, 1970). They studied populations living on tailings from copper or lead-zinc mines and in immediately adjacent pastures uncontaminated by the heavy metals. The populations living on the tailings were tolerant of the heavy metals but those from uncontaminated soils a few meters away soon died if grown on contaminated soil. Copper-tolerant ecotypes were not found on the uncontaminated soil near the mines (McNeilly, 1968), which was interpreted to mean that there was selection against the copper-tolerant type living on normal soil although it was weaker than the selection favoring it on the tailings. The extremes of tolerance and intolerance were found within a continuous range of distribution forming a cline over distances as small as 50 to 100 meters (Jain and Bradshaw, 1966). Moreover, there was evidence of considerable gene flow over these short distances (McNeilly, 1968; Antonovics and Bradshaw, 1970). It is worth noting that these examples provide still another illustration of the rapidity with which evolutionary change can occur under strong selection (and in spite of gene flow), for intensive working of the mines began only a little more than a century ago. Similar sorts of genetic differences between coastal and inland populations of the grass *Agrostis stolonifera* were found within an area of only two hectares (about 5 acres) by Aston and Bradshaw (1966).

Ehrlich and Raven (1969) and Levin and Kerster (1974) argued against the evolutionary importance of gene flow primarily because their work indicated rather low levels of dispersal in the populations they studied. These studies in British plants indicate that even if considerable gene flow is occurring, it may have relatively little impact, and they suggest that the importance of gene flow cannot be assessed independently of other factors such as selection. However, other authors (e.g., Wolfenbarger, 1946, 1959; Bateman, 1950; Dingle, 1972) have reviewed dispersal in a wide variety of species, leaving a rather different impression of the dispersal capabilities of these species compared to those studied by Ehrlich and Raven or Levin and Kerster. Bateman made the additional observation that gene dispersion was leptokurtic rather than normal in that there were higher proportions of short- and long-range dispersal than expected with a normal distribution. In contrast to the work with the checker-spot butterfly, Camin and Ehrlich (1958) interpreted their observations on the distribution of unbanded water snakes, *Natrix sipedon*, on islands in Lake Erie in terms of strong selection favoring the unbanded snakes on the islands and considerable immigration of banded snakes from the mainland. The widespread distribution at a uniform, low frequency of the unspotted Burnsi dominant gene in leopard frog populations of the Upper Midwest has been attributed to migration and the vagility of this species (Merrell, 1970; 1977). Thus, perhaps the safest conclusion to draw is that the importance of dispersal and gene flow differs from species to species or from one population to another of the same species or even from time to time in the same population.

MacArthur and Wilson in *The Theory of Island Biogeography* (1967) attempted to develop a theory to account for the numbers and distribution of species on islands. They worked with the following variables: the area of the islands; their distance from a source of immigrant species; the numbers of species on the islands; and, finally, the rate of colonization of new species to the islands and the rate of extinction of existing species on the islands. Thus, they treated colonization and extinction as recurrent phenomena and postulated that an equilibrium would be achieved dependent on the rates of colonization and extinction. They further postulated that the rate of colonization would be higher, the closer the island is to a source of immigrant species, and that the rate of extinction would be greater, the smaller the island.

Such a model is very crude. The idea of an immigration rate or a colonization rate may be applicable to islands near a source of immigrants, but some cases, such as Darwin's finches in the Galapagos Islands or the Hawaiian honeycreepers, may have involved unique events. It would seem desirable to know something about not only the successes but also the failures, the immigrants that arrive but fail to become established; however, such data would be very difficult to obtain. Similarly, it may be difficult to prove extinction. Finding members of a species on an island is good evidence of their presence, but failure to find them is not proof of extinction and may merely indicate an inadequate search. A further factor not taken into account in the model is that colonizing populations, upon their arrival, will start to adapt to their new environment. Thus, early arrivals to a newly formed island will have a better chance to become established and a longer time to adapt to conditions there than later arrivals. The idea of a dynamic equilibrium between colonization and extinction may not reflect the actual situation, for the island may act more like a sponge, soaking up additional species with the passage of time. Extinctions and success in colonization may be more closely related to habitat diversity or the lack thereof than to area alone. It should be added that endemic island populations have frequently been overwhelmed by species introduced by humans, either accidentally or intentionally. Such introductions, most of which would never have occurred without human intervention, and which have greatly increased the rate of colonization, have sometimes greatly disrupted the ecological balance on the islands.

Colonization is dependent on the phenomenon of dispersal. Wolfenbarger (1946, 1959) has shown from data for many species that dispersal generally decreases more or less exponentially with distance. The actual distances involved may differ greatly for different species. For example, the range of dispersal of the plum curculio beetle is from about 10 to 100 meters, but the range of dispersal of the beet leafhopper is from about 1000 to 1 million meters; however, in both cases, relatively few individuals disperse the greater distance. Dispersal may not be simply dependent on the vagility of a species; it may also depend on the population density at the source.

Successful colonization will depend not only on the probability of reaching an area but also on the probability of becoming established there. In some species a single gravid female may be sufficient to establish a colony, although in others, some minimum number of immigrants may be necessary, in which case arrivals below this number might be irrelevant to colonization. The probability of becoming established will further depend on the characteristics of the colonizing species, the characteristics of the habitat they are invading, and the characteristics of the species already there. An area already occupied by a similar species may be much less open to successful colonization than if that species were not present.

Therefore, dispersal is important not only to the study of gene flow, but also to the study of colonization or the invasion of previously unoccupied habitat. The great difficulty in dealing with dispersal is that, to treat it quantitatively, one considers it a recurrent phenomenon. However, some dispersals may be unique and thus do not lend themselves to quantitative treatment or permit probabilistic predictions to be made. Yet these unique dispersal events, like the other kinds of unique events cited earlier, events of virtually complete indeterminacy, may have great evolutionary significance.

As a footnote, it might be added that emigration of individuals from a population under study may fit under the rubric of selection in that population rather than migration. If there is any sort of differential emigration by different genotypes, in a formal sense that is a type of selection.

Considerable work on dispersal has been done with *Drosophila*, starting in the 1940s with the work of the Timoféeff-Ressovskys (1940) and Dobzhansky and Wright (1943, 1947). Various methods for identifying the migrating flies have been used. In some cases, mutant flies reared in the laboratory have been released and their dispersal among natural populations was detected by their unusual appearance (Timoféeff-Ressovsky and Timoféef-Ressovsky, 1940; Dobzhansky and Wright, 1943; Burla et al., 1950; Wallace, 1970b). The hazard in using laboratory-reared mutant flies is that they may not behave like the wild flies either because of their mutant genotype or because of the laboratory environment in which they were reared or both. Powell and Dobzhansky (1976) reported much greater vagility of wild flies of *D. pseudoobscura* than of the orange-eyed flies tested in the same place many years previously by Dobzhansky and Wright.

If a long time has elapsed after release, the recessive mutant may be detected by test-crossing flies from the wild populations to homozygous recessive flies (Dobzhansky and Wright, 1947). Dubinin and Tiniakov (1946a) released 100,000 flies homozygous for an inversion rare in the local population and subsequently checked the larval salivary gland chromosomes of descendants of these flies to find how far the inversion had dispersed.

Another more indirect approach is to make tests for the allelism of lethals within and between populations or from sites separated by different distances; and from the results of these tests, one can infer the amount of gene flow between them (Wright, Dobzhansky, and Hovanitz, 1942; Wallace, 1966b). A drawback is that this method is indirect, that interpopulation allelism of lethals may have other explanations than migration or gene flow between the localities involved. More recently (e.g., Selander, 1970), differences in gene frequency for biochemical polymorphisms have been used to measure gene flow or lack of it between adjacent populations.

Other approaches involve capture-mark-release-recapture of wild flies in the locality under study. Toda (1974), for example, marked them lightly with a few dots of colored lacquer. Others have used radioactive phosphorus (Yerington and Warner, 1961) or micronized dust that fluoresces under ultraviolet light (Stern and Mueller, 1968) or rare earths (Richardson et al., 1969; Richardson, 1969, 1970; Dobzhansky and Powell, 1974).

These studies have produced some wildly different results. For example, the Timoféeff-Ressovskys (1940), Dobzhansky and Wright (1943), and Wallace (1966b, 1968b) all reported that *D. melanogaster* is a rather sedentary species, but Yerington and Warner (1961) found that flies tagged with P^{32} had moved upwind nearly 5 miles in 24 hours. Similarly, the flies in some cases appear to disperse more or less at random (Dobzhansky and Wright, 1943; Wallace, 1966a; Wright, 1968a), yet there is also evidence that such factors as food sources, temperature, crowding, and wind velocity all influence rates of dispersal so that a Brownian motion-diffusion model of dispersion must be far too simple. The flies tend to disperse from densely crowded areas; however, they are attracted to food, and they do not fly if the wind velocity is too high or the temperature too low. Thus, their movements seem governed by environmental stimuli. Further discussion of dispersion in *Drosophila* is found in Crumpacker and Williams (1973) and Dobzhansky (1974). Johnson (1969) has reviewed a massive amount of information on insect migration and dispersal in general, although not from the viewpoint of population genetics, a viewpoint best summed up, perhaps, by the notable quote from Spieth (1974), "In terms of gene flow, the distinction between absolutely none and almost none is enormous."

HYBRIDIZATION

Earlier it was mentioned that hybridization fell under the rubric of migration and gene flow. Now the role of hybridization in evolution must be considered. Unfortunately, the word hybrid has been used in so many contexts that it is necessary to restrict its meaning here to discuss the relevance of hybridization to evolution. In a monohybrid or dihybrid cross, F_1 hybrid individuals heterozygous at one or two loci are produced. Within a single cross-fertilizing variable deme, individuals heterozygous at many loci are present; they too may be called hybrids. *Primary intergradation* is used to refer to the hybridization between members of adjacent populations that have never been isolated, whereas *secondary intergradation* refers to hybridization between members of populations that have been isolated for a time but have once again come in contact. Secondary intergradation is used to refer to hybridization between different ecotypes or different subspecies or even, in some cases, different species. In the context of gene flow, we shall use the last meaning of hybridization, the crossing between members of different populations that have undergone some degree of genetic divergence. The effect of hybridization is to reverse this process of evolutionary divergence. The questions we need to address are How often does natural hybridization of this sort occur? What is its outcome? and How important is it to evolution? On this, opinions have differed greatly. Some have considered hybridization to have been of negligible significance to evolution. Others, such as Lotsy (1916), who wrote *Evolution by Means of Hybridization*, considered hybridization to play a major role in evolution. More recently, Edgar Anderson, who developed the concept of introgressive hybridization, about which more will be said later, has been accused of seeing a hybrid under every bush, and even of thinking the bush is a hybrid, too. In general, zoologists have dismissed hybridization as unimportant to evolution, whereas botanists have considered it important because there can be little doubt that hybridization has been of much greater significance to the evolution of plants than animals. The phylogenies of most groups of higher plants are often reticulate owing to hybridization and allopolyploidy in contrast to the dichotomous, branching phylogenies of most groups of animals.

Hybridization between different species is more common in plants than animals for several reasons. One is that sexual isolation stemming from behavioral differences during courtship, which may prevent mating between males and females of different species of animals, is not a factor in plants. Even though the mating may be fertile and the progeny vigorous and fertile, behavioral or ethological isolation may be sufficient to prevent such matings from occurring in nature between members of different animal species. Only in the case of higher plants that are pollinated by insects may the behavior of the insects secondarily become involved in the reproductive isolation between different species of flowering plants.

Another factor promoting the frequency of hybridization in plants is the greater life span of individual plants compared to animals. Perennials, and especially trees, may live a long time compared to animals, and the adaptive advantages of the hybrids may be sufficiently great to outweigh the burden of low fertility, especially if asexual means of propagation are also available. In any case, semi-sterility should not be as great a handicap to long-lived individuals as to short-lived ones, and thus, hybrids should be found more often among perennial plants than among annuals. In addition, development in plants is somewhat simpler than the complex pattern of differentiation in animals so that a wide cross in plants is more apt to result in viable offspring than a wide cross in animals.

Finally, natural hybrids may appear to be more common in plants than in animals because plant hybrids are easier to detect in the field. In plants, large populations can be examined and the unusual individuals singled out to determine if they are hybrids. It is ordinarily far more difficult to examine equally large numbers of animals to detect hybrids among them. Therefore, failure to find species hybrids in animals, in some cases, may simply be due to lack of a comprehensive and effective search.

The literature on natural hybridization is extensive. Plant hybridization was reviewed by Roberts (1929), Heiser (1949, 1973), Anderson (1949, 1953), Stebbins (1950, 1959), and Grant (1971); animal hybridization was reviewed by Dobzhansky (1951), Cockrum (1952), Gray (1954, 1958), Hubbs (1955), Mayr (1963), Remington (1968), White (1973), and Moore (1977).

INTROGRESSIVE HYBRIDIZATION

When hybridization between different species or subspecies occurs in nature, F_1 individuals will be as uniform in phenotype as either parental population, and in general, the F_1 phenotype will be intermediate between the phenotypes of the two parental populations. If the F_1 hybrids interbreed, the F_2 and subsequent generations will be quite variable because of Mendelian segregation and recombination. If the differences between the parental populations are due to differences at relatively few gene loci, individuals resembling the parents may even be recovered in the F_2. However, if, as is usually the case, numerous gene loci are involved, a variety of recombinants will be recovered, but they will not cover the entire spectrum of variation from one population to the other. This restriction in the range of phenotypes is probabilistic, for the chances of recovering the extremes of the parental populations are vanishingly small if many loci are involved, and there are other restrictions to the types of phenotypes recovered as well. Genetic linkage, of course, will limit complete, random assortment of the gene differences involved; however, the recombinant individuals observed will not represent a random sample of combinations of the phenotypic traits observed in the parental groups. Developmental correlations and restrictions also exist that will limit the kinds of phenotypes observed among the hybrid progeny.

Under normal conditions, naturally occurring hybrids are not apt to be F_2 or F_3 individuals. If an F_1 hybrid is formed, it is apt to be rare compared to the numbers of individuals in the parental populations that generated it. Thus, most of its potential mates will be members of the parental populations so that backcrossing to these populations will occur rather than the formation of F_2 and F_3 generations. If the environment is undisturbed and the parental groups are well-adapted to their niches within the existing environmental conditions, the backcross individuals most closely resembling one or the other of the parents will be better adapted than the others, and natural selection will tend to eliminate the more aberrant recombinant backcross types. Thus, even though hybridization may be occurring in natural populations, it may not be easy to detect. For this reason, Anderson (1949) developed methods to detect the subtle morphological differences that would indicate that some gene flow was occurring. More recently (Ayala, 1976, Manwell and Baker, 1976), studies of biochemical variability have been used for a similar purpose.

Where the habitat has been disturbed, the environmental conditions favoring the parents may be disrupted, and some of the hybrids may actually be better-adapted to the changed conditions than either of the parental types. Anderson (1949) cited a case of this sort in *Iris fulva* and *I. hexagona* var. *giganti-caerulea*, which grow sympatrically on the lower Mississippi Delta. Their ecological requirements are rather different. *I. fulva* is a wide-ranging species found in the Mississippi Valley as far north as Illinois and grows in wet clay soil along rivers and ditches in partial shade. *I. hexagona* var. *g.-c.*, on the other hand, is found in the lower delta, never far from the sea and grows in the mucky alkaline soil of tidal marshes in full sun. Because of these differences, they normally remain fairly well isolated ecologically from one another. They are not as closely related to each other as to other *Iris* species, but they can cross to form hybrids whose fertility is somewhat lower than the parents'. The species are strikingly different in appearance—*I. fulva* has numerous, small, brick-red flowers whereas *I. hexagona* var. *g.-c.* has few, large, colorful, variegated blue, yellow, and white flowers.

Occasionally, hybrids may have formed in the past where a natural levee ran into a tidal marsh; however, settlement by the French greatly changed the lower delta. Small French farms were established along the rivers and bayous, with a narrow river frontage and property lines at right angles to the river. Each farm was long and narrow, and each farmer treated his land somewhat differently, clearing some land for crops, some for pasture, and leaving some as woodlands. This disruption of the habitat brought the two *Iris* species into contact and led to extensive hybridization between them. In one place where the land had been cleared and then heavily overgrazed, a hybrid swarm grew; however, it did not extend beyond the fence lines into the adjacent farms where the land had been less disturbed. The plants in this hybrid swarm were a colorful mixture ranging from some much like *hexagona* to a few that resembled *fulva*. Because there were more hybrids in this disturbed habitat than anywhere else in the vicinity, Anderson (1948, 1949) argued that "hybridization of the habitat" was responsible for the success of these hybrid plants. The idea that hybrids will persist in nature in environments where they are more fit than the parental populations has also been supported by Stebbins (1959), Grant (1971), and Moore (1977).

A second hybrid population growing on the same farm was rather similar to *hexagona*, but was somewhat more variable than normal *hexagona* populations, and the variability was in the direction of *fulva*—that is, the flowers tended to be slightly smaller, redder, and more numerous than usual. The habitat where this second population grew was far less disturbed by humans and their animals than in the other hybrid colony. This type of effect, where genes from one population have entered the gene pool of another, different group, has been termed *introgressive hybridization* by Anderson (1949).

Hybridization between *fulva* and *hexagona* followed by backcrossing to *hexagona* has permitted the transfer of genes from *fulva* into *hexagona* or the introgression of genetic material from one species to another.

Introgression is often difficult to detect. However, it can occur only where members of the populations concerned are living sympatrically so that hybridization can occur. Since a disturbed habitat is apt to provide an environment favorable to hybridization and to the resulting hybrid and backcross offspring, such habitats in areas of sympatry are prime targets to search for introgression. If introgression is occurring, the recipient species should be more variable in the region of overlap than where it is geographically isolated from the other species, and the increased variation should be in the direction of the introgressing species, as in the case of introgression from *fulva* into *hexagona* cited above. Further details on methods of developing a *hybrid index* to detect cryptic introgression can be found in Anderson (1949). If protein differences between the parental groups can be detected by electrophoresis, then hybridization can be detected in zones of overlap by this method, as has been done, for example, in the *Rana pipiens* species complex (Salthe, 1969; Platz, 1972; Kruse and Dunlap, 1976).

Introgression may be a significant factor in evolution. Mutation and genetic recombination within a population generate variation upon which natural selection can act to bring about adaptation. With introgression, new genes enter the germ plasm of the population, not by mutation, but across a reproductive barrier of some sort, and blocks of linked genes rather than single genes are involved. This infusion of germ plasm may lead to convergence but in disturbed environments, recombination and selection may cause the population to move rapidly to new adaptive peaks.

POLYPLOIDY

The high frequency of allopolyploidy in plants (see Chapter 9) is sufficient evidence that hybridization has played a significant role in their evolution. Roughly half of the angiosperms, for instance, are allopolyploids. Typically, an allopolyploid is formed when hybridization between two species produces a relatively infertile F_1 diploid hybrid. Chromosome doubling then may give rise to a fertile F_2 tetraploid, which is reproductively isolated from the two parental species, and thus behaves as a new species. This sequence of events is the only way known by which new species can originate in a single step. Müntzing (1930, 1932) was the first to resynthesize a naturally occurring allotetraploid plant, *Galeopsis tetrahit*, from its putative diploid ancestors, *G. pubescens* and *G. speciosa*. The artifical allotetraploid resembled the wild *G. tetrahit* in morphology and chromosome number, and crossed readily with it to produce fertile F_1 offspring. But it was, like the wild *G. tetrahit*, isolated by a sterility barrier from *G. pubescens* and *G. speciosa*.

It is not possible to estimate how often hybrids have been formed in nature from crosses between members of different taxonomic units. Of those that are formed, most probably fail to be perpetuated either because of sterility or because they are less fit than the parental populations amidst which they live. Chromosome doubling in nature seems likely to be a rare, fortuitous event. Despite this combination of improbabilities, polyploidy is common among the higher plants, which suggests that it must confer certain adaptive advantages that explain its evolutionary success.

Natural polyploids are usually allopolyploids—that is, they are of hybrid origin. Hybrids between different ecotypes or geographic races or species often manifest hybrid vigor or heterosis and may also show increased physiological or developmental homeostasis. These phenotypic traits of the hybrids must confer added fitness to these individuals; however, as hybrids, they will not breed true for these traits if they reproduce sexually, and may even be sterile if the genomes making up the hybrid are too different. Allopolyploidy circumvents these problems. The allopolyploid, in contrast to the diploid hybrid, is fertile and can reproduce sexually, yet it breeds true for the hybrid genotype and hence for the hybrid vigor. Unless the parental genomes are very different, a certain limited amount of segregation and recombination is possible so that some further adaptive changes may occur. However, in general, because of their polysomic nature, polyploids are much slower to release their stored genetic variability than diploids. The segregation ratio for homozygous recessives, for example, is much lower for polyploids than for diploids. Although polyploidy is common in higher plants, it is not uniformly distributed among them (Stebbins, 1971a). Polyploidy is predominantly found among perennial herbaceous plants. Such plants have a long life span and also usually have some form of vegetative propagation. These factors presumably enhance the chances that somatic chromosome doubling will occur in the hybrid plant to produce a fertile allopolyploid. Where comparisons have been made on the incidence of polyploidy among perennial herbs, annual or biennial herbs, and woody plants, polyploidy is more common among herbs than trees and more common among perennials than annuals. Woody plants are long lived; however, they seldom reproduce vegetatively. Polyploidy is especially rare among conifers, but *Sequoia sempervirens*, one of the few natural coniferous polyploids, is also unusual in being able to regenerate from suckers.

Polyploidy is most common in groups in which speciation at the diploid level involves chromosome rearrangements. If chromosome repatterning does not occur, polyploidy tends to be rare or absent in a group even though natural hybridization is frequent between its members and it has the perennial growth habit (Grant, 1971). Apparently all three requirements must be met, for surveys show that polyploidy is common only in those groups which hybridize naturally, are perennials, and undergo chromosome repatterning during speciation as diploids. This divergence in chromosome structure helps to ensure the fertility of the allopolyploid.

Efforts have also been made to associate the incidence of polyploidy with ecological factors as well as with the breeding system. For example, it has often been shown that a higher frequency of polyploidy is associated with both higher latitudes and higher altitudes; however the interpretation of this is still in doubt. Efforts to relate polyploidy to temperature or moisture or disturbed habitats have generally foundered because there are exceptions. Nevertheless, the idea persists that polyploids may be better able to withstand severe climatic conditions or to colonize newly available habitats than their diploid ancestors.

At present, it appears that a number of conditions, both internal and external, must be met if polyploidy is to become established in a given group. These conditions have frequently been met in the higher plants but only rarely in animals (White, 1973). Therefore, even though polyploidy has been of great significance in plant evolution, it cannot be universally significant as an evolutionary mechanism, for evolution has occurred in many groups in the absence of polyploidy. On the other hand, if allopolyploidy is considered as one manifestation of gene flow, as it clearly seems to be, then gene flow must be a significant factor in evolution, for it has played a major role in the evolution of the higher plants.

CHAPTER 12

The Origin of Races

GEOGRAPHICAL VARIATION

We have discussed variation in natural populations, primarily in terms of polymorphic variation within a single breeding population or deme. In the previous chapter, we dealt with population structure and gene flow between populations, but our primary concern was still the effect of subdivision and gene flow on gene frequencies within a single deme. Now we shall consider geographical or polytypic variation in an effort to understand the nature and origin of the genetic differences between different populations of the same species.

When two populations of the same species are compared, the observed differences may be due to genetic differences, or they may simply reflect the effects of different environmental conditions. Usually differences are due to a combination of genetic and environmental effects. To determine the nature of such differences requires more than simple inspection. One approach is through transplantation experiments that may involve reciprocal transplantation, growing members of each population in the environment of the other, or transplanting members of both populations to a common environment where they can be compared while growing side by side.

Another approach is to analyze the differences through genetic crosses. If genetic differences exist, the study of segregation and recombination in the F_2 and backcross generations from hybrids between the two populations will give some indication of the nature and extent of their genetic differences.

Still another approach is through electrophoretic analysis of protein variation. It is usually thought, because the path from gene to molecule is so short, that this technique is a means of assessing genetic differences between populations. However, even at the molecular level, gene expression can be influenced by environmental conditions as in the case of temperature-sensitive mutants, enzyme induction or repression, and other forms of regulation. Hence, some thought should be given to making such analyses on organisms that have been subjected to similar environmental regimes before analysis.

Some of the classical studies of geographic variation have been done with plants (Turesson, 1922b, 1925, 1930; Turrill, 1938, 1946; Clausen, Keck, and Hiesey, 1940; Clausen, 1951; Clausen and Hiesey, 1958). The great advantage of plants for such studies is that, lacking the mobility of animals, they grow only where conditions are favorable for their existence, and it is thus possible to identify their ecological requirements with considerable precision. Animals, on the other hand, have the ability to move about within an area as the environmental conditions or their needs change, and they may seek out favorable conditions or seek to avoid unfavorable conditions. Thus, it is generally considerably more difficult to specify the ecological requirements of a population of animals than it is for a population of plants.

We have already discussed several kinds of intrademic variation; several types of interdemic or polytypic variation have also been studied. Usually external morphological differences are the first to be observed and studied. These studies often lead to further work on physiological differences in temperature tolerance, edaphic requirements, moisture demands, developmental rates, and the like.

As genetic techniques have become available, studies of interdemic differences in lethal frequencies and allelism, in the nature and frequency of chromosomal rearrangements, and, most recently, of biochemical variants have added to our store of information about the differences between populations of the same species.

The simplest kind of geographical variation is that with a simple genetic basis. In industrial melanism in moths, for example, the dominant genes for melanism have high frequencies in industrial areas, but the frequencies decrease the farther removed the populations are from these areas until in rural, unpolluted areas they reach zero (Kettlewell, 1958). Human blood-group alleles show considerable variation in frequency in different human populations as do the alleles governing various other human enzyme and protein polymorphisms (Cavalli-Sforza and Bodmer, 1971). The Burnsi and Kandiyohi dominant genes are found in populations of leopard frogs, *Rana pipiens,* at polymorphic frequencies in Minnesota and adjacent areas (Merrell, 1965c), but are rare or absent from populations in other parts of the species range. We have already discussed the geographic variation in the frequency of chromosome inversion types; these also fall in the category of simply inherited, geographical variation (see Chapter 9). Molecular genetics in recent years has provided still another approach to the study of simply inherited, geographical variation through the identification of protein variants or electromorphs with electrophoretic techniques (e.g., Ayala et al., 1972). The great problem with most of this geographical variation is that, apart from industrial melanism, its significance remains a puzzle. The adaptive or evolutionary significance of the differences between populations of the same species remains obscure despite many efforts at clarification. In fact, the great debate over protein variations is whether they have any adaptive significance at all—that is, Are they influenced by natural selection or are they adaptively neutral?

QUANTITATIVE TRAITS

Thus far, most of the discussion has been directed toward simply inherited, *discontinuous* traits, traits that show clear-cut, easily categorized differences resulting from the segregation of alleles at a single locus. For example, a flower color may be white or purple, or a human blood type may be A, B, O, or AB, or a mouse may be albino or pigmented. These are discrete classes, without intermediates, and the mode of inheritance is simple. Mendel deliberately chose this type of variation for study in the garden pea (round vs. wrinkled or green vs. yellow seeds, tall vs. dwarf plants, etc.) because he realized that this type of discontinuous variation offered him the best opportunity to unravel the mechanism of inheritance. Similarly, population genetic theory has largely focused on events at a single locus in an effort to determine the effects of mutation, selection, migration, and drift on gene-frequency change in populations. Again, the hope and expectation were that if events at a single locus could be understood, by extension it might be possible to comprehend the nature of such events involving many loci. Theoretical treatments for two, and more, loci have been developed; however, even for two loci, the models rapidly become very complex.

The truth of the matter is, of course, that not all biological variation is discontinuous, and not all of it is simply inherited. Many traits such as height or weight are quantitative and are measured on a continuous scale. In fact, *continuous variation* is more common than discontinuous variation, and quantitative or metric characters are important both in evolution and in plant and animal breeding. (Recently they have become a subject of interest and controversy in human genetics.) Quantitative traits include not only such characters as height or weight, but growth rate, milk production, yield in corn, mental abilities as measured by IQ tests, flower length in tobacco, and the like. In each, it is not possible to classify individuals into a few distinct classes. Instead, they form a continuous series from the smallest to the largest, with relatively few individuals at the extremes of the distribution and most of them clustered near the center. The frequency distribution for quantitative biological variation thus approximates the normal bell-shaped curve; this frequency distribution is characteristic not only of morphological variation but of physiological and behavioral variation as well.

The biological determination of a quantitative trait is generally far more complex than the determination of a qualitative trait such as blood type. For example, yield in corn is dependent on the size of the plant, the number of ears per plant, the size of the ears, the number of rows per ear, the number of kernels per row, and the size of the kernels. Each of these traits is apt to be influenced by genes at a number of loci so that many genes must be involved in the determination of yield in corn. In addition, the yield is also influenced by such environmental factors as temperature, light, moisture, soil type, and available nutrients. Therefore, this and other quantitative traits are influenced by a number of genetic and environmental factors.

The above traits are all continuously variable. If enough individuals could be accurately measured and plotted, the frequency distribution would form a smooth curve. Usually, however, the measurements are grouped into equally spaced classes, either because of limitations in the accuracy of measurement or for convenience in plotting the data. Thus, a frequency histogram is formed rather than a smooth curve.

Meristic variation involves traits that can be expressed only in whole numbers, such as litter size, number of eggs laid, number of fin rays or scales in fish, bristle number in *Drosophila,* and so on. Since they are enumerated rather than measured, variation in these traits is, strictly speaking, discontinuous rather than continuous. However, the frequency distribution for meristic traits usually approximates a normal distribution if the number of classes is not too small, and thus it closely resembles the frequency distribution for continuously variable traits whose measurements are grouped. It is generally assumed that meristic traits reflect some underlying tendency that is continuously variable but can be expressed only in whole numbers. For example, fecundity may be regarded as a continuously variable trait that can be expressed only as the number of eggs laid. Similarly, the number of scales in fish or bristles in *Drosophila* may reflect a continuous developmental tendency to produce scales or bristles that can be realized only by the production of a complete scale or bristle. For each unit to be produced, it can be postulated that some *threshold* must be crossed. Then the development of scales or bristles would involve a succession of thresholds.

Although, in practice, meristic traits are treated in the same manner as truly continuous variation, it is not inconceivable that the developmental mechanisms underlying the formation of bristles, for instance, may differ in significant ways from those responsible for weight in *Drosophila*. Until more is known, this cautionary word is about all that can be said.

In addition to continuous variation and meristic variation, still another type of variation, in the so-called *threshold characters,* is also considered to be a type of quantitative variation. Examples of threshold traits are disease resistance, expressed as either survival or death, and congenital malformations such as cleft palate or anencephaly, which are either present or absent. In such *all-or-none* situations, it may seem surprising that this type of variation is regarded as a form of quantitative variation. However, these traits are not inherited in a simple Mendelian fashion. The fundamental similarity in continuous traits, meristic traits, and threshold traits lies in the fact that their expression is influenced by many gene loci and by the environment. Thus, even though the phenotypic expression is discontinuous, dead or alive, affected or unaffected, the mode of inheritance and development for threshold traits resembles that for continuously varying characters. The difference lies in the concept that there is an underlying continuity in development, but the phenotypic expression changes abruptly if the threshold point is crossed. In this way, the developmental continuity is reconciled with the phenotypic discontinuity. All individuals anywhere along the developmental axis below the threshold will appear in one phenotypic class; all those above the threshold will appear in the other. The similarity to meristic characters, which appear to involve a series of thresholds rather than just one, should be obvious.

The reason for this digression into the nature of quantitative traits is that most geographical variation involves quantitative rather than qualitative differences, as shall now be seen.

THE ECOTYPE CONCEPT

Turesson (1922a) coined the word *ecotype* to characterize populations of a species well adapted to life in a particular habitat, for he
found that common, widespread species of plants were differentiated
into a number of ecotypes, each adapted to a different set of ecological conditions. Before this work, students of intraspecific, polytypic
variation had been primarily concerned with morphological differences. Following the work of Turesson and others, it was realized
that the significant evolutionary unit was the ecological race or ecotype
rather than the morphological race or subspecies. The classic studies
of Clausen, Keck, and Hiesey on *Potentilla* and other plants helped
to clarify the nature of geographical variation and the concept of the
ecotype (Clausen, Keck, and Hiesey, 1940, 1945, 1948; Clausen,
1951; Clausen and Hiesey, 1958).

Potentilla glandulosa, a perennial herb of the rose family, Rosaceae,
is found in California from the Coast Ranges up into the higher elevations of the Sierra Nevada Mountains. In central California four
morphologically and ecologically distinct subspecies were recognized:
P. g. subsp. *typica* is found primarily in the Coast Ranges; subsp.
reflexa occupies warm sunny slopes from about 900 to 6,000 feet in
the foothills and mid-elevations of the Sierra Nevadas; subsp. *Hanseni*
is found in mountain meadows between 4,000 and 8,000 feet; subsp.
nevadensis occurs even higher, in meadows or on slopes ranging from
8,000 to 11,000 feet. Two of the subspecies, *reflexa* and *Hanseni,*
overlap in altitudinal distribution, but remain separate because
reflexa occurs on dry slopes and *Hanseni* in moist meadows. Even
though four morphologically distinct subspecies could be identified
in the field along the central California transect, and each was also
ecologically distinct, further careful study revealed that each contains
at least two ecotypes. *Nevadensis* consists of a dwarf, early-flowering
alpine form and a subalpine form that is somewhat larger and flowers
later. Because size and flowering time are variables that are strongly influenced by environmental conditions, their hereditary nature had to be
established by growing the plants together under the same conditions.
Reflexa also consists of at least two ecotypes of climatic races, one
that grows higher than 3,000 feet and is winter-dormant, and another
that grows continuously throughout the winter in the lower foothills.

To follow the response of the different subspecies to different environmental conditions, representative plants from each were cloned

and members of each clone were reared at three stations, each under very different climatic conditions owing to differences in elevation: Stanford, at 100 feet, Mather, at 4,600 feet, and Timberline, at 10,000 feet. These experiments revealed, for example, that if *typica* from the Coastal Range were grown in the alpine environment at Timberline, it failed to survive the winter. Conversely, *nevadensis* from the high elevations managed to survive near sea level in the Stanford garden, but it did not thrive there. Even though conditions were favorable for continuous year-round growth, *nevadensis* retained its winter dormancy, which was, however, shortened to two or three months in contrast to nine months in its normal alpine habitat.

Therefore, the transplant experiments revealed that even though the phenotypic traits of these subspecies were modified in different environments, the subspecies nonetheless retained their distinctiveness, and the differences between them had an underlying genetic basis. The experiments also revealed that within the subspecies, which were originally described primarily on the basis of morphological differences, more than one ecotype might exist. The ecotypes were distinguished primarily on the basis of their physiological adaptations to particular environments, which may or may not be reflected in morphological differences. Thus, Clausen, Keck, and Hiesey did not consider the ecotype or ecological race and the subspecies or geographical race to be synonymous. They recognized several levels of variability (Clausen and Hiesey, 1958) the first of which is that within a single local population. Next is the variability between isolated local populations living under similar climatic and edaphic conditions. Although subjected to similar selection pressures, such populations reared together in a common environment still may differ significantly from one another. The third level involves the differences observed between populations living under different environmental conditions. These are the physiological differences that give rise to distinct ecotypes, which become evident when the plants are grown in a common garden or cloned and tested in several contrasting environments. As noted above, the different ecotypes may or may not be detectable morphologically. Finally, a fourth level of intraspecific variation, the taxonomic subspecies, is morphologically as well as ecologically distinct.

These distinctions hold reasonably well for *P. glandulosa.* However, in other species of plants, these different levels of variation may be overlapping and blurred, and hybridization may further obscure such distinctions. Zoologists have apparently not found the ecotype concept useful, for the word seldom appears in zoological literature. The importance of the ecotype concept lay in its emphasis on the possible adaptive significance of intraspecific variation among different populations belonging to the same species. It drew attention away from the purely morphological differences toward a more holistic study of the differences between demes. Botanists have found the concept useful because plants are sedentary and it is much easier to study plants in relation to their environments than it is to study animals. This is not to say that animals may not be as finely attuned to their environments by natural selection as are plants, but merely that it is far more difficult to determine in animals, given their mobility, what the relevant environmental selection pressures may be.

In addition to their work on ecotypes in *Potentilla* and other groups, Clausen, Keck, and Hiesey (1947; Clausen, 1951) studied two geographically isolated races or subspecies of the coast tarweed, *Hemizonia angustifolia,* that appear to occupy ecologically similar environments. This species is a California member of the sunflower family that occupies a narrow strip of land on the coastal side of the outer Coast Range, extending inland only as far as the coastal fog belt. One of the races ranges along a 275-mile strip of coast from northern California to south of Monterey Bay. There, a 40-mile gap in the distribution of the species is created by the Santa Lucia Mountains, which rise so precipitously from the ocean that there is no coastal plain. The other race occupies a 40-mile coastal strip south of the Santa Lucia Mountains. Therefore, the two races are effectively isolated geographically, but occupy ecologically similar habitats.

Despite the fact that these two populations presumably belong to the same ecotype, they manifest slight but consistent morphological differences. The plants from the northern race have a low, broad growth habit, slender, open branching, and rather small flower heads. The plants from the southern race are more erect and robust in their growth habit and have larger flowers. Although these differences are rather small, they are so consistent that sometimes the two races (northern subsp. = *typica;* southern subsp. = *macrocephala*) have been called separate species.

In this case, Clausen, Keck, and Hiesey (1947) analyzed the nature of the differences between the two forms genetically. The two types could be easily crossed, and the F_1 hybrids were fully fertile and of uniform phenotype intermediate between the parents. An F_2 of 1,152 individual plants was grown, among which no two plants were exactly alike and none completely resembled either one of the parents. These results indicated that the rather slight differences between the parents depended on the action of genes at a number of loci, each of relatively small effect. The recombination of these genes produced the wide range of variation in the F_2. Even though no F_2 plants were recovered that were identical to one of the parents, some were found that matched a parent in a single trait. For this reason it was thought that the differences between the parents were determined by a moderate number of genes. Clausen and Hiesey (1958) later estimated the minimum number of loci controlling a given character difference between the two races to be approximately four or five. Therefore, multiple-gene differences were involved in the phenotypic differences between the northern and southern subspecies rather than simply inherited single-locus differences.

Among the 1,152 F_2 plants, 57% were of normal size and vigor compared with the parents, and 43% were smaller and had a reduced rate of growth. There was a continuous gradation from normal size to extreme dwarfs that were only 1/1000 the volume of a normal plant. These small plants seemed healthy and continued to flower for the 7 or 8-month growing season, but failed to produce any significant growth. Therefore, genetic recombination between these two races in some cases gave rise to F_2 gene combinations with detrimental effects on development and growth. Thus, the genomes of the two subspecies are apparently no longer completely compatible and interchangeable. Nevertheless, a sufficient proportion of the hybrid offspring was of normal vigor and fertility for Clausen, Keck, and Hiesey to continue to regard the northern and southern populations as geographic subspecies rather than as separate species. They felt that these two groups are in a transitional stage between subspecies and full-fledged distinct species. Two further points to note are that the two geographic subspecies appear to represent just one ecotype and that even though a number of gene loci were involved in the differentiation of the two races, neither fertility nor chromosome structure was affected.

Finally, Clausen and Hiesey (1958) state that the difference between the two subspecies is "of a kind that could have arisen simply through isolation and genetic drift," and that "the morphological separation in this instance was not accompanied by ecological differentiation." Although these statements may be true in a general way, it hardly seems appropriate to assume that these differences arose simply by chance and lack any adaptive significance in the absence of any evidence one way or the other. No two different geographical areas are likely to be ecological replicates of one another. Therefore, even though the northern and southern populations occupy seemingly similar coastal habitats, they are unlikely to be identical, and the observed differences between the northern and southern races may reflect the subtle differences in the selection pressures in the two areas rather than chance events. Here again we encounter one of the central questions in evolutionary biology: To what extent are the observed variations adaptive and to what extent are they selectively neutral?

CLINES

Thus far, the examples of geographical variation considered in *Potentilla* and *Hemizonia* have involved discontinuities so that the species were separable into such discrete intraspecific entities as ecotypes or subspecies. Not all geographical variation falls into such neat patterns, however, for frequently there is a tendency for characters to change gradually and continuously over large areas. These character gradients have been called *clines* by Huxley (1939, 1943).

Clinal variation appears to be rather common and comes in a variety of forms. Clines have been described for size, color, and other morphological traits. In addition, clines have been found for other, less visible traits such as the frequency of the blood-group genes in humans or of inversion types in *Drosophila* and for various physiological characteristics such as temperature tolerance, flowering time, moisture requirements, and the like. It should be noted that clines, which are continuous character gradients, may involve either quantitative, continuously variable traits such as size, or discontinuous polymorphic traits such as blood type. In one case it is the size itself that changes gradually as samples are taken along a geographical transect. In the other, it is the frequencies of the blood-group genes that change gradually along the transect.

The explanation for clinal variation is that environmental gradients exist for many factors such as temperature, moisture, day length, solar intensity, and so on. Therefore, those phenotypic characters, morphological or physiological, that are influenced by these environmental selection pressures will also show gradients in their expression. Most ecological conditions vary gradually; only a few such as soil type will change abruptly. Hence, it is not surprising that Huxley (1943) and Endler (1977) could cite so many examples of various types of clines. Gene flow between adjacent populations is another factor that tends to smooth out any discontinuities among them. The nature of the geographical variation within a species is dependent then on the selection pressures tending to make each breeding population uniquely adapted to its own local environmental conditions, on chance events related to mutation or drift, which would also contribute to local divergence, and on the amount of gene flow, which tends to prevent the local differentiation of populations.

The concept of clinal variation has been useful in pointing up the possibility that geographical variation in a species may be gradual and continuous and that species cannot always be divided into discrete subspecific units such as ecotypes or geographical races. However, studies of clines also have certain drawbacks. A cline refers to a gradient in a particular character, usually along a geographical transect, and does not refer to a population. If several characters are studied simultaneously, there may be a strong correlation in the changes among them. However, in other cases, the clines may vary quite independently of one another. Therefore, there may be as many different clines as there are traits under study, and a cline does not coincide with any natural biological entity. Moreover, there may be a cline for one trait and random variation, or no variation at all, for another. Thus, clines have not been particularly useful in taxonomy; rather, they are useful abstractions for dealing with geographical variation in individual traits.

A cline suggests a continuously varying character gradient, but most of the evidence for them is based on inadequate sampling. If samples of a few widely separated populations indicate a gradient in variation, there is no way to know whether this is a truly continuous character gradient or represents a series of discontinuous steps in variation along a transect. Only more thorough sampling can answer this question. Furthermore, most observations on clines have not been made in a uniform environment where environmental modifications could be distinguished from hereditary variations.

Clines and ecotypes are different ways of dealing with geographic variation within a species. However, they are not necessarily mutually exclusive. Both tend to emphasize the adaptive significance of the observed variation. The ecotype concept stresses the discontinuities in geographic variation whereas the clinal concept stresses the gradual changes.

ECOGEOGRAPHICAL RULES

The regularities of geographical variation in relation to climate have led to the formulation of certain climatic or ecogeographical rules (Rensch, 1960). Just as clines can usually be related to gradients in the environmental conditions, these rules related geographical variation in morphological traits to broad-scale climatic gradients. These rules are empirical generalizations to which there may be a number of exceptions; they are by no means inexorable biological laws. Also, they apply only to intraspecific geographic variation and not to interspecific variation.

Perhaps the best known is *Bergmann's rule,* which states that in the warm-blooded vertebrates (i.e., birds and mammals) the races of a species living in the colder parts of the range have larger body size than the races living in the warmer parts. That means there is an inverse relationship between body size and temperature — the lower the mean temperature of the habitat, the larger the average body size. The usual explanation for Bergmann's rule is that an increase in body size decreases the surface-to-volume ratio since the volume increases as the cube and the surface only as the square of the linear dimensions. Thus, an increase in body size will result in a relative reduction of the body surface and will have an adaptive advantage in cold climates because it will tend to minimize heat loss. Conversely, in hot climates small body size and relatively large surface areas will be at a premium to facilitate cooling.

Allen's rule is similar to Bergmann's rule in that it also deals with the surface-to-volume ratio in warm-blooded vertebrates. Allen's rule is that the protruding body parts (ears, legs, tail, bill) are shorter in the races of a species inhabiting cool areas than in the races living in warmer parts of the species range. I sometimes think of this as the roly-poly rule, for a sphere has the lowest surface-to-volume ratio of any body shape, and the closer an organism approximates a sphere, the more efficiently it can retain heat. Thus, the reduction in size of the protruding body parts in cold regions gives the animals their roly-poly appearance, which is adaptive in cold climates. Observation of cottontail rabbits in Minnesota with frost-bitten ears suggests the reality of this type of natural selection. It is worth noting that Bergmann's and Allen's rules appear to apply to humans as well as to other species (Dobzhansky, 1962a), for the stature-to-weight ratio is considerably higher among peoples living in hot climates than cold.

Gloger's rule applies to pigmentation, also in warm-blooded vertebrates. Here it has been observed that the races living in the warmer, more humid parts of the range are more heavily pigmented than those living in the cooler, dryer parts of the range. In this case both temperature and humidity are factors in the amount of pigmentation. Races living in hot, humid climates are the darkest and are characterized by black eumelanins; those living in hot, dry habitats are somewhat lighter and are characterized by the reddish-brown phaeomelanins. Because pigmentation appears to be related to two environmental variables, the situation appears to be more complex than for Bergmann's or Allen's rules even though there are relatively few exceptions to Gloger's rule.

Gloger's rule also appears to hold among such insect species as butterflies, wasps, beetles, and flies. In insects humidity is apparently more important than temperature, for the amount of pigmentation increases in humid, cool areas and decreases in hot, dry climates in certain species (Dobzhansky, 1970). Some authors profess to be at a loss to explain the adaptive significance of Gloger's rule (Mayr, 1963, 1970; Dobzhansky et al., 1977). However, it should be open to experimental test. The significant environmental variables appear to be temperature and humidity, and the most promising adaptive possibilities are protective coloration and temperature regulation. It is difficult to test for adaptive value under the best of circumstances. The difficulty here lies in the number of variables that seem to be involved.

These three ecogeographical rules, and others (Rensch, 1960), represent attempts to describe some of the regularities that were observed in geographical variation. They resemble clines because they relate character gradients to environmental gradients. Their continued use is evidence for the reality of the phenomena they describe and evidence for the continuous gradual nature of some geographic variation in contrast to the discontinuities represented by ecotypes or subspecies.

A number of terms have been used to characterize the geographical variation within a species: variety, race, subspecies, biotype, ecotype, cline, geographical race, ecological race, *Rassenkreis,* semispecies, and so on. The same term may be used with different meanings by different authors. For our purposes now, the important thing to realize is that this welter of words is a reflection of the wealth of geographic variation to be found in almost any species and of the efforts by biologists to describe it in an orderly fashion. Whatever it may be called, the variation is there, and one of the basic problems in evolutionary biology is to explain the nature and origin of the differences between different populations of the same species.

QUANTITATIVE INHERITANCE

Earlier we discussed the nature of quantitative variation, and it should now be clear that most geographic variation is quantitative. The crosses between the northern and southern subspecies of *Hemizonia angustifolia* showed that the differences between them were quantitative and that genes at a number of different loci (indeterminate in number, although minimum estimates were made) were involved. Similarly,

Clausen and Hiesey (1958) reported on an extensive genetic analysis of the differences between ecological races of *Potentilla glandulosa* in which they found that multiple gene loci were involved in all 19 of the traits studied. Again, estimates of the minimum number of loci influencing each trait were made. In nearly all cases, the minimum number of loci segregating was estimated to be less than 10. It should be emphasized that the number of loci segregating for a given trait may be less than the number of loci involved in the determination of that trait. If the two ecological races are homallelic at some loci affecting the trait, these loci will not show up in this type of analysis. However, the analysis suggested that the trait differences between these ecological races were regulated by finite and moderate numbers of gene loci. Even so, the minimum estimate for segregating loci for the 19 traits studied was over 100. Since these are minimum estimates, and since the races undoubtedly differ in other unmeasured traits, it is clear that the differentiation of these ecological races has entailed genetic changes at a considerable number of gene loci.

We cannot treat the inheritance of quantitative traits extensively. However, given the fact that so much geographic variation is quantitative in nature, it seems desirable to consider quantitative inheritance in somewhat more detail. For this purpose the inheritance of corolla-tube length in the flowers of tobacco, *Nicotiana longiflora* (East, 1915), is informative. When two varieties that differ in the average length of the corolla tube were crossed, the F_1 flowers had an average length intermediate between the means of the parents. The frequency distribution of all three groups (P_A, P_B, and F_1) approximated a normal curve (Figure 12-1). Furthermore, the F_1 had about the same amount of variability as the parental populations, as indicated by their comparable values for the *coefficients of variation* (C.V. = standard deviation $\times$ 100/mean). The F_2 mean did not differ significantly from the F_1 mean, but the F_2 was considerably more variable than the F_1 or the parents, as shown in the figure and also by the larger coefficient of variation. An F_3 was also raised, which had variability intermediate between the F_1 and the F_2. In addition, the means of the different F_3 individuals were correlated with the phenotypes of their F_2 parents—that is, if the F_2 parents had smaller corolla tubes than the average for the F_2, their F_3 progeny had smaller corolla tubes than the average for the F_3, as seen in the figure. Results such as these are more or less typical of the results obtained in the study of the inheritance of quantitative traits.

At first glance these results appear quite different from the results ordinarily obtained from a cross involving Mendelian segregation and recombination. In fact, in the early days of this century a bitter

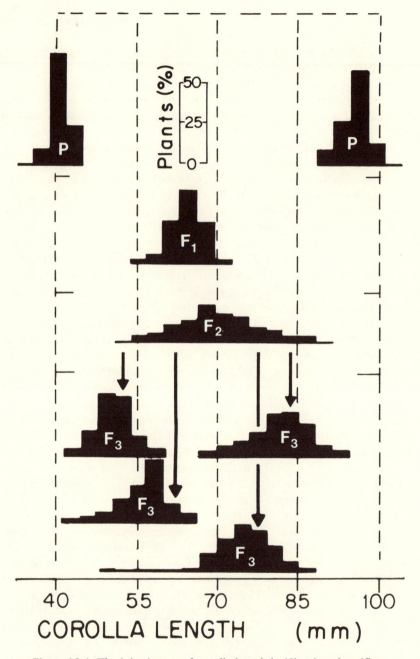

Figure 12-1. The inheritance of corolla length in *Nicotiana longiflora*. See text for details. Reprinted from Kenneth Mather and John L. Jinks: *Biometrical Genetics*. Copyright © 1971 by Kenneth Mather and John L. Jinks. Used by permission of the publishers, Cornell University Press, and Associated Book Publishers Ltd. (Based on data from East, 1915.)

controversy developed in England between the biometricians led by Karl Pearson and the Mendelians led by William Bateson over whether a Mendelian explanation was possible for the inheritance of quantitative traits. This controversy was finally laid to rest by R. A. Fisher's classic paper (1918) on "The correlation between relatives on the supposition of Mendelian inheritance." Earlier Nilsson-Ehle (1909) and East (1910) had independently formulated the *multiple-factor hypothesis,* which became the basis for the interpretation of the inheritance of quantitative traits in Mendelian terms. The essence of the theory is that quantitative traits are influenced by genes at a number of different loci, each of relatively small effect, and also by non-genetic environmental effects. Even though the alleles at each locus segregate in a discrete, discontinuous Mendelian fashion, if there are many such loci segregrating with similar and cumulative effects, many different classes for the trait will be formed, each differing slightly from the next. Moreover, the intermediate phenotypic classes will be the most frequent so that the frequency distribution approximates a normal curve. The differences between the different genotypic classes are so slight that it is difficult to distinguish between them. The slight genetic discontinuities are obliterated by environmental effects, which cause the phenotypes generated by the various genotypes to overlap to the point where the variation becomes truly continuous and quantitative.

As originally formulated, the multiple-factor hypothesis of Nilsson-Ehle and East was based on genes at a number of different loci, each of similar, slight, additive effect. They postulated independent inheritance and the absence of dominance for these loci—assumptions that are obviously oversimplified. R. A. Fisher (1918) took into account the possibility that the factors might differ in the magnitude of their effects, in their degree of dominance, and, furthermore, might be linked. He also considered the possible effects of multiple allelism, epistasis, assortative mating, and the environment. Kempthorne (1977a) stated that Fisher's paper has been the basis for and has dominated thought about quantitative inheritance ever since it was published.

Some idea of the more recent developments in the study of quantitative inheritance can be gained from Falconer (1960), Wright (1968b), Mather and Jinks (1971), and Pollak, Kempthorne, and Bailey (1977). From them it can be seen that the approach is heavily mathematical and statistical. Furthermore, most of the research has been focused on plant and animal breeding, in other words, on applied quantitative genetics; and it is here that the greatest successes have been registered in the development of improved types of domesticated plants and animals. It can also be seen that while the approach to quantitative inheritance is mathematically and statistically rather sophisticated, biologically and genetically it is quite crude and simplistic. It is a remarkable fact, for example, that Falconer's book (1960) contains no reference at all to DNA and DNA is barely mentioned in Wright (1968b). Apart from Lewontin's brief paper on "The relevance of molecular biology to plant and animal breeding," the same is true of Pollak, Kempthorne, and Bailey (1977). Thus it appears that the revolution in molecular biology of the last few decades, which has contributed so much to our understanding of the nature, action, and regulation of genes, has had little or no impact on quantitative genetic theory. The usefulness of the present theory lies in its predictive value as an aid in the genetic improvement of domesticated animals and plants. However, it seems inevitable that future developments will include a more realistic model of the genetic basis for quantitative traits, which will incorporate some of the discoveries in molecular biology. The most surprising aspect of the present model, perhaps, is that it appears to work as well as it does in predicting selective advance.

On the other hand, despite the fact that most intraspecific geographical variation appears to involve differences in quantitative traits, surprisingly little attention is paid to quantitative inheritance in some of the best known works on evolution such as Mayr (1963, 1970), Dobzhansky (1970), Grant (1971), Ford (1975), and Dobzhansky et al. (1977). Their approach is based on the concepts of population genetics set forth in Crow and Kimura (1970) and Li (1976) rather than the concepts of quantitative genetics set forth in Falconer (1960) or Mather and Jinks (1971), and thus bears out the perceptive comments of Kempthorne (1977b) about the distinction between population genetics and quantitative genetics. He argued that even though the two areas start from the same basic Mendelian ideas, they are quite divergent and that "an expert in population genetics may not be at all an expert in quantitative genetics and vice versa." This difference in outlook may help to explain why evolutionists have been so preoccupied with the biochemical polymorphisms revealed by electrophoresis over the past decade. Because the theory of evolution developed by Fisher, Haldane, and Wright is a population genetics theory, the biochemical variation among different populations of the same or different species is more suitable for analysis with this theory than the quantitative variation of the sort studied by Clausen, Keck, and Hiesey. We shall return later to the findings with respect to intra- and interspecific biochemical variation; first we must deal further with quantitative inheritance.

Earlier we saw that estimates of the number of loci segregating for a quantitative trait are possible even though the loci cannot be individually identified. One method is to determine what proportion of the F_2 resembles one or the other of the parents. If only one locus is segregating, 1/4 of the F_2 will resemble one of the parents; with two loci, 1/16 of the F_2 will resemble one of the parents; for three loci, 1/64, and so on, as shown below:

Number of Gene Loci Segregating	Proportion of F_2 Resembling One of the Parents
1	1/4
2	1/16
3	1/64
4	1/256
.	.
.	.
.	.
n	$1/4^n$

Even if none of the F_2 resembles one of the parents, the number of F_2 individuals will permit an estimate of sorts. For example, if, among 256 F_2 individuals, none resembles one of the parents, it seems likely that more than four loci are segregating for the trait.

The above method of estimating number of loci is not very useful if many loci are involved because the size of the F_2 population required to recover one individual like one of the parents increases so rapidly with each additional locus. Therefore, a different method is often used which involves a comparison of the *variance* in the F_1 and the F_2. Ordinarily, the F_1 variance in such crosses is primarily environmental, but the F_2 variance is both environmental and genetic. If it is assumed that the environmental variance is about the same in the F_2 as in the F_1, then subtraction of the F_1 variance from the total F_2 variance should provide an estimate of the F_2 genetic variance.

$$VG(F_2) = VP(F_2) - VP(F_1)$$

where

$VG(F_2) = F_2$ genetic variance,

$VP(F_2) = F_2$ phenotypic variance, and

$VP(F_1) = F_1$ phenotypic variance.

The F_2 genetic variance can be estimated as

$$VG(F_2) = \frac{Ne^2}{2}$$

where

N = the number of loci segregating and
e = the contribution of each effective allele.

The value of e can be estimated from

$$e = \frac{D}{2N}$$

where D = the difference between the means of the parental lines. Substitution for e in the equation for the F_2 genetic variance then gives

$$VG(F_2) = \frac{D^2}{8N}$$

and

$$N = \frac{D^2}{8[VP(F_2) - VP(F_1)]}.$$

This estimate is rather crude because it assumes equal and additive effects for the genes, no dominance, epistasis, or linkage, and also that all the positive factors are in one parent and all the negative factors are in the other. One of the more intriguing points to note about these relationships is that the greater the number of loci segregating in the F_2, the smaller the F_2 variance. Again, it is worth noting that the number of loci segregating may be fewer than the number controling the development of the trait if the parents are homallelic at some loci. Further details on the estimation of the number of loci involved in a quantitative trait can be found in Wright (1968b) and Mather and Jinks (1971).

Although these formulas estimate the number of genes responsible for a quantitative trait, the loci are not individually identified. In most of the work on quantitative inheritance, the number and location of the genes involved are unknown. The early work of Sax (1923) on the garden bean and Rasmusson (1935) with the garden pea was directed toward demonstrating the existence of Mendelizing genes influencing quantitative traits such as size and flowering time through their linkage to *major* genes affecting qualitative traits such as seed-coat color and flower color. Castle (1919), in his study of the hooded trait in rats, demonstrated the existence of *modifying factors*, which can be defined as multiple factors detectable only through their quantitative effects on the expression of a major gene, in this case, hooded. In none of these cases, however, were the multiple genes studied individually. More recently, some attempts have been made to locate and study the effects of multiple factors, or *polygenes,* as they are sometimes called (Thoday, 1961; Wehrhahn and Allard, 1965; Robertson, 1967; McMillan and Robertson, 1974; and Thoday, 1977). These studies provide some insight into the nature and action of those previously more or less hypothetical entities, the genes influencing quantitative traits.

Polygene is often used synonymously for multiple factor, but Mather (1944) gave polygene a special meaning when he suggested that the genes governing quantitative traits were located in the heterochromatin. Therefore, since polygene has become associated with heterochromatin, and multiple factor with equal and additive effects, the most appropriate expression for such genes may be *multiple genes.* However, these three terms, plus *polymeric genes,* are usually used interchangeably. If a more restricted meaning is intended, it can usually be inferred from the context.

HETEROSIS

Sometimes, when a cross is made between individuals differing in a quantitative trait, the F_1 progeny, rather than being intermediate between the parents for the trait, exceed the average of the better parent. This hybrid vigor was called *heterosis* by G. H. Shull (1948) with no implications as to its genetic mechanism. Heterosis is trait specific—that is, if it is found in relation to one quantitative trait difference, it may or may not be found with respect to other quantitative characters in F_1 progeny from the same set of parents.

Because heterosis has been so useful in agriculture, it is usually associated with larger size or yield in the F_1 compared to the better parent. However, if developmental rate or early flowering is the trait of interest, heterosis implies that the F_1 will develop faster or will blossom sooner than either parent. In such cases heterosis would involve an F_1 with a lower value than the lower parent. Therefore, it should be recognized that for a given trait, the F_1 mean may be intermediate to the means of the parents, it may exceed the mean of the greater parent, or it may fall below the mean of the lower parent. The essential point about heterosis is that the F_1 hybrid means fall outside the limits set by the means of the parents.

Because larger size or yield is usually regarded as favorable, especially in an economic context, these traits are defined as heterotic when the F_1 mean exceeds that of the better parent. Shull (1948) certainly thought in these terms, for he wrote, "By definition, heterosis is the *increase* of size, yield, vigor, etc. If there is no such *increase*, there is no heterosis." Nevertheless, as we have seen earlier, there are instances such as flowering time where a lower value in the F_1 than in either parent would be regarded as favorable, and therefore heterotic. When the F_1 mean lies in a direction assumed to be favorable, this is considered to be *positive heterosis*. If the F_1 mean lies in a direction considered to be unfavorable, this has been called *negative heterosis*, although Shull (1948) considered the expression a misuse of the word he coined.

As long as the heterosis concept is used in an economic context, it is possible to define what is favorable or unfavorable fairly easily in terms of maximizing yield. When this mode of thinking is carried over to natural populations, however, it creates problems. The largest, or fastest growing individuals may not be the most favorable in an adaptive sense. The fittest individuals may be intermediate in size, for example. Therefore, heterosis, as defined by Shull, and fitness are not necessarily related concepts.

Dobzhansky's solution (1952a) was to redefine heterosis in terms of fitness. However, this definition created as many problems as it solved: a classical example of heterosis such as the mule has zero fitness because it is sterile. Hence, Dobzhansky had to resort to calling such cases *luxuriance* or *pseudoheterosis*. Another difficulty with trying to tie heterosis to fitness is that heterosis is trait specific while fitness is characteristic of an entire phenotype. Therefore, it seems wise to keep the two concepts separate. Even though heterozygotes may be more fit in many cases than their more inbred relatives, this is not invariably true, any more than it is necessarily true that the largest, most vigorous individuals are better adapted than their punier relatives.

As mentioned above, the F_1 from parents differing in a quantitative trait does not always show heterosis. Analysis has shown that the appearance of heterosis in an F_1 depends on the degree of genetic divergence between the parents (Grant, 1975). If they are closely related — e.g., sibs from the same inbred line — there will be no heterosis. In general, the more remote the relationship, the greater the amount of heterosis, up to a point. Thus, intervarietal, interracial, and even some interspecific crosses will frequently produce heterotic F_1 individuals. However, when the genetic differences become too great, the F_1 may show reduced vigor, stunted growth, sterility, or even lethality. The dwarfed F_1 plants from the cross between the northern and southern races of *Hemizonia angustifolia* are an example of this type of effect.

The genetic mechanism responsible for heterosis has been the subject of speculation and controversy for years. One major theory is Jones's (1917) theory of *linked favorable dominant genes*. He assumed that there are numerous gene loci on all the chromosomes influencing such quantitative traits as size and yield. The genes responsible for normal growth and development were assumed to be dominant or at least partially dominant while the more deleterious genes were recessive. There are so many loci involved that it is highly unlikely that one chromosome would ever carry nothing but favorable dominants. Therefore, when inbred lines are formed in normally outcrossing species, they will be homozygous for both dominant and recessive genes, and the homozygous recessives will reduce the vigor of the inbred lines below that of the original parental population. However, each inbred line will tend to be homozygous for a different set of recessives so that, when two inbred lines are crossed, the dominants in one line will cover up the recessives in the other, and vice versa. Therefore, there will be a favorable dominant allele at virtually every locus, and the F_1 hybrids will show maximum vigor. The decline in vigor in the F_2 and subsequent generations with inbreeding is due to segregation, which results in an increasing proportion of loci homozygous for deleterious recessives.

A second major theory to explain heterosis is the concept of allelic interaction, first presented in detail by East (1936). It was earlier called superdominance by Fisher, Immer, and Tedin (1932) and is now best known as the *overdominance* theory of heterosis, a term coined by Hull (1945, 1946, 1952). In this case, the heterozygote was postulated to be superior to either homozygote owing to the interaction of alleles such that $a_1 a_1 < a_1 a_2 > a_2 a_2$. Here, heterosis could result from the interaction of alleles at a single locus, which was not the case for the dominance theory where at any one locus, $AA = Aa > aa$.

A crucial test to distinguish between the dominance and overdominance theories of heterosis then requires a comparison of the heterozygote at a given locus with its two corresponding homozygotes, with the rest of the genotype held constant. A satisfactory test is extremely difficult to perform because of the practical difficulties in generating the three genotypes, $a_1 a_1$, $a_1 a_2$, $a_2 a_2$, on a common genetic background. Since in most cases, the dominance and over-dominance theories lead to the same expectations (Allard, 1960) and since it has proved so difficult to distinguish between them experimentally, one may wonder at the great efforts over many years to make the distinction. However, the implications of the two theories for the practical animal and plant breeder are quite different. If the dominance theory is correct, the breeder should strive to accumulate the maximum number of favorable genes for size, yield, hardiness, etc. in the inbreds, with his ultimate goal being an inbred free of deleterious recessives. However, if the overdominance theory is correct, the primary concern of the breeder in the development of inbred lines should be their combining ability with other inbreds. Above some minimum level of viability and fertility, the breeder should select for improved combining ability in the inbreds rather than for improved quality of the inbred line itself.

Although overdominance has always had strong advocates as an explanation of heterosis, it sometimes seems that the more carefully the experiments are conducted, the smaller the role overdominance appears to play in the observed heterosis (Wright, 1977). For example, if, after several generations to allow for recombination, the apparent overdominance is dissipated, it is probable that the observed heterosis was due, not to single-locus heterosis, but to the retention of heterozygosity over short chromosome segments, with the favorable genes in repulsion. With the passage of generations these linkages are broken up and the apparent single-locus heterosis disappears. Gardner (1963) has summarized the information from maize on the effect of linkage in causing apparent overdominant effects in genes controling yield in maize and concluded (1977) that "The overdominance hypothesis advanced by Hull as the primary cause of heterosis in corn has been refuted." Similarly, Frydenberg (1964) reported that the apparent overdominance at the ebony locus in *D. melanogaster,* which initially led to an apparent balanced equilibrium between *e* and +, disappeared if the populations were run more than 1,000 days and that ebony was then eliminated. (For further discussion, see Merrell, 1965d.) This phenomenon of heterosis related to linkage disequilibrium has been called *pseudo-overdominance* by Mangelsdorf (1952), *multilocus heterosis* by Hexter (1955), *associative overdominance* by Frydenberg (1963), and *apparent overdominance by repulsion linkage* by Wright (1977).

Grant (1975) has cited a number of reports of single-gene heterosis in both animals and plants, and the possibility should certainly not be excluded. However, it is probably not as important a factor in heterosis or in balanced polymorphism as has frequently been assumed. The most likely situation favoring true overdominance would be where isoalleles, each with positive, favorable, but somewhat different codominant effects, are involved. In this case, the heterozygote would have somewhat greater physiological or biochemical versatility than either homozygote, which is expressed as heterosis. It seems less probable that overdominant heterosis would develop when one of the alleles is clearly deleterious. In other words, *heterotic lethals* or heterosis in heterozygotes for albinism, for instance, seem to be likely candidates for examples of pseudo-overdominance. However, the case of sickle-cell anemia is sufficient to show that even seriously detrimental genes may produce a heterozygous advantage (Cavalli-Sforza and Bodmer, 1971). In this case, the Hb^S allele causes the production of hemoglobin S, which differs from normal adult hemoglobin A by a single amino acid substitution. Hemoglobin A, under the influence of the normal allele Hb^A, contains a sequence of 146 amino acids in its beta polypeptide chain, with glutamic acid in position 6. In hemoglobin S, the amino acid in position 6 is valine. This seemingly trivial difference has far-reaching consequences, for the Hb^S/Hb^S homozygotes suffer from a severe form of anemia, which is often fatal. The alleles Hb^A and Hb^S are codominant in heterozygotes whose red cells contain both hemoglobin A and hemoglobin S. The heterozygotes' red cells show the *sickling trait* under reduced oxygen tension, but otherwise, the heterozygotes do not seem to be seriously affected. In fact, in areas where subtertian malaria caused by *Plasmodium falciparum* is prevalent, the heterozygous Hb^A/Hb^S individuals seem to be more resistant to malarial infection than are the normal Hb^A/Hb^A homozygotes (Allison, 1964). Therefore, in these areas, single-locus heterosis or overdominance appears to be responsible for balanced polymorphism between Hb^A and Hb^S and the relatively high frequencies of the otherwise detrimental Hb^S allele. However, the very frequency with which the sickle-cell case is cited may be indicative of the rarity of well-based examples of single-gene heterosis or overdominance.

In the literature on population genetics and population biology, overdominance is sometimes used synonymously for heterosis — in other words, it is used for those cases where the heterozygotes are superior to both homozygotes without regard to the mechanism producing this effect. Since many of these cases result from heterozygosity for chromosome segments containing complexes of linked genes, or *supergenes* as they are sometimes called, this usage undermines the usefulness of the term overdominance. As originally defined by Hull, it refers to heterosis due to interactions between alleles at a single locus. Using the word in cases where the mechanism of heterosis is unknown implies greater knowledge of the mechanism than is actually the case.

Dominance and overdominance are not the only theories of heterosis. Powers (1944), in a study on tomatoes, showed that *complementary gene interaction* provided the best explanation for his data. When he crossed a line with a few large fruits with one that had numerous small fruits, the F_1 means were lower than the arithmetic means of the parental values for both fruit size and fruit number. Thus, there was no heterosis for either fruit size or fruit number and not even any indication of dominance for larger numbers or size of the fruit. Nevertheless, the yield or total weight of fruit produced by the F_1 far exceeded the yield of either parent so that there was heterosis for yield. Neither dominance nor allelic interaction can adequately account for these results. The most reasonable interpretation is that the heterosis is caused by complementary interactions between nonalleles, between the genes affecting size on the one hand and number on the other. In this case, heterosis was caused by epistatic interactions in the broad sense of the word. This cross also illustrates the point made earlier that heterosis is trait specific, for only one of the three traits studied showed heterosis in the F_1. Cytoplasmic theories of heterosis have also been suggested (e.g., Michaelis, 1951), but even though favorable nucleocytoplasmic interactions may be involved in heterosis, cytoplasmic inheritance is ordinarily exclusively maternal and can hardly account for the typical effects of inbreeding and crossbreeding, which depend equally on both parents.

It seems unlikely that heterosis or hybrid vigor can be explained by one theory to the exclusion of all others. Each of the following may contribute to heterosis to a greater or lesser degree: (1) dominance, the masking of deleterious recessives; (2) overdominance, or allelic interaction; and (3) epistatic interactions, or complementary gene actions among genes at different loci. In natural populations, it appears that some of the more unusual genetic systems have evolved as a means to maximize or perpetuate heterosis. Among them are translocation heterozygosity (in *Oenothera*), inversion polymorphism (in *Drosophila*), allopolyploidy in plants, balanced lethal systems, certain types of asexual reproduction that perpetuate heterozygotes, and incompatibility systems and other mechanisms that necessitate or encourage outcrossing. The particular genetic mechanisms involved may depend on the circumstances under which selection favoring heterosis occurred; however, it is noteworthy how often differentiated chromosome segments are being maintained.

HERITABILITY

The effects of individual genes cannot ordinarily be identified and studied for a quantitative trait. Instead, a phenotypic value is obtained by measuring the trait in appropriate metric units in each individual. From these phenotypic values, the mean, the variance, and the covariance for the population can be calculated. From these estimates, some knowledge of the genetic properties of the population can be obtained. The analysis, in essence, partitions the variance into the components responsible for the observed phenotypic variation.

The *variance* is a numerical estimate of the variability of the phenotypic values about the arithmetic mean. The numerator is the sum of squares of the deviations of the phenotypic values from the arithmetic mean, and the denominator is simply one less than the number of observations. Therefore the phenotypic variance is

$$V_P = \frac{\Sigma(X - \bar{X})^2}{N - 1}$$

where

X = a phenotypic value,

$\bar{X}$ = the arithmetic mean of the X's,

Σ = symbol for summation, and

N = number of observations.

The total phenotypic variance has two major components or causes for the observed variation, genetic and environmental. Therefore,

$$V_P = V_G + V_E$$

where

V_G = genetic variance and

V_E = environmental variance.

The genetic variance can be still further subdivided:

$$V_G = V_A + V_D + V_I$$

where

V_A = additive genetic variance,

V_D = dominance variance, and

V_I = variance due to epistatic interactions.

If a given environmental change does not produce a comparable effect in all genotypes, the phenotypic variance should also include a V_{GE} term to take this genotype-environment interaction into account. However, it is usually assumed in this type of analysis that there are no genotype-environment interactions. In a practical sense, it would be necessary to know the expression of the different geno-types in different environments, and this type of information is seldom available, especially in human populations. However, the fact that such interactions are possible and even probable should lead to con-siderable caution in the interpretation of the data. Thus, a more com-plete expression for the components of phenotypic variance would be

$$V_P = V_A + V_D + V_I + V_E + V_{GE}.$$

The ratio V_G/V_P is a measure of that proportion of the phenotypic variance that is genetic in origin. This ratio is known as the *degree of genetic determination* or as *heritability in the broad sense.* Ignoring V_{GE}, we can write the more complete equation

$$H^2 = \frac{V_G}{V_P} = \frac{V_A + V_D + V_I}{V_A + V_D + V_I + V_E}.$$

Therefore, H^2, heritability in the broad sense, gives the proportion of the total phenotypic variance attributable to differences among the genotypes.

The ratio of the additive genetic variance of the total phenotypic variance, V_A/V_P, is known as *heritability in the narrow sense,* which can be written as

$$h^2 = \frac{V_A}{V_P} = \frac{V_A}{V_A + V_D + V_I + V_E}.$$

Heritability in the narrow sense or h^2 is used much more than H^2, and when the word heritability is used in the literature without further explanation, h^2 is usually meant. Heritability in the narrow sense is of particular interest because the additive genetic variance is the chief cause of the resemblances between relatives and can provide a mea-sure of the breeding value of individuals and of the response of popu-lations to selection. Moreover, the additive genetic variance is much easier to estimate than the other variance components. Further in-formation about heritability can be obtained from Lerner (1958), Falconer (1960), Crow and Kimura (1970), Cavalli-Sforza and Bodmer (1971), and Spiess (1977).

Like heterosis, a heritability estimate is trait specific—heritability for one trait in a population may be high; for another in the same population at the same time, it may be quite low. Moreover, herita-bility is not just a property of a particular trait, but is a property of that particular trait in that particular population in that particular environment. The reason for this statement can be readily seen in the equation for h^2, above. All of the genetic components are influenced by gene frequency, by gene action, and by assortative mating, about which little or nothing is known with respect to quantitative traits. Since different populations tend to differ in allele frequencies, genic interactions, and degree of assortative mating, the heritability for the same trait in different populations will tend to differ. Moreover, if there were, for example, a marked change in environmental conditions from generation to generation, the heritability estimate within a given population might even vary. Variable environmental conditions will increase the environmental variance and thus reduce the heritability, but uniform environmental conditions will increase it. Moreover, a relatively homozygous population will show lower heritability for the same trait than a relatively more heterozygous population.

Heritability estimates have been most useful to animal and plant breeders in predicting the response to selection of quantitative traits. The selection response is dependent on the intensity of selection and on the heritability of the trait under selection. The intensity of selection is measured by the selection differential, which is the difference between the mean of the individuals selected for breeding and the mean for the entire population from which these parents were drawn. If h^2 = 1.0, the observed variation among phenotypes being due solely to additive genetic variance, the mean of the progeny would equal the mean of the selected parents. If h^2 = 0, as it would in a homozygous inbred population, no progress would be made. However, h^2 is seldom 0 or 1, but somewhere between these values. If a heritability estimate (h^2) and the selection differential (S) are available for a population, it is possible to predict the expected response to selection, which is simply

$R = h^2 S.$

Conversely, it is possible to obtain a heritability estimate for a trait if the selection differential and the response to selection are known because $h^2 = R/S$.

When used in this fashion, heritability estimates have been a useful tool for practical animal and plant breeders. Usually, they are working with a particular population in which they have some control over or at least knowledge of both the genetic composition and the environmental conditions. From the discussion above, it should be clear by now that the use of heritability estimates should be approached with extreme caution. The tendency to give average or representative values for heritabilities (e.g., Falconer, 1960, Table 10.1) in domestic or laboratory animals has, perhaps, been misleading to some not quite so aware of the pitfalls and limitations in their use.

Heritability estimates have not been used to any great extent in studies of natural populations for reasons that should be obvious from the discussion above. An exception is the controversy over IQ and race in human populations (which, I suppose I should add, meet my criteria for natural or perhaps wild populations). A sampling of the literature includes Jensen (1969), Scarr-Salapatek (1971, 1974), Herrnstein (1971), Lerner (1972), Loehlin, Lindzey, and Spuhler (1975), and Lewontin (1975). The results of IQ tests fit a normal distribution and thus can be treated as a quantitative trait in humans. Furthermore, the results are widely used as predictors of academic success and are correlated with "success" in a broader sense so that they tend to be taken very seriously. Given this combination of circumstances, it was perhaps inevitable that the methodology of quantitative genetics would be applied to studies of IQ. Much of this work exemplifies the inappropriate use of heritability estimates, and the rest points out the errors that have been made. One common error is the assumption that the heritability of IQ is the same in all populations. Other difficulties are the lack of information about the genetic parameters in different populations, the inability to determine the nature or effects of different environmental conditions in different populations, and the absence of any information about possible genotypic-environmental interactions. Twin data have been used to estimate the degree of genetic determination of quantitative traits in humans or heritability in the broad sense, but these studies are equally open to question. More complete critiques on estimates of heritability or degree of genetic determination in humans can be found in Cavalli-Sforza and Bodmer (1971) and Lewontin (1975). Perhaps the greatest value of this controversy is that it has clarified and dramatized the shortcomings of our present approach to and understanding of the inheritance of quantiative characters.

RACE

In discussing the origin of races, we must reach some understanding of the meaning of the word *race*. A number of definitions have been proposed with varying degrees of refinement. They all refer to biological units below the species level. I shall use the terms *race* and *subspecies* interchangeably even though some authors have drawn a distinction between them.

First, in an effort to get at the essence of the concept, consider some of the definitions that have been proposed.

1. Mayr (1942) wrote, "The subspecies, or geographic race, is a geographically localized subdivision of the species, which differs genetically and taxonomically from other subdivisions of the species."

2. Dobzhansky (1951) wrote, "Races may be defined as Mendelian populations of a species which differ in the frequencies of one or more genetic variants, gene alleles, or chromosomal structures."

3. Later, Mayr (1970) wrote, "A subspecies is an aggregate of phenotypically similar populations of a species inhabiting a geographic subdivision of the range of the species and differing taxonomically from other populations of the species."

4. Dobzhansky's (1970) later definition also changed somewhat. "A race is a cluster of local populations that differs from other clusters in the frequencies of some gene alleles or chromosomal structures," or, more briefly, "Races are populations which differ in the incidences of some genes."

5. W. S. Laughlin (quoted in Lerner, 1968) defined races "as groups between which gene flow has been restricted."

This brief sampling of the definitions of race or subspecies—and it is by no means exhaustive—should suffice to indicate the diversity and lack of clarity in such definitions. For example, some of the definitions state that races replace one another geographically, but this is not mentioned in other definitions. Some stress that the populations should differ phenotypically or taxonomically without specifying how much they should differ in order to merit subspecific status. Taxonomists sometimes use the so-called 75% rule, one interpretation of which is that subspecific separation is warranted if 75% of a group of specimens can be placed in one subspecies or the other. There is, however, disagreement about the interpretation of this rule (Mayr, Linsley, and Usinger, 1953), which helps to point up the arbitrary, subjective nature of such a rule.

Similarly, the definitions dependent on gene-frequency differences do not specify how numerous or how great the differences should be. Dobzhansky's first definition specified "one or more" but this was later dropped, presumably because it is doubtful that anyone would distinguish subspecies on the basis of gene-frequency differences at a single locus. Laughlin's definition, which takes an entirely different tack, depending not on phenotypic or gene-frequency differences, but on the restriction of gene flow, also fails to specify how much restriction is required for subspecies to be recognized. Presumably, some degree of genetic differentiation is also involved.

From this brief discussion, it should be clear that there is little agreement about the nature of races or subspecies or even about the criteria used in trying to define them. One author (Livingstone, 1964) went so far as to deny the existence of human races. Unfortunately, these nonexistent entities have been the source of much controversy and many misconceptions. As long as the discussion can be confined to intraspecific differences in leopard frogs or song sparrows, people seem able to deal with it fairly objectively. The matter of intraspecific differences in humans, however, often seems to evoke a much less rational attitude.

The essence of these definitions is that races or subspecies of a species are ordinarily geographically distinct aggregates of breeding populations that differ genetically from one another and between which gene flow is restricted. The last requirement is perhaps the most essential, for if gene flow is unrestricted, the gene-frequency differences between the subspecies will tend to be swamped out by hybridization. If a cline exists, the problem of delimiting separate subspecies is very difficult, if not impossible, for it is very hard to determine where one subspecies stops and the next begins. However, if the clines are steeper in some places than in others, for example in the transition from forest to prairie or from the mountains to the plains, then the problem for the taxonomist is somewhat simpler. The paradox in the situation is that geographic races or subspecies are never clear cut unless they are isolated from one another; however, if they are isolated from one another so that the possibility of inter-breeding is remote, it becomes a matter of subjective judgment as to whether these similar but somewhat different groups should be called subspecies of a single species or separate species. It should also be noted that the presence of a transition zone or ecotone may not necessarily mark a boundary between subspecies. Wright and Wright (1949) assumed that the boundary between what they considered the eastern subspecies of the leopard frog, *Rana pipiens pipiens,* and the western subspecies, *R. p. brachycephala,* lay in the transition zone between forest and prairie in Minnesota. However, more adequate sampling (Merrell, 1977) revealed no differences between frogs from either side of this supposed boundary. In fact, the transition zone it-self appears to provide the most favorable habitat for the species in that area.

The most reasonable conclusion from this discussion is that the subspecies category can be a useful tool in some cases so long as its limitations are recognized. Since scientific names seem to be held in veneration and often take on a life all their own in the literature, great caution should be used before trinomials are conferred, lest further sampling and study reveal them to be misnomers. Once the damage is done, the literature seems to be cluttered forever with these discarded, erroneous names. It is far better to try to understand the biological situation thoroughly first before any attempt is made to confer formal taxonomic designations.

The discussion of geographic variation thus far should make clear that a species is composed of a number of individuals dispersed over a geographical area. Unless this area is very small, these individuals do not constitute a single, freely interbreeding Mendelian population. Instead, the population of the species is subdivided into a number of demes, which are isolated from one another to varying degrees, either by unfavorable habitat or simply by distance.

Given some degree of isolation, it becomes highly probable, indeed almost inevitable, that some genetic divergence among the demes will occur. The reasons are fairly obvious: No two areas are exactly alike; each has its own unique set of physical and biological characteristics. This means that even if two populations had identical gene pools at the outset, they would diverge because of the different selection pressures imposed on them in the two different areas. Furthermore, actual populations are finite and may be quite small, either all of the time or effectively so because of periodic bottlenecks. Because of limited population size, the spectrum of mutations occurring and becoming established in one population will tend to differ from that in others. Added to this, of course, is the possibility of random genetic drift if the effective population size becomes small enough. Thus, all three factors—selection, mutation, and drift—will tend to promote genetic divergence between different populations of the same species. If this process is carried far enough, it may lead to the formation of a polytypic species composed of a number of recognizable subgroups, which may be formally recognized as subspecies. The origin of races, then, can be due to the combined effects of selection, mutation, and drift. The differences observed between different populations of the same species may, therefore, be adaptive, owing to natural selection, and they may also be due to chance, related to mutation or drift. There is also the ever-present possibility that the observed differences are nongenetic, owing simply to the different environmental conditions under which the populations exist. Simple inspection is rarely adequate to determine which factors are involved. The one factor opposing genetic divergence between different populations of the same species is gene flow. Gene flow may prevent divergence from occurring or may reverse the process if it has already occurred. Thus, the gene pool of a species is in a state of flux, the particular form it takes depending on the distribution and breeding structure of the populations of the species at the moment.

BIOCHEMICAL GEOGRAPHIC VARIATION

In Chapter 3 it was seen that gel electrophoresis permits an estimate of the amount of genetic variation within a population in terms of $\bar{H}$, the average frequency of heterozygous individuals per locus, and P, the proportion of polymorphic loci in the population. This type of information can also be used to compare the amount of genetic variation in different populations and to estimate the degree of genetic differentiation between them. The advantages and limitations discussed there still apply and are worth reviewing.

The discovery of the colinearity of genes and proteins meant that it was possible to study the genetic variation of populations through the study of protein variation. The existence of a gene could now be detected even when there were no alternative alleles. Thus, it became possible for the first time to estimate what proportion of loci was invariant and what proportion was variable. A number of techniques have been developed in recent years to study protein variation, but most, like amino-acid-sequencing, are time-consuming and costly. Gel electrophoresis is more efficient, but has certain drawbacks. The proteins are placed in an electrical field through which they migrate at a rate dependent on net electrical charge and molecular size and shape. After electrophoresis, the gel is treated so that the enzyme or other protein is stained and its position can be visualized. The choice of the proteins for study is dependent on the availability of suitable assay methods, which means that either a specific enzyme stain is available or else the protein has a high enough concentration to be detected by a general protein stain. Because the proteins studied are chosen for the availability of a suitable assay, it is assumed that these proteins and the loci they represent are a random and representative sample of all the loci in the genome. However, they are, in fact, representative only of the loci coding for enzymes and other soluble proteins and not of the structural genes coding for nonsoluble proteins or of the entire category of regulatory genes that do not code for proteins at all. Furthermore, not all allelic variation is detectable by gel electrophoresis. If two soluble proteins differ in their mobility in an electric field, it is safe to assume that they differ in at least one amino-acid substitution. However, if two such proteins have identical electrophoretic mobilities, they may or may not be identical. The reason is that many amino-acid substitutions are electrophoretically neutral in the pH range normally used. In addition, the electrophoretic technique will fail to detect the genetic variation related to redundant codons, for in this case, the proteins will be identical, but the DNA coding for these proteins will not. Thus, the amount of protein and genetic variation is underestimated to an unknown degree by this technique. Some efforts to determine the extent of this undetected variation by isoelectric focusing and other techniques have been made, but the cost and effort involved have limited such efforts, and most of the available data are subject to the limitations cited above.

The fact that an enzyme might occur in a multiplicity of different forms had been known for some time when Markert and Møller (1959) coined the term *isozymes* to describe the multiple forms of a single enzyme. Isozymes are different molecular forms of an enzyme arising from any cause, including control by genes at different loci. Prakash, Lewontin, and Hubby (1969) coined the term *allozymes* to refer to different enzyme forms produced by different alleles at the same locus. These terms frequently appear in the literature on gel electrophoresis, for many of the proteins studied are thought to be allozymes although the genetic tests necessary to prove it are not always carried out.

Once the genetic variation in a population has been measured, the information can be used to estimate the degree of genetic similarity between that population and others or the amount of genetic differentiation between them. Many methods have been developed to express the genetic differences between two different populations in a single statistic (Rogers, 1972; Crow and Denniston, 1974; Nei, 1976). The most widely used have been the measures of *genetic identity, I,* and *genetic distance, D,* of Nei (1972).

The *genetic identity, I,* is calculated as follows: Let X and Y be two different, randomly mating diploid populations in which multiple alleles are segregating at a given locus, K. Let x_i and y_i be the frequencies of the ith alleles at the K locus in populations X and Y, respectively. The probability of identity of two randomly chosen alleles in population X is $j_X = \Sigma x_i^2$ while it is $j_Y = \Sigma y_i^2$ in population Y. The probability of identity of an allele from X and an allele from Y chosen at random is $j_{XY} = \Sigma x_i y_i$. If there is no selection and each allele originates from a single mutation in an ancestral generation, the expected values of j_X and j_Y equal Wright's inbreeding coefficient (1922) in X and Y, and the expected value of j_{XY} equals Malécot's (1948) coefficient of kinship. The normalized probability that two alleles at the K locus, one from each population, are identical is

$$I_K = \frac{\Sigma x_i y_i}{\sqrt{\Sigma x_i^2 \Sigma y_i^2}} \quad \frac{j_{XY}}{\sqrt{j_X j_Y}}.$$

I_K will equal one when populations X and Y have the same alleles with identical frequencies, and will equal zero when they have no common alleles. The index is normalized by dividing j_{XY} by the geometric mean of the frequencies of the identical homozygotes.

When all loci are considered, including monomorphic loci, the normalized mean genetic identity of populations X and Y is defined as

$$I = \frac{J_{XY}}{\sqrt{J_X J_Y}}$$

where J_X, J_Y, and J_{XY} are the arithmetic means of j_X, j_Y, and j_{XY}, respectively, for all gene loci sampled, including the monomorphic loci. The value of I can range from one (complete genetic identity) to zero (complete genetic differentiation).

The *mean genetic distance* between two populations is defined as

$$D = - \log_e I.$$

The value of D can range from zero for no genetic difference between the populations to infinity.

This definition of genetic distance assumes that the rate of base substitutions per locus is the same for all loci, that these nucleotide substitutions within a locus occur independently of one another, and that the number of nucleotide substitutions per locus fits a Poisson distribution. If so, then D, the genetic distance, can be interpreted as the average number of electrophoretically detectable allelic substitutions per locus that have accumulated in the two populations since they diverged from a common ancestral population. If not, D tends to underestimate the number of substitutions per locus. As noted above, it was also assumed that the allelic substitutions are adaptively neutral, but the numerical values of I and D are not dependent on these assumptions.

It should be noted that when I is close to zero, D becomes very large and also that even though I and D purport to estimate two different parameters, genetic identity and genetic distance, both estimates are based on the same information and thus are simply two different ways of presenting that information.

To provide some idea of the type of information derived from comparisons of populations at different levels of evolutionary divergence, values of the average genetic identity and genetic distance in the *D. willistoni* group are shown in Table 12-1 from the work of Ayala, Tracey, Hedgecock, and Richmond (1974). There it can be seen that different local populations of the same species are genetically quite similar. In different subspecies of the same species, however,

Table 12 - 1. Average Genetic Identity, *I*, and Genetic Distance, *D*,
Between Populations at Different Levels of Evolutionary
Divergence in the *Drosophila willistoni* Group

Taxonomic Level	I	D
Local Populations	0.970 ± 0.006	0.031 ± 0.007
Subspecies	0.795 ± 0.013	0.230 ± 0.016
Semispecies	0.798 ± 0.026	0.226 ± 0.033
Sibling Species	0.563 ± 0.023	0.581 ± 0.039
Non-sibling Species	0.352 ± 0.023	1.056 ± 0.068

Source: Ayala, Tracey, Hedgecock, and Richmond, 1974.

23 electrophoretically detectable allelic substitutions have occurred on the average per 100 loci tested in a given pair of subspecies since their divergence. *Semispecies* are considered to be borderline cases between species and subspecies. Semispecies generally show somewhat greater morphological divergence and particularly more sexual isolation than subspecies. In this study, even though sexual isolation was well developed between most semispecies, the average genetic distance was no greater than that between subspecies. This finding was interpreted to suggest that relatively few genes are involved in the development of sexual isolation. *Sibling species* are reproductively isolated true species that are still morphologically very similar. Despite this similarity, they are genetically quite different, for on the average, 58 electrophoretically detectable allelic substitutions per 100 loci tested have accumulated since the sibling species started to diverge. Even so, the sibling species have diverged less than the morphologically distinct, non-sibling species; in that case, about one allelic substitution per locus has occurred since they diverged from a common ancestral population. The value for *I* in non-sibling species is, however, 0.352, which means that on the average, 35 of 100 loci tested remained identical — or, if they had changed, the change was not detectable.

These figures are rather remarkable, for they indicate an amount of genetic change at the lowest levels of evolutionary divergence far greater than had previously been suspected. If the loci tested are truly representative of all loci, the genome undergoes a rather complete transformation even at this evolutionary level. It should be remembered that these figures apply to groups so similar that they are placed within a single genus, *Drosophila,* and furthermore, within that genus they are members of a single species group, the *D. willistoni* group.

In addition, these estimates are based on electrophoretically detectable differences, and thus are apt to be underestimates of the actual amount of genetic differentiation.

Ayala, Tracey, Hedgecock, and Richmond (1974) made further use of the genetic distances to construct a dendrogram showing the phylogenetic relationships within the *D. willistoni* group. The agreement between the phylogeny derived in this manner and one based on studies of reproductive isolation, chromosomal polymorphisms, morphological traits, sexual behavior, ecology, and geographical distribution (Spassky et al., 1971) is essentially complete. Ayala (1975a) reviewed comparable studies of genetic differentiation during speciation based on electrophoretic variants, or *electromorphs* as they are sometimes called, in other *Drosophila* species, a number of other invertebrate groups, fishes, salamanders, lizards, mammals, and plants. In general, although there is some heterogeneity, the values for genetic identity and genetic distance for this diverse assemblage of organisms seemed to parallel those seen in the *D. willistoni* group for comparable levels of evolutionary divergence.

Lewontin (1972), using a different measure of the magnitude of genetic diversity known as the Shannon information measure, studied the genetic diversity within human populations, between different populations of the same race, and between different human races. He reported that the mean proportion of the total species diversity found among individuals within populations is 85.4 %, the proportion between populations within a race is 8.3%, and hence, the proportion of the total genetic diversity found between races is only 6.3%. Despite the different measures of diversity used, this result is so at odds with the conclusions reached in Ayala's review that a fundamental difference appears to exist. One factor of unknown significance is Lewontin's dependence on the published data of others. It is possible, for example, that the supposed within-population samples are not actually individuals drawn from the same human breeding population in the way they presumably were in the other species reviewed by Ayala. In other words, the differences may reflect differences in the sampling procedures used in humans as compared to the other species. Lewontin also used a conservative racial classification that included only seven human races. Nei and Roychoudbury (1974; Nei, 1975) subdivided *Homo sapiens* into just three races, and reached conclusions similar to Lewontin's. In contrast, Dobzhansky (1962a), for example, distinguished 34 human races. These differences help to point up how elusive and difficult is the definition of race, as was stressed earlier. It would be interesting to see whether the results would change if the analysis were made using Dobzhansky's classification. Given the world-wide distribution and the large numbers of humans, the larger number of races seems at least as reasonable as the smaller numbers used by Lewontin and by Nei and Roychoudbury. In any case, one reasonable conclusion appears to be that the human races as defined by Lewontin are not comparable to the races or ecotypes we have been discussing in other species.

However, the problem is even more serious, for the neat, consistent pattern of intra- and interspecific differences outlined in Ayala's review has begun to unravel. A major conclusion from the work reviewed by Ayala (1975a) was that an extensive reorganization of the genome occurred even at such a relatively low level of genetic divergence as the origin of species within a genus. As Throckmorton (1977) has pointed out, other studies have shown (e.g., in the *D. virilis* group, and in a study of six sibling species pairs each with a close non-sibling relative) that, in terms of electrophoretic differences, subspecies appear to have diverged in about the same way and to the same degree as full species. Furthermore, in the well-studied sibling species, *D. pseudoobscura* and *D. persimilis,* the genetic distance, D, is only about 0.05. Contrast this value with the 0.58 for sibling species in the *D. willistoni* group, the 0.23 for subspecies and semispecies, and the value of 0.03 for local populations in Table 12-1. Obviously the genetic distance for the loci tested is small for *D. pseudoobscura* and *D. persimilis* and comparable to the value for local populations in the *D. willistoni* group. Similar findings have been reported for the sibling species *D. melanogaster* and *D. simulans* and for the *western* and *eastern A* forms of *D. athabasca,* which are sympatric and behave as separate species without interbreeding. In all these sibling species pairs, the value of genetic identity, I, is of the order of 0.9 or higher. Therefore, an extensive reorganization of the genome does not appear to be necessary for the origin of species. In fact, very few changes at these loci appear necessary for speciation. In studies of this sort, the minimum value is the most significant, for it indicates the minimum amount of change necessary for speciation. Larger values of D may include the consequences of phyletic evolution as well—that is, of evolutionary change unrelated to speciation. Of course, even the minimum values of D may include changes related to phyletic evolution as well as speciation, and to an unknown degree.

Therefore, there appears to be little relationship between biochemical divergence and speciation. From this conclusion still others flow—namely that the dendrograms based on genetic distances may not be reliable guides to phylogeny. Moreover, a premise of some students of biochemical evolution is that it occurs at a constant rate. However, Throckmorton (1977) points out that in the *D. virilis* group, for example, molecular evolution has occurred at different rates in different phylads and that rates may differ at different loci within the same group and appear dependent on the evolutionary opportunities available.

One further recent development is that methods have been developed to detect cryptic variation—that is, genetic variants not revealed by the electrophoresis of proteins. For example, with the use of heat denaturation at the xanthine dehydrogenase locus (XDH) of the *D. virilis* group, the number of alleles identified rose from 11 to 32 (Bernstein et al., 1973), and similar results have been reported in other cases. G. B. Johnson (1977a) has reviewed the problem of hidden heterogeneity and suggested methods to detect the hidden alleles not revealed by standard electrophoretic techniques. In addition to the thermal variants revealed by heat-denaturation studies, other variants have been revealed by varying the pore size of the electrophoretic gels. Still another group has been shown to vary in isoelectric point; although at one pH they have the same net charge and migrate similarly on gels, when the pH is changed, their mobilities differ. Alternatively, as mentioned earlier, the isoelectric points can be determined directly. Johnson offered a protocol for determining the physical and catalytic properties of allozymes. With the realization that cryptic variation is not just hypothetical but real, it seems desirable to reassess some of the earlier work based solely on gel electrophoresis with the use of such a protocol. Throckmorton (1977) does not feel that the conclusions reached to date will be changed markedly if this is done, but it is clear that the early euphoria that greeted this new approach to the study of evolution and speciation has become somewhat moderated. Nevertheless, the study of biochemical variation in natural populations has provided a new tool for the study of geographic variation and for the study of speciation and evolution. Now the question is, what is the role of this biochemical variation in natural populations?

CHAPTER 13

Neutralist vs. Selectionist

CLASSICAL AND BALANCE THEORIES

The first analyses of electrophoretic variation in enzymes and soluble proteins in natural populations were conducted by Harris (1966) in humans, by Hubby and Lewontin (Hubby and Lewontin, 1966; Lewontin and Hubby, 1966) in *D. pseudoobscura,* and by Johnson et al. (1966) in *D. ananassae.* As pointed out in Chapter 3, they reported what was at that time considered to be a remarkably high level of genetic variation in natural populations. The studies in humans and in *Drosophila* agreed in the finding that, on the average, about a third of all loci studied were polymorphic and that the average proportion of heterozygous loci per individual was about 10%. The realization that electrophoresis was capable of detecting only a fraction, probably less than half, of the genetic variation actually present made these findings seem even more remarkable. For one thing, they seemed to lay to rest the differences over what are known as the *classical* and the *balance* theories of genetic population structure (Dobzhansky, 1955). The classical scheme, usually identified with Muller (1950), in its most extreme form postulated that every individual is homozygous at nearly all loci for the wild-type allele at that locus except for a small number of loci at which it is heterozygous for rare deleterious alleles. The balance theory of genetic population structure, usually associated with Dobzhansky and his associates, postulated in its most extreme form (Wallace, 1958b) that individuals in a cross-breeding diploid species will be heterozygous at nearly all of their loci. The balance concept does not permit the identification of a normal or wild-type allele at each locus because the normal individuals in the population are heterozygotes. (For further ramifications of these models, see Dobzhansky, 1970 and Lewontin, 1974). It seemed clear, from the amount of genetic variation revealed by electrophoretic techniques, that the balance theory must be correct.

This wealth of genetic variation, however, posed a problem in that it created what appeared to be an intolerable genetic load. Lewontin and Hubby (1966) raised but dismissed the possibility of selective neutrality for these alleles or of a balance between selection and mutation in which the selection coefficient and the mutation rate were of the same order of magnitude. The third possibility raised was that selection favored the heterozygotes. In keeping with the times, they envisioned this selection in terms of heterosis, of a selective advantage of the heterozygotes over the homozygotes, and for this reason felt that the genetic load would be intolerable with so many loci segregating. Lewontin and Hubby (1966), for example, calculated that if polymorphism at 2,000 loci were being maintained by heterosis, with each homozygote 98% as fit as the heterozygote, the population's reproductive potential would be cut to 10^{-9} that of the most fit multiple heterozygote. This implies that a female *Drosophila* of optimum fitness could lay a billion eggs, which seems a bit absurd. Their argument about the magnitude of the genetic load was refuted in three papers the following year (Sved et al., 1967; King, 1967; and Milkman, 1967). It is noteworthy that all three accepted the premise that the high degree of polymorphism was due to heterosis and then argued that a large number of polymorphisms could be maintained in a population by heterosis without generating an excessive genetic load. The essence of their argument was that even though individuals with the optimum genotype are of primary importance in defining the genetic load, these hypothetical multiple heterozygotes would be so rare in actual populations that they would play little part in determining the average selective advantage at individual loci. If, rather than a multiplicative model of fitness, it is assumed that fitness reaches an upper asymptote or plateau not far above the average level of heterozygosity, then it was estimated that about 1,000 polymorphic loci could be maintained in *Drosophila* if the heterozygote advantage at each locus was of the order of 1%. Moreover, selection involves individuals, and if a fitness threshold is visualized above which an individual survives and reproduces and below which it does not, and if the probability of exceeding the threshold increases with increasing heterozygosity, then when selection occurs, it acts, not against individual genes, but against those individuals carrying an excessive number of deleterious genes. Thus with this *truncation-selection* model, each *genetic death* results in the elimination of a number of deleterious genes from the population in a single selective event. With an upper asymptote for fitness and a selective threshold, the heterotic explanation for polymorphism became more credible. However, one difficulty remaining was that the amount of inbreeding depression expected with the heterotic model was much greater than the amount actually observed.

THE NEUTRALIST THEORY

The following year Kimura (1968) challenged the idea that protein polymorphisms were maintained by some form of balancing selection and argued, instead, that the alternative alleles were adaptively neutral. This paper was followed by others, notably Kimura (1969) and Kimura and Ohta (1971a, 1971b, 1972, 1974), and was supported by an influential paper by King and Jukes (1969). This concept is called the *neutral mutation-random drift theory* by Kimura, which is reasonably descriptive, and it is also known as *non-Darwinian evolution*, a phrase first used in this connection by King and Jukes. It is also sometimes called simply the *neutralist* theory of protein polymorphism in contrast to the *selectionist* theory. Lewontin (1974) has termed it the *neoclassical* theory, for he regards it as a phoenix-like resurrection of the dead and discarded *classical* theory from its own ashes. In other words, he supposes that the former controversy over the *classical* and *balance* theories of the genetic structure of populations, seemingly settled in favor of the balance theory by the discovery of so much protein polymorphism, has been resurrected in a new guise, the dispute between the *neutralists* and the *selectionists*.

Lewontin (1974) also argues that the positions taken on the *classical-balance* or *neutralist-selectionist* controversy reflect the political and socioeconomic ideology of the scientists involved. It is certainly true that it is possible to predict, with rather disturbing accuracy, which side of the controversy a particular paper will favor, merely by knowing the names of the authors and before the experimental results are known. However, the most obvious correlation is not with the political background of the scientists, which is usually unknown, but with their scientific background. Those whose background is mathematical, statistical, or theoretical tend to be neutralists; those whose background is more biological and who have worked with experimental or natural populations tend to be selectionists.

The neutralist theory appears to have been proposed because the amount of protein polymorphism observed seemed too great to be maintained by selection. It was widely accepted at first that the form of balancing selection responsible involved heterosis and, hence, generated a considerable genetic load. However, Dobzhansky, who with E. B. Ford had long promoted heterosis as the mechanism that maintained balanced polymorphisms, was soon (1970) suggesting frequency-dependent selection and diversifying selection—that is, selection in heterogeneous environments—as alternative forms of balancing selection because they do not generate genetic loads at the equilibrium point. The past decade has been notable in population genetics for the attempts to resolve the controversy over the neutralist and selectionist positions. Much of the research on biochemical polymorphism has been directed toward shedding light on this question.

VARIATIONS IN PROTEIN POLYMORPHISM

After the initial discoveries of the high level of protein polymorphism in natural populations, a number of different groups were surveyed to determine how widespread the phenomenon was and whether any pattern of differences might emerge. It was discovered that there was variation in the amount of genetic polymorphism among the major taxonomic groups. As shown to some extent in Table 3-7, the insects and most other small invertebrates are highly polymorphic. Among the vertebrates, the mammals, reptiles, and birds are only about half as variable as the invertebrates, as indicated by the values of H and P. The amphibia and fish appear to lie between the invertebrates and the amniote vertebrates in variability although the estimates for fish are quite heterogeneous. The results with plants were also heterogeneous, but the average level of polymorphism was high.

A few cases were reported in which no genetic variation was found, for example, in U.S. populations of the introduced, self-fertilizing snail, *Rumina decollata*, (Selander and Kaufmann, 1973a) and the Northern elephant seal, *Mirounga angustirostris*, (Bonnell and Selander, 1974). The elephant seal was close to extinction at the turn of the century, and it is thought that the lack of genetic variability in these cases is attributable to the fact that these populations went through a recent reduction in population size, leading to fixation of alleles due to random genetic drift. If nothing else, such species show that high levels of genetic variability are not essential to the survival of wild populations.

Because the lack of allozyme variability seemed to be associated with small population size and random drift in these and a few other cases (Selander and Johnson, 1973; Selander, 1976), other situations were studied in which it was expected that a reduction in heterozygosity and polymorphism might occur. These included studies of cave animals and their surface-dwelling relatives, fossorial animals, island-dwelling and mainland populations or marginal and central populations, populations inhabiting different life zones, and species using such breeding systems as parthenogenesis, self-fertilization, and haplodiploidy. The results of these studies tended to be ambiguous. For example, Avise and Selander (1972) found a marked reduction in genetic polymorphism in cave populations of the fish *Astyanax mexicanus,* as compared to surface-dwelling populations of the same species. The surface populations were eyed and pigmented. Two cave populations were eyeless and unpigmented, and a third contained a range of phenotypes from eyed and pigmented to eyeless and pale. Despite these morphological differences and coefficients of genetic similarity (Rogers, 1972) of 0.96 for the six surface populations, only 0.83 for the three cave populations and 0.82 for cave versus surface populations, all the populations were placed in the same species. The low level of genic polymorphism in the cave populations seems most reasonably attributable to random drift in effectively small populations. The lack of eyes and pigment raises other questions, specifically about the mechanism of regressive evolution.

Similarly, it was thought that fossorial rodents might have unusually low levels of heterozygosity either because of their small N_e or else because selection in their uniform subterranean environment would tend to act in this direction. However, no striking differences were found between fossorial and other types of rodents (Selander, 1976). Similar arguments about marine fish from thermally stable environments or the fauna from the depths of the sea, also considered to be a homogeneous, stable environment, were not supported by the data, which showed such populations not to differ significantly in genic polymorphism from populations living in less stable environments (Selander, 1976; Valentine, 1976). Powell (1975a) attempted to relate the amount of genetic variability to the different life zones occupied. In this survey the consistent difference in heterozygosity (H) between vertebrates and invertebrates held up, but no meaningful differences were found among the various life zones (tropical, temperate, terrestrial, marine, fresh water, continental, island).

Another expectation was that marginal or peripheral populations of a species would be less variable than centrally located populations and that island populations would be less variable than continental populations. As pointed out in Chapter 9, this is true of chromosomal polymorphism in *Drosophila*, but it is not true of protein polymorphisms. Despite the low level of chromosomal polymorphism in island and peripheral populations of *Drosophila*, there is no significant difference in the level of genic biochemical polymorphism between island and continental populations or between peripheral and central populations. However, a study of deer mice of the genus *Peromyscus* (Selander et al., 1971) revealed lower levels of heterozygosity and polymorphism in island populations than on the mainland in Florida, and similar results were obtained for such populations in the Gulf of California (Avise et al., 1974). The reduced genetic variation seems attributable to genetic drift in small populations, but the consistent pattern was spoiled by the discovery that some mainland populations of *Peromyscus* also had very low levels of variation.

Another theory to account for the variation in levels of genic heterozygosity relates the variation to Levins's concept (1968) of environmental grain. According to this concept, large, mobile organisms with some degree of physiological and behavioral homeostatic control tend to experience their environment as *fine-grained.* Their optimum adaptive strategy is toward a single phenotype able to cope with the usual range of environmental conditions. Smaller, less mobile organisms with poor homeostatic control encounter a *coarse-grained* environment. Their optimum adaptive strategy is toward a variety of phenotypes, each adapted to a particular subniche in the patchy environment (Selander and Kaufmann, 1973b). This homeostasis theory then would postulate more genic polymorphism in smaller, less homeostatic species, and thus might account for the differences found between invertebrates and vertebrates.

It might be anticipated that animals reproducing asexually by parthenogenesis, or *thelytoky,* as it is sometimes called, would show low levels of heterozygosity and polymorphism. On the contrary, such species, in which males are absent or else extremely rare, may show very high levels of genic variation (Selander, 1976). Even though the populations may be phenotypically quite uniform (White, 1978), many parthenogenetic species have a hybrid origin, and the mode of reproduction preserves and promotes heterozygosity. Furthermore, a fairly large proportion of thelytokous species are polyploid, which is also favored by those genetic systems that favor heterozygosity. Relatively few cases of thelytoky are known in which the genetic system promotes homozygosity. Therefore, given that so many parthenogenetic species have a hybrid origin and a genetic system promoting heterozygosity, it is not surprising that the genic variability in these species equals and sometimes exceeds the variation found in their sexually reproducing diploid relatives.

In the Hymenoptera, males develop from unfertilized eggs and are haploid; the females, from fertilized eggs, are diploid. This genetic system is known as *haplodiploidy* or *arrhenotoky*. It might be expected that knowledge of the levels of polymorphism in the Hymenoptera would help to resolve the controversy over the mechanism maintaining genetic polymorphism. When one sex is haploid, the genes carried by that sex are all exposed to selection unless their expression is sex-limited to the opposite sex. However, the results have been sufficiently heterogeneous that no clear picture emerges. Snyder (1974), for instance, found no allozyme variability in three species of bees, and Selander (1976) showed it to be greatly reduced in six species of wasps. Nevertheless, other studies of Hymenoptera with haploid males have found significant levels of allozyme polymorphism in ants (Johnson et al., 1969; Crozier, 1973). Even though genetic variation was reduced in some of these haplodiploid species, the reasons are not clear. Whereas selection in the haploid males might have been responsible, random drift is equally plausible because the species of wasps and bees studied have been solitary or primitively social and, thus, effective population sizes were small.

The results in plants (Allard and Kahler, 1971) and snails (Selander, 1976) agree in showing that allozyme variation in strongly self-fertilizing species equals the amount of allozyme variation in their outcrossing relatives. The main difference between the inbreeding and outbreeding species is that more of the genetic variability is carried as differences between the homozygotes in the self-fertilizing populations.

A final theory to explain allozyme variation related the level of variability to the metabolic function of the enzymes. The first to suggest this possibility were Gillespie and Kojima (1968), who postulated that glucose-metabolizing enzymes would be less variable than non-glucose-metabolizing enzymes. This prediction has been confirmed in a number of studies in a variety of different groups (Powell, 1975a). Kojima et al. (1970) argued that the enzymes involved in glucose metabolism act on a single substrate of relatively constant concentration whereas the non-glycolytic enzymes act on a variety of substrates that may vary considerably in concentration. Some of these substrates may lie outside the organism and thus reflect environmental variation. The argument, then, is that the greater allozyme variability is needed to handle the greater variability in the substrates. Johnson's work (1973a) supported the idea that genetic polymorphism was far greater for enzymes utilizing substrates from the external environment than for enzymes utilizing internal metabolites as substrates. Subsequently, Johnson (1973b, 1974) drew up a different classification of enzymes: variable substrate, regulatory, and nonregulatory. In *Drosophila* (13 species), small vertebrates (22 species), and humans, a consistent pattern emerged. The highest values of H, the average heterozygosity, were associated with variable substrate enzymes, the next highest with regulatory enzymes, and the lowest with nonregulatory enzymes. A later, more inclusive survey by Powell (1975a), however, did not produce results nearly as clear-cut as Johnson's. The major difference was that the regulatory enzymes appeared to be about as polymorphic as those acting on variable substrates. The nonregulatory enzymes, however, averaged less than half as variable as the other two groups.

The interpretation of the association between enzyme polymorphism and metabolic function was, of course, that the enzymes must play an adaptive role in the populations (Johnson, 1975, 1976a). However, Selander's critical analysis (1976) of such studies together with Powell's survey (1975a) suggest that even in this case, the results are not nearly as clear-cut as one might wish.

The upshot of all of these surveys—of cave animals and their surface-dwelling relatives, of fossorial and non-fossorial rodents, of island and mainland populations, of marginal and central populations, of populations in different life zones, of species with different types of breeding systems, and, finally, of enzymes with different metabolic functions—was that patterns of association sometimes appeared, but they were never so clear-cut and unambiguous that they won unreserved acceptance. An inherent assumption in these surveys is that the groups being lumped together, all fossorial rodents, say, or all marginal populations, or all haplodiploid Hymenoptera, or for that matter, all glycolytic enzymes, will behave in a similar fashion and therefore can properly be grouped. If this assumption is wrong, it is small wonder that the survey approach has produced such ambiguous results.

THE CONCEPT OF NEUTRALITY

Where does all of this material leave us with respect to the *neutralist* versus *selectionist* controversy? The truth of the matter is, a long way from a resolution of the question. The problem stems, in part, from the difficulty in devising an adequate way to discriminate between the two concepts. The neutralist position is so adaptable, resting on several virtually untestable assumptions about mutation rates, migration rates, population size, and the number of gene loci, that just about any set of data can be accommodated by the theory if the right values of u, m, N_e, etc., are chosen. Often the product of two unknowns is involved, which permits considerable latitude in the assumptions made about them.

An even more fundamental problem stems from the nature of scientific proof. It is not possible to prove that a gene is adaptively neutral. In an experiment to test whether a given allele has an effect on fitness, for example, the null hypothesis is tested (i.e., that the allele has no effect on fitness). If the experiment reveals that this particular allele does have a statistically significant effect on fitness, it is regarded as evidence that the allele is influenced by natural selection—as proof, in other words, of an effect on fitness. If, on the other hand, the experiment reveals no statistically significant effect on fitness, this is not proof that the allele is adaptively neutral. It merely shows that in this particular experiment, the adaptive value of this allele did not differ significantly from neutrality. Even if 100 experiments failed to reveal a significant effect on fitness, it would not constitute proof of adaptive neutrality, for the 101st experiment, run, perhaps, under slightly different conditions, still might reveal a significant effect on fitness.

This fundamental difficulty seems responsible for the approach to validating the mutation-drift hypothesis suggested by Fuerst, Chakraborty, and Nei (1977). They wrote, "Because of this nature of the hypothesis, it is obvious that proper tests of its validity must be statistical," and they carried out a number of statistical analyses of gene-frequency data to test the null hypothesis of neutral mutations. The rationale is that if enough statistical tests are run and all are in reasonable agreement with predictions from the neutral mutation theory (i.e., the null hypothesis is not rejected), then the probability that the hypothesis is correct will be enhanced. This approach is far less satisfactory than a well-designed experiment to test the validity of the hypothesis, but that avenue is not open to the neutralists.

As one reads the literature pertaining to the neutralist-selectionist controversy, one sometimes wonders if the two camps are actually as far apart in their thinking as it sometimes seems. The selectionists seem to have difficulty in accepting the idea that any gene, however trivial its effects may appear, can be completely neutral. Yet Kimura, even in his original 1968 paper, wrote of *neutral* and *nearly neutral* mutations in relation to his theory. Crow (1972) stated that "the definition of neutrality is dependent on the effective population number. A gene is effectively neutral if its selective advantage or disadvantage is small relative to the reciprocal of the effective population number." In other words, if $s < 1/N_e$, the gene acts in the population as if it were neutral. Thus, the neutralists were talking not of genes that were neutral under all circumstances necessarily, but rather of genes that behaved as if they were neutral because, under the existing conditions, the effects of selection were too slight to be detected relative to drift. This may seem to be a trivial distinction, but it is not, for it makes the neutralist position far more palatable to the selectionist if it is defined in this way. The idea that finely tuned organisms, which are also finely attuned to their environments, are to any great extent the products of random evolutionary processes seems to most biologists a highly questionable assumption. To postulate that genes may range in fitness from advantageous through neutral to detrimental, depending on the conditions, seems far more reasonable than to separate them into two categories—neutral and selected.

The expression *nearly neutral* deserves comment. In some ways it seems comparable to the expression *slightly pregnant.* Even though neither situation may be immediately apparent, conceptually there is all the difference in the world between *not pregnant* and *slightly pregnant* just as there is between *neutral* and *nearly neutral.*

A few words are warranted about the meaning of *nearly neutral.* The usual assumption seems to be that nearly neutral genes will have a small but constant selection coefficient of the order of 0.01 or 0.001, which ordinarily would be very difficult to detect. There is another possibility, which can best be illustrated by an analogy. Near the center of a large urban area there is a neighborhood notable for its many trees, including numerous oaks. It supports a large, successful population of gray squirrels, not just because of the oaks but also because of the many bird-feeders. As numerous cartoonists have testified, urban squirrels are notoriously tough and intelligent, as evidenced by their ability to master squirrel-proof bird-feeders and to evade dogs and cats, their urban predators. If one were to study this more or less isolated population of gray squirrels over a period of several years, one would undoubtedly conclude that, whatever the limiting factors to the growth of this population, they would not include the great horned owl. In this one would be wrong, for one winter when the squirrels had become superabundant, a great horned owl moved into the neighborhood. By spring, where a dozen or more squirrels could be seen at once the previous fall, one was lucky to see one.

That the owl was a major contributor to their demise was evidenced by the pellets containing gray fur and squirrel teeth at the foot of the owl's favorite roosts. For that brief period, the owl became a major selective force in that population with effects that could extend over many generations even though owls were no longer preying on the squirrels and it would no longer be possible to detect their selective role. Similarly, a nearly neutral gene could be one that manifested its effects on fitness only sporadically under particular circumstances and appeared neutral most of the time. The adaptive role of this gene might be even more difficult to detect than that of a gene with a small but constant effect on fitness. For example, the effect on fitness of genes involved in resistance to infectious diseases would be very difficult to determine in the absence of the disease, but might be crucial in an epidemic.

THE NEUTRAL MUTATION-RANDOM DRIFT THEORY

Consider now the nature of the neutral mutation-random drift theory of evolution that has been set forth in detail in a number of publications (Kimura, 1968, 1969; King and Jukes, 1969; Kimura and Ohta, 1971a, 1971b, 1971c, 1972, 1973, 1974; Crow, 1972; Nei, 1975) and has also been subjected to scrutiny and criticism in a number of others (Richmond, 1970; Ayala, 1972b, 1974, 1975b; Allard and Kahler, 1972; Stebbins and Lewontin, 1972; Johnson, 1973b, 1975, 1976a; Lewontin, 1974; Powell, 1975b; Ewens and Feldman, 1976).

In his early papers, Kimura gave several reasons for proposing the neutral mutation-random drift theory of molecular evolution. One was the high rate of amino-acid substitutions in the protein molecules studied. For hemoglobin, for example, he estimated the rate to be one amino-acid substitution in a chain of about 140 amino acids in 10^7 years. By extrapolation, he estimated that if the rate observed for hemoglobin applied to all of the DNA, the rate of base substitution in mammals would be about one nucleotide substitution every two years. If this rate of substitution were due to selection, it would, according to Haldane's idea of the cost of natural selection, create an intolerable substitutional load. Haldane (1957) had calculated that the rate of nucleotide substitution due to selection could only be about one nucleotide substitution every 300 generations at best. Kimura resolved this problem by suggesting that most of the nucleotide changes were brought about, not by selection, but by random drift and, hence, there would be no load. By 1973, as Ewens pointed out, the idea of an intolerable, substitutional load with selection was no longer being used to support the neutral theory by Kimura and Ohta (1973).

A second argument favoring the neutral theory was the apparent uniformity in the rate of amino-acid substitutions in each protein molecule. This observation was contrary to expectation if selection were involved but was in accord with the neutral hypothesis. If selection were primarily responsible, the rate of amino acid substitutions would be expected to vary as the selection pressures changed. If the substitutions were adaptively neutral, the rate of substitution should be constant. However, as Stebbins and Lewontin (1972) pointed out, the seeming constancy of rates of substitution over millions of years is more apparent than real, for over such spans of time, the rates being examined must be average rates rather than constant rates. Moreover, Throckmorton (1977) recently reviewed the evidence from *Drosophila* and concluded that rates of molecular evolution in this group were not constant but rather were related most closely to the opportunities open to the evolving populations. Another point to note is that the neutral theory predicts a constant rate of substitution per generation, but the apparent constancy is a constant rate per year. Since the generation length in some of the species compared differs by one or two orders of magnitude, this discrepancy is too great to be ignored.

On a much broader scale, in an earlier study of four proteins in vertebrates ranging from fish to humans, Fitch (1975) also concluded that the rates of molecular evolution were far from constant, too much so to be due to stochastic processes.

Another argument was that the average amino-acid frequencies, calculated for a number of vertebrate proteins, agree so well with those expected from random permutations of the nucleotides. This predictability of amino-acid composition from nucleotide frequencies and the genetic code is interpreted as evidence for a random pattern of base substitutions. Here, too, however, taking an average for a heterogeneous sample of 53 proteins from a number of different species is more apt to conceal significant differences in amino-acid composition than to reveal them, and, in fact, significant deviations from expectations based on the code are known (Stebbins and Lewontin, 1972).

Despite the fact that all the major reasons for proposing the neutral theory have been challenged, the theory has persisted. Let us examine briefly the essence of the theory to appreciate why it has been so difficult to refute or to authenticate.

In the first place, the neutralists accept the idea that the evolution of morphological, physiological, and behavioral traits is primarily adaptive and governed by natural selection. Even though these traits are ultimately governed by DNA and, hence, by adaptive changes in DNA, nevertheless they argue that much of the evolutionary change in DNA and in the proteins coded for by DNA is due to chance — to random genetic changes at the molecular level that have no effect on fitness.

It is assumed that a large proportion of all new mutations are detrimental, and these are either rapidly eliminated or else kept at low frequency by natural selection. In other words, selection against these harmful alleles is directional and *purifying*. Another group of mutations are functional and are adaptively equivalent to one another. These mutations will be favored by natural selection over the deleterious mutations, but, relative to one another, they have no effect on fitness, and hence, their frequencies will not be determined by natural selection. These are the *neutral* mutations. The possibility of favorable mutations is recognized (Crow, 1972), but it is assumed that they are so rare that their effect on molecular evolution is negligible. Crow shows (Figure 1) a frequency distribution for deleterious, neutral, and slightly advantageous alleles that reflects the usual assumptions about their relative frequencies. The assumption is that the less deleterious mutants are more frequent than the more severely deleterious mutants while the nearly neutral and neutral mutations are the most frequent of all. The slightly beneficial mutations (no highly beneficial mutations are shown), on the other hand, are shown as being very rare. The usual argument for the rarity of beneficial mutations is that existing alleles are the products of many generations of natural selection and, hence, it is improbable that any new mutations will confer greater fitness than the existing alleles. Thus, the curve reflects a selectionist interpretation of the frequency distribution of new mutations in their effects on fitness. An implicit assumption in a curve such as this is that it is possible to assign a fixed fitness value to each new mutation whereby some will be detrimental and others will be neutral. Given the constant flux in the environmental conditions of most species, this is, at best, a simplifying assumption.

The fate of virtually all new mutations, as was pointed out in Chapter 4, is that they are lost from the population within a few generations through the random sampling of gametes in reproduction. This applies not just to the deleterious mutants (which are also sped on their way by natural selection) and the neutral mutations, but even to the beneficial mutations unless their selective advantage is very large.

We discussed the definition of neutral alleles earlier and pointed out that they were defined not as being adaptively identical, but rather as having such slight differences in their effects on fitness that their frequencies change primarily owing to accidents of sampling rather than natural selection. The alleles will be *effectively* neutral when $s < 1/N_e$. Thus, whether they behave as neutral alleles depends not on just the selection coefficient but on the effective size of the breeding population as well. We have already discussed some of the difficulties in ascertaining s and N_e. However, if s were 0.005 in a population with $N_e = 50$, random drift would predominate as a cause of gene-frequency change; in a population with $N_e = 1000$, selection would predominate.

The mutations considered are base substitutions, the most common type at the molecular level. Because each gene or cistron normally consists of hundreds of nucleotides, at any one of which a base substitution could occur, the number of possible alleles is enormous. The usual model of mutation in population genetics, with two alleles, A and a, mutating at comparable rates from one allele to the other, is obviously unrealistic. Rather than two alleles, there is essentially an infinite number (Kimura and Ohta, 1973). Rather than reversible mutations between two allelic states, mutation is essentially irreversible because the probability of further mutation at any one of a number of sites is far greater than the probability of an exact reversal of the original base substitution.

Kimura and Ohta (1973) illustrated these points with the cistron coding for the alpha chain of human hemoglobin. This chain consists of 141 amino acids so that the cistron has 3 × 141 or 423 nucleotide sites. Each site could be occupied by any one of the four bases (A, T, G, C) so that the possible number of alleles is 4^{423} or about 10^{254}. Just by single base substitutions, each allele can mutate to 3 × 423 or 1,269 different alleles. If a second mutation were to occur, the chance that it would be an exact reverse mutation back to the original allelic state would be only one in 1,269, provided mutation rates are equal in all directions. Therefore, these figures suggest that at the molecular level in the cistron, a model of irreversible mutations to an effectively infinite number of alleles is more realistic than the usual model of reversible mutations between just two allelic states. Furthermore, the mutation rate per site is of the order of 10^{-8} or 10^{-9} per site per generation rather than the rate usually given of 10^{-5} or 10^{-6} per locus per generation. With such low rates of mutation and the

finite, limited size of breeding populations, each new mutation is probably unique to that population rather than being a recurrent event.

Now consider the probability that a newly arisen neutral mutant
will be fixed. In a population of N diploid individuals, there will be
$2N$ alleles at that locus, one of which will be the new mutant. All of
the alleles are assumed to have an equal probability of being fixed so
that the probability of the new mutant being fixed simply equals its
frequency, or the probability of fixation is

$$p_n = \frac{1}{2N}.$$

The probability of loss is $1 - p_n$. For any neutral allele, its probability
of fixation, then, is equal to its frequency.

If it is assumed that each new mutation is unique, as seems reasonable, and that the mutation rate per cistron per gamete per generation
equals u, then the number of new alleles in a population of size N
will equal $2Nu$.

Now calculate k, the rate of neutral allelic substitutions in the
population. This value simply equals the rate of origin of new alleles
in the population times their probability of fixation, or

$$k_n = 2Nu \left(\frac{1}{2N}\right) = u.$$

In other words, the rate of substitution for neutral alleles simply
equals the mutation rate and is independent of population size. It
means that if the mutation rate remains constant, the rate of molecular evolution will remain constant. Conversely, an estimate of the
evolutionary rate of neutral allelic substitutions provides a direct
estimate of the rate of neutral mutations.

For each neutral allele that reaches fixation, it will take on the
average $4N_e$ generations from the time of the original mutation for
it to reach fixation, where N_e is the effective size of the breeding
population (Kimura and Ohta, 1969). Since the time to fixation is
dependent only on N_e, it should be the same for all loci within a given
population even though their mutation rates may differ. If N_e is
large, allelic substitution will take a long time. For example, if $N_e =$
10,000, the fixation time is 40,000 generations. For humans, if the
generation time is estimated to be 25 years, the fixation time for a
neutral allele is a million years. During this period a transient polymorphism will exist in the population, but in a practical sense, one
would be hard put to detect it as transient. Such a time span is
apparently much longer than has been required for racial divergence
in *Homo sapiens*. Not all alleles are destined to reach fixation; some

will increase in frequency for a time and then decline by chance to extinction, but they, too, will contribute to the transient polymorphism.

At equilibrium between the production of new neutral alleles by mutation and their loss by random drift, the average homozygosity of an individual will be $1/(4N_eu + 1)$. The reciprocal of this number $(4N_eu + 1)$, as discussed in Chapter 3, is known as the *effective number of alleles* (Kimura and Crow, 1964), and is usually smaller than the actual number of alleles in the population. The average proportion of heterozygotes for this locus will equal

$$\bar{H} = \frac{4N_eu}{4N_eu + 1}.$$

Now consider the rate of allelic substitution for alleles with a slight selective advantage. If a new mutant, A_2, has an advantage over the existing allele, A_1, without dominance so that $W_{A_1A_1} = 1$, $W_{A_1A_2} = 1 + s$, and $W_{A_2A_2} = 1 + 2s$, and s is much smaller than 1, then the probability of fixation of the new allele is approximately (Kimura and Ohta, 1972)

$$p_s = \frac{2N_es}{N}.$$

As given before, if $s = 0$, the probability of fixation of a new allele equals $1/(2N)$. If the value of N_es is not far from zero, the probability of fixation will not differ greatly from $1/(2N)$. However, if the product N_es is greater than +1 or less than -1, the probability of fixation or elimination will be determined primarily by selection.

The rate of allelic substitution for a slightly advantageous allele without dominance, as shown in the example above, is given by the equation

$$k_s = 2Nu\left(\frac{2N_es}{N}\right) = 4N_esu.$$

The rate of substitution of selected alleles again equals the rate of origin by mutation in the population times their probability of fixation. However, where the rate of substitution for neutral alleles was dependent solely on the mutation rate, the rate of substitution for selected alleles is dependent on three factors: the effective size of the population, the selection coefficient, and the mutation rate. What this equation indicates is that the rate of substitution for favorable alleles is faster than for neutral alleles, which is merely restating the obvious. Moreover, it should be noted that this is a special case of selection, without dominance, and that s is assumed to be a constant. Other forms of selection are known that could greatly modify the predicted rate of substitution.

THE NATURE OF NEUTRAL MUTATIONS

Recall that the neutral mutation-random drift theory accepts the idea that most new mutations will be deleterious and eliminated by natural selection. The mutations that persist and are the basis for the observed biochemical polymorphisms are postulated to be adaptively neutral or nearly so. However, they are all functional alleles, not null alleles, and are adaptively equivalent to one another. Thus, the non-Darwinian concept does not exclude natural selection from the evolutionary process anymore than the Darwinian concept of evolution excludes the possible importance of random events. The differences lie in the relative importance the two schools of thought attribute to selection, mutation, and random genetic drift in the course of evolution.

When a base substitution occurs in DNA, it may have one of several effects on its corresponding polypeptide. First, no change occurs because of the redundancy of the genetic code. The mutation is *synonymous* because the new codon codes for the same amino acid as the old codon. Second, a *missense* mutation occurs—that is, the new codon codes for a different amino acid so that an amino acid is replaced in the polypeptide. Third, the base triplet coding for an amino acid is converted into a chain-terminating or *nonsense* codon, and synthesis of the polypeptide molecule is prematurely terminated.

The chain-terminating mutations are thought to represent about 5% of the mutations, and since their effects are so drastic, they are thought to be rapidly eliminated by natural selection. The synonymous mutations, in which there is no change in the amino-acid sequence, constitute about one-quarter to one-third of the base substitutions, and since there is no change in the polypeptide, King and Jukes (1969) and Kimura and Ohta (1972) assumed that they must be selectively neutral. The remainder, the missense mutations, which lead to an amino-acid substitution, were also assumed to be neutral unless they occurred at an active site essential to the function of the molecule. This *ad hoc* assumption was made because, when different proteins were compared, it was found that some amino-acid sites were constant and others were variable. It was assumed that the constant or conservative sites contain amino acids that are essential to the proper functioning of the molecule, say an enzyme, and cannot change without adversely affecting this function. Therefore, this constancy is assumed to be maintained by natural selection. The variable sites, on the other hand, are free to change because they can undergo amino-acid substitutions without seriously affecting the function of the molecule. The codons that are variable at a given time are known as the *concomitantly variable codons* or *covarions*

(Fitch and Markowitz, 1970). It is the covarions, then, that are assumed to undergo non-Darwinian evolution.

Here is, then, a situation in which not all base substitutions are thought to undergo neutral evolution—just those that lead to synonymous mutations or else to missense mutations, and not all missense mutations, only those in covarions. In effect, neutrality is being postulated where selection has not been observed or is thought to be unlikely. Of course, what is being assumed is actually what should be tested. The question really is, Are synonymous mutations, and missense mutations in covarions, neutral in their effects on fitness?

However, the neutralist position on these types of mutations has been challenged (Richmond, 1970; Stebbins and Lewontin, 1972; Wills, 1973; Powell, 1975b). These authors suggest that even synonymous mutations can have differential effects on fitness if there are different transfer RNA's for different codons for the same amino acid or (shades of Goldschmidt) if different rates of synthesis of the proteins prevail. Kimura and Ohta (1971a) dubbed as "naïve pan-selectionists" those who believe that any change at the molecular level is apt to have an effect on fitness. However, the neutralists' conception of natural selection can be rather puzzling. Ohta (1974), for example, wrote, "Selection pressure must, in general, be very weak and indefinite for mutations that can only be detected at the molecular level, compared with mutations that have a visible effect. As neo-Darwinism is based on observations at the visible phenotypic level, it should be reexamined."

This passage represents a very simplistic and misleading view of natural selection. The neo-Darwinian conception of natural selection rests on the differential reproductive fitness of individuals with different genotypes. Reproductive fitness is not a visible phenotypic trait in the usual sense of the word. Sterility mutants and behavioral mutants are not normally classed as "mutations that have a visible effect," but they can have major effects on fitness. The neutralists seem to have erected an artificial dichotomy between mutations with visible and invisible effects, however one wishes to define the word visible. Nevertheless, it can be argued that all mutations have visible phenotypic effects, though some are manifested only as amino-acid substitutions in proteins or as base substitutions in DNA, and obviously some are more visible than others. Similarly, it can be argued that all mutations may be subjected to natural selection, but some will be more strongly selected than others. However, the neutralists appear to believe that genetic effects manifested at higher levels of organization—physiological, morphological, or behavioral, for example—are subject to natural selection, but that changes at the molecular level are, for the most part, somehow exempt from the effects of selection. Given what little is known at present about the evolution of chromosome structure and function, the evolution of genetic systems of reproduction, and the evolution of complex metabolic systems, there is no reason to suppose that adaptive evolution has not occurred at the molecular level as well as at higher levels of organization.

The neutralists, of course, accept the idea that some adaptive evolution occurs at the molecular level. For example, they agree that the constant amino-acid sites in enzymes are under the control of natural selection. They would also agree that such molecular diseases as sickle-cell anemia and phenylketonuria are subjected to strong selection pressures. These conditions provide an example of how artificial is the dichotomy between molecular and higher traits. The basic defect in both PKU and sickle-cell anemia is molecular. However, among other things, PKU affects head shape and hair color and causes a heightened level of phenylalanine in the blood and a reduction in IQ, whereas the sickle-cell gene, when homozygous, causes an often fatal anemia as well as the sickling of red cells visible under the microscope. Thus, the effects of these mutations range from the molecular to the physiological, the morphological, and the behavioral. The association between blood type and certain diseases (Bodmer and Cavalli-Sforza, 1976), particularly in the ABO system, provides another example in which differential selection occurs at the molecular level, and in this case there are no visible phenotypic effects.

Perhaps the basic question should be whether molecular variation differs in any significant way from morphological, physiological, or behavioral variation in its relation to fitness. Is adaptive neutrality any more probable at the molecular level than at any other level of organization? Fitness is difficult to determine under the best of circumstances. What is needed is more information about whether synonymous mutations and missense mutations, the ones thought most likely to be neutral, have any effect on fitness. In this connection, note that a demonstration that the same enzyme taken from two different species behaves similarly *in vitro* hardly constitutes evidence for the adaptive neutrality of the amino-acid differences between the enzymes. The substitutions in the covarions, usually assumed to be neutral, may, in fact, provide the fine tuning that enables the enzyme to function optimally in each species.

VARIATION IN DIFFERENT POPULATIONS

The neutral mutation-random drift theory postulates that a large number of neutral alleles can exist at each locus. Although not all of these alleles are detectable by the usual electrophoretic techniques, nonetheless the number of neutral electromorphs must also be quite large. Because they are selectively neutral, the frequencies of these alleles, and electromorphs, will change through time owing to sampling drift. These random changes would then lead to a gradual divergence in the pattern of allelic frequencies in different populations of the same species in the course of time. The longer the time since their derivation from a common ancestral population, the more the populations should differ in their allelic frequencies.

Contrary to this expectation, however, where different populations of the same species have been studied, electromorph frequencies have tended to be similar throughout the range of the species. The electromorph frequencies not only do not show random local differentiation, but they do not indicate differential adaptation to different habitats or differences attributable to different population size (King and Ohta, 1975). Results of this sort have been reported by Prakash, Lewontin, and Hubby (1969) in *D. pseudoobscura*, by Ayala et al. (1971), and Ayala (1972b, 1974, 1975b) in the *D. willistoni* group, and also in a variety of other groups reviewed by Ayala (1975a).

This deviation from expectations under the theory of neutrality has been explained in two ways by the neutralists. One is that too little time has elapsed since the populations separated from their common ancestral population for them to have diverged from one another by random drift. Although not a great deal is ordinarily known about the history of such populations, their age can usually be inferred to be too great for recent divergence to be a very convincing explanation for their genetic similarity.

The other explanation is that the observed similarities between different populations of the same species are due to gene flow, that migration of individuals between populations is sufficient to ensure that all populations of the species form, in effect, one large panmictic breeding population with similar allelic frequencies throughout the range of the species (Kimura and Ohta, 1971a). The difficulty with this explanation is that the amount of migration sufficient to prevent the differentiation of local populations by random genetic drift of neutral genes is so small that it would be virtually undetectable (Nei, 1975). Lewontin (1974), for example, calculated that a migration rate as low as one individual per thousand per generation is sufficient to prevent differentiation between populations of moderate size. Maruyama (Kimura and Ohta, 1971b) showed that for a two-dimensional stepping-stone model, marked local differentiation is possible only if the number of individuals exchanged between adjacent populations averages less than one per generation no matter what the size of the populations may be. To detect such low rates of migration of individuals between populations wuld be difficult enough; to detect whether the migrants were, in fact, contributing genetically to the population they invaded would be even more challenging.

Ayala, Tracey, Barr et al. (1974) have presented the evidence against migration as a plausible explanation for the similarity in electromorph frequencies among different local populations of the same species. A variety of deviations from predictions based on the neutral theory are given. However, the most difficult bit of evidence to explain under the neutral theory is that populations with essentially the same allelic frequencies at most loci have some quite different allelic frequencies at a few other loci. If migration were responsible for the similarities observed at most loci in the different populations for which the genetic similarities fall in the range from 0.95 to 1.00, then, because the migrating flies carry complete sets of genes from one locality to another, all loci should show similar patterns of allelic frequencies. In fact, however, a few loci have genetic similarities close to zero, and the neutrality theory plus migration fails to account for the observed patterns of genetic variation in different populations of the same species.

In some cases, correlations have been reported between variations in electromorph frequency and certain environmental variables (e.g. Koehn et al., 1971; Johnson, 1971; Clegg and Allard, 1972; Hamrick and Allard, 1972; Kojima et al., 1972; Vigue and Johnson, 1973; Johnson and Schaffer, 1973). The correlations observed were between the patterns of gene-frequency distribution and such environmental variables as temperature, latitude, altitude, and humidity. The neutralists argued, however, that there was no evidence that the enzyme loci themselves were being directly influenced by selection. Instead, they proposed that the neutral polymorphic enzyme loci were linked to other loci that were undergoing selection and that the neutral protein polymorphisms were actually maintained by selection at these other linked loci. Moreover, if selection at the selected loci correlates with the environmental variables, then the polymorphic neutral loci will appear to correlate with these variables also. Thus, linkage of a neutral allele at one locus to a selected allele at a nearby linked locus may lead to linkage disequilibrium and the development of what has been termed *associative overdominance* (Frydenberg, 1963; Ohta and Kimura, 1970). This phenomenon has also been called the *hitchhiker* effect (Maynard Smith and Haigh, 1974; Thomson, 1977). Johnson's (1975) opinion of this concept is worth quoting: "Such 'associative overdominance' is not unlike invoking divine intervention, as it is an argument that can explain practically any pattern of field data."

VARIATION IN DIFFERENT SPECIES

The neutralist theory predicts that different species should have different patterns of electromorph frequencies. Since different species are reproductively isolated from one another, gene flow between them is no longer possible, and genetic divergence between them due to random sampling of neutral alleles should proceed independently in different species. The longer two species have been isolated from one another and from their common ancestor, the greater should be the degree of genetic divergence between them. The magnitude of the differentiation in allelic frequencies is directly dependent on the length of time involved and inversely related to population size. Over long periods of time, the amount of genetic divergence should be about the same for all loci, and if a number of loci are analyzed, the pattern of distribution for the degree of divergence for each locus should fit a normal distribution curve about the mean. This expectation holds whether all alleles can be distinguished or not (Ayala, 1975b). However, a study of the sibling species in the *D. willistoni* group (Ayala, Tracey, Barr et al., 1974) showed an entirely different frequency distribution of loci than that expected under the neutrality theory. Rather than a normal curve, the frequency distribution for degree of genetic similarity formed a bimodal, U-shaped curve. In other words, at most loci, the alleles carried by two sibling species were either virtually identical or completely different. Most of the loci were found at the two extremes of the distribution. None of the possible explanations under the neutrality theory is adequate to account for this bimodal curve. It has been suggested, however, that the populations are not in equilibrium, that they may, for example, have gone through a bottleneck in population size, which would affect the amount and distribution of the genetic variability. Nei (comment following Latter, 1976) pointed out that it would take the reciprocal of the mutation rate, usually assumed to be 10^{-7} per year for electrophoretically detectable alleles, for the genetic variability to recover to its original level following a bottleneck. If this is so, it would take ten million years for equilibrium to be restored. Again, an impasse is reached because this sort of explanation is virtually untestable. It is next to impossible to determine whether a population has gone through a bottleneck in numbers in the past; and an equilibrium that requires ten million years to achieve is not really open to study.

Another observation is that different species of *Drosophila* have similar levels of electromorph heterozygosity despite the probability that the population sizes have varied greatly in the different species (King and Ohta, 1975). Furthermore, the levels of electromorph heterozygosity fall within a relatively narrow range in all kinds of species. This observation, again, is contrary to expectations under the neutral theory, which predicts a continuing increase in heterozygosity and the effective number of alleles as the population size increases. As pointed out earlier, Kimura and Crow (1964) calculated the expected effective number of unique neutral alleles (n_e) in a finite population as

$$n_e = 4N_e u + 1$$

where N_e is the effective population size and u is the mutation rate to unique alleles. Ohta and Kimura (1973) developed an equation appropriate for neutral electromorph mutations in which

$$n_e = \sqrt{8N_e u + 1}.$$

Here, n_e is the effective number of electromorphic variants and u is the mutation rate to electrophoretically detectable neutral mutations. Although the relation is no longer linear, there is still a direct relation between the effective number of variants (and also the amount of heterozygosity for these electromorphs) and the effective size of the population. The study of Ayala and Gilpin (1974) showed that, in a comparison of several species of *Drosophila,* any two species have very similar frequencies of electromorphic alleles at a substantial proportion of all loci, but at many other loci, the species have very different frequencies of electromorphic alleles. Again, this bimodal pattern of frequency distributions is not in accord with the predictions of the neutrality theory. Ayala and Gilpin also asserted that their arguments are valid even if each electrophoretic "allele" actually consists of several alleles that may or may not be different in different species.

TESTS FOR SELECTION AT POLYMORPHIC ENZYME LOCI

Thus far, we have been primarily concerned with the neutral mutation-random drift theory as an explanation for the high levels of biochemical polymorphism and heterozygosity observed in natural populations and with how well the observations agree with the predictions made under the theory. Where predictions and theory do not agree, the findings have been taken not only as evidence against the neutral theory, but also as evidence favoring a selectionist interpretation of the data. However, such evidence is by no means conclusive. As pointed out earlier, correlations have been reported between variations in electromorph frequency and environmental variables. In such cases there is always the problem of identifying the relevant environmental factor(s), and even then the neutralist argument that the neutral electromorph locus is linked to a nearby locus undergoing selection may be difficult if not impossible to refute.

Similarly, clines in gene frequencies are usually taken as evidence for natural selection along some sort of environmental gradient, but can also be explained as the result of hybridization between previously isolated populations that had diverged as the result of random genetic drift. Thus, a single cline is not very convincing evidence for selection, but if similar clines are found in several independent groups of populations, selection becomes more plausible.

Still another approach is to seek correlations between gene frequencies at homologous polymorphic loci in closely related sympatric species between which no hybridization is occurring. Such correlations certainly suggest some form of selection, but do not indicate its nature. Finding suitable species may be a problem, although one such study indicating parallel selective forces in sympatric species of mussels was reported by Koehn and Mitton (1972).

Other studies of allozyme polymorphisms in natural populations have been interpreted as supporting the operation of selection on these loci. In studies such as Koehn (1970), Johnson (1971), Allard and Kahler (1972), Ayala (1972b), Ayala, Tracey, Barr et al. (1974), Schaffer and Johnson (1974), Koehn, Milkman, and Mitton (1976), McKechnie, Ehrlich, and White (1975), Mitton and Koehn (1975), and Watt (1977), the data seem most compatible with a selectionist interpretation and the selective factors have sometimes seemed fairly obvious. However, the alternative interpretations offered by the neutralists seldom can be ruled out entirely. Therefore, the approach advocated cogently by Clarke (1975) is to seek experimental evidence that natural selection is acting directly on loci controling protein polymorphisms.

One of the first such studies (Gibson, 1970) reported that the frequency of the alcohol dehydrogenase fast allele ($AdhF$) increased from 0.5 to 0.73 and 0.82 in two experimental populations of *D. melanogaster* to which 6% ethanol had been added, but that no comparable gene-frequency change occurred in the controls lacking the added ethanol. Gibson also showed that the $AdhF$ enzyme from larvae had a higher level of enzymatic activity than the $AdhS$ enzyme, but that $AdhS$ was more heat stable at $40°C$ than $AdhF$, thus suggesting a possible balancing mechanism. Some other studies of the effects of *Adh* alleles on fitness in experimental populations are: Bijlsma-Meeles and Van Delden, 1974; Johnson and Powell, 1974; Briscoe, Robertson, and Malpica, 1975; Clarke, 1975; Morgan, 1975; Van Delden, Kamping and Van Dijk, 1975; Oakeshott, 1976; Thompson and Kaiser, 1977; and Van Delden, Boerema, and Kamping, 1978. In general, these reports confirm and expand the results reported by Gibson. In essence, Gibson and most of the others sought to demonstrate the action of natural selection at the *Adh* locus by challenging the polymorphic populations with environmental additives acting as specific substrates for the enzyme under study. From a knowledge of the properties of the different allozymes and a manipulation of the environmental conditions, including the substrate of the enzyme, it was possible to predict which allele would be favored under a given set of conditions. The precision with which these predictions came true suggested that selection was acting directly on the ADH phenotypes and that the results were not due to indirect selection on closely linked loci. In other words, the results of these experiments make it very difficult to continue to regard the *Adh* alleles are neutral alleles.

For example, Vigue and Johnson (1973) showed a cline in the frequency of *Adh* alleles in the eastern United States with *Adh^6* decreasing from 0.9 in Florida to about 0.5 in Maine in wild populations of *D. melanogaster*. Johnson and Powell (1974) showed that in polymorphic laboratory populations exposed to heat or to cold shock (41°C or 0°C for sufficient time to cause 90% mortality), the frequency of the *Adh^6/Adh^6* type decreased under cold stress and the frequency of the *Adh^4/Adh^4* type decreased under heat stress. They also demonstrated that the *Adh^6* enzyme molecule was more stable and reactive at elevated temperatures than the *Adh^4* molecule, and that the *Adh^6* allele decreased in frequency during the winter in North Carolina. All these results lead to the conclusion that the *Adh^6* allozyme confers greater fitness than *Adh^4* under warm conditions but that *Adh^4* is more fit than *Adh^6* under cool conditions.

Briscoe, Robertson, and Malpica (1975) showed that on food with 12.5% ethanol added, the survival relationships were as follows: *Adh^F/Adh^F* = *Adh^F/Adh^S* > *Adh^S/Adh^S*. They also found that *Adh^F* had a high frequency in Spanish wine cellars where a concentration of 12% to 15% ethanol is not uncommon, but that the *Adh^S* allele was more frequent elsewhere. Again the field and laboratory observations agree in pointing to direct selection on the *Adh* locus.

Morgan (1975) tested *Adh^F* and *Adh^S* not only on ethanol but on six other alcohols as well and found that the *Adh^F* allozyme had a greater specific activity on these alcohols than did Adh^S. He also tested fitness in the presence of these alcohols and found that when the *FF* flies broke down the alcohol to a nontoxic product, they survived better than *SS* flies. However, when 1-penten-3-ol is broken down by ADH, a toxic compound, ethyl vinyl ketone, is produced. In the presence of 1-penten-3-ol, the *SS* flies survived better than *FF* because less of the toxic compound was being produced. The results of these experiments suggest that alcohol dehydrogenase permits the flies to cope with high and possibly toxic amounts of alcohols in their environments.

Thompson and Kaiser (1977) tested two Adh^S alleles with identical electrophoretic mobility, one of which produced only one-half as many ADH enzyme molecules as the other. When egg to adult viability was tested on alcohol-supplemented media, the viability of flies with the allele producing less enzyme was lower than that of flies with the more active allele. Again the results support direct selection at the enzyme locus.

These studies, which involve observations both in the field and in the laboratory, typify work in ecological genetics and have provided some of the best evidence thus far bearing on the neutralist-selectionist controversy. These results, however, deal with one locus in one species, and, even though they are very suggestive, they can hardly be regarded as conclusive evidence in the controversy.

A number of other studies seeking evidence of selective effects on individual enzyme loci have also been reported. Koehn (1969), for example, was one of the first to report a latitudinal cline in the frequency of alleles for serum esterase in the sucker *Catostomus clarkii.* Moreover, he showed that the activity of the enzyme produced by the most frequent allele in southern populations increased with increasing temperature from 0° to 37°C whereas the activity of the enzyme for the most frequent northern allele increased with decreasing temperature. Berger (1971), in addition to reporting relative constancy of allelic frequencies for seven enzyme loci over both space and time in natural populations of *D. melanogaster,* also observed systematic changes in allelic frequencies at several dehydrogenase loci in laboratory populations toward the frequencies observed in natural populations. Both observations suggested selection of some sort, but not much about its nature.

Powell (1971) demonstrated greater numbers of alleles per locus and higher levels of heterozygosity per individual at 22 protein loci in populations of *D. willistoni* reared in varied environments than in populations reared in more constant environments. A similar result was obtained by McDonald and Ayala (1974) with populations of *D. pseudoobscura.* Van Delden, Boerema, and Kamping (1978) also studied selection at the *Adh* locus in populations of *D. melanogaster* in different environments and found evidence indicating selection. Powell also (1973) reported on studies of three enzyme loci in laboratory populations of *D. willistoni* and found some evidence of allelic frequency changes that suggested selection. However, because the alleles used were nonrandomly associated with chromosome inversions, the results were ambiguous. It was impossible to attribute the observed change in allelic frequencies to direct selection at the enzyme loci. They may merely have been neutral markers for the larger blocks of genes under selection.

More recently, a number of studies have attempted to study the action of selection at a number of enzyme loci in different species of *Drosophila* (e.g., Marinković and Ayala, 1975a, 1975b; Wills, Phelps, and Ferguson, 1975; DeJong and Scharloo, 1976; Hickey, 1977; Bijlsma and Van Delden, 1977; Kamping and Van Delden, 1978).

Not all authors have found selective differences at enzyme loci. For example, Yamazaki (1971) found no evidence for selection at the X-linked esterase-5 locus of *D. pseudoobscura,* nor did Powell (1973) at the only X-linked esterase locus in *D. willistoni,* nor did Cobbs and Prakash (1977a) for esterase-5 in *D. pseudoobscura, D. persimilis,* and *D. miranda.* Similarly, Yardley, Anderson, and Schaffer (1977) found no evidence of selection at the *a*-amylase locus in *D. pseudoobscura,* but De Jong and Scharloo (1976) and Hickey (1977) did find selection acting on amylase alleles in *D. melanogaster.*

In general, where evidence has been sought for the direct action of selection on individual enzyme loci, it seems to have been found more often than not. It also appears that the more that is known about the biological function of the locus, the better the chances of revealing selective differences between different alleles, the *Adh* locus being a case in point.

THE MUTATION-EQUILIBRIUM THEORY

When it appeared that the data on electrophoretic variants failed to agree with predictions based on the neutralist theory, a new theory was proposed by the neutralists, which was called the mutation-equilibrium theory, also known as the mutation-selection theory or purifying selection (Ohta, 1973, 1974, 1976; King, 1974; King and Ohta, 1975; Ohta and Kimura, 1975; Kimura, 1979a, 1979b). This theory deals with mutations that are postulated to have very slight deleterious effects—so slight that their selective disadvantage is of the order of the mutation rate. These mutations were envisioned solely in terms of single amino-acid substitutions. Since these slightly detrimental mutations may persist in the population, they are subject to further mutation so that, in addition to the original allele, the population will contain a heterogeneous array of alleles one mutant step removed from the original allele, another array two steps removed, and so on. The fitness of the original or type allele was postulated to equal 1, the fitness of the one-step mutations $1-s$, that of two-step mutations $1-2s$, and that of the nth mutant array $1-ns$.

King and Ohta (1975) then interpreted their model in terms of electromorph frequencies rather than amino-acid substitutions. Dayhoff et al. (1972) showed that 70% of the amino-acid substitutions result in no change in the net charge of the protein molecule, 15% in an increase of one unit of net charge, 15% in a decrease of one unit of net charge, and a negligible number in two-unit changes in charge. Thus, it was possible to estimate what proportion of amino-acid substitutions would give rise to detectable electromorphic variation. The assumptions about fitness were then transferred from amino-acid substitutions to electromorphs. In other words, the greater the unit charge difference between the type electromorph and the mutant electromorphs, the lower the fitness of the mutant electromorphs. In support of this idea, King and Ohta (1975) stated that the frequency distribution of electromorphs at polymorphic loci is symmetrical with the most common or type electromorph flanked by progressively less frequent electromorphs.

However, fitness estimates based on electromorphs are two steps removed from the actual allelic differences whose fitnesses should be of primary concern. When a mutation occurs, it affects the DNA. A base substitution, for example, may not result in an amino-acid substitution, but, nonetheless, may have an effect on fitness. If a base substitution does lead to an amino-acid substitution, it may not result in a change in net charge of the protein molecule, but it does not necessarily follow that similar proteins with the same net charge have similar effects on fitness. Thus, changes in net charge are rather far removed from the allelic differences resulting from mutation. Nevertheless, King and Ohta (1975) assumed that net charge is directly related to fitness and that the most common electromorph has the highest average fitness and that "electromorphs removed by one, two or more steps from the type electromorph are considered to have progressively reduced mean fitness values." Although a disclaimer was added that a given electromorph may represent a heterogeneous array of structural alleles with different fitnesses and that net charge may be only indirectly related to fitness, nonetheless, net charge and fitness were obviously assumed to be closely related. Given how far removed changes in net charge are from the primary mutational event and how difficult it is to estimate fitness in any case, such an assumption seems tenuous at best. The necessities of mathematical modeling seem to require simplistic assumptions about the operation of selection in natural populations, but, unfortunately, the way selection acts is unlikely to be simple. As R. A. Fisher (1954) wrote in "Retrospect of the criticisms of the theory of natural selection" with regard to difficulties in explaining the evolution of such structures as the bat's wing or the vertebrate eye: "They are all, in somewhat different ways, difficulties less of the reason than of the imagination." The same may be true for evolution at the molecular level.

Thus, in addition to the neutral mutation-random drift theory and the theory of balancing selection, there is now a third theory, the mutation-selection theory, to account for the high levels of genetic variability in natural populations. This theory postulates the presence of a wild-type allele of fairly high frequency and a number of other very slightly deleterious alleles whose frequencies are determined by the equilibrium between mutation and natural selection. Although this theory involves selection, it is purifying selection rather than balancing selection. E. B. Ford's (1940b) original definition of genetic polymorphism excluded purifying selection—that is, the balance between mutation and selection, from the definition. Instead, he postulated that true balanced polymorphism is maintained by some form of balancing selection that ensures that more than one type of allele is actively maintained in the population by selection. Ford, of course, favored heterozygote advantage as the primary form of balancing selection. Subsequently, especially after the implications of heterozygous advantage for genetic load were realized, frequency-dependent and density-dependent selection and selection based on spatial or temporal heterogeneity in the environment and the like have been advanced as other possible types of balancing selection that would not generate such onerous genetic loads.

Although it involves selection, the mutation-selection theory is not a selectionist theory or a compromise between the neutralist and selectionist positions. Instead, it represents a new version of the non-Darwinian or neutralist position. The selectionists postulate that molecular variation is maintained in populations by balancing selection and that directional selection tends toward the fixation of favored alleles. The neutralists (e.g., Ohta, 1974) believe that molecular evolution proceeds essentially by mutational pressure rather than by positive Darwinian selection, [and that] once the structure and function of a molecule are determined in the course of evolution, natural selection acts mainly to maintain them, because all later evolutionary changes proceed under selective constraints. Natural selection then becomes mostly "negative" and positive Darwinian selection is only a minor part of both total selection and the total number of mutant substitutions. If selection pressure is very weak, a mutant allele, even if slightly deleterious, can occasionally replace the original allele by random drift, particularly in small populations, and such chance events are apt to be important in molecular evolution. [Finally they feel that] random genetic drift and mutational pressure play a much bigger part (in molecular evolution) than has been believed.

Thus, the mutation-equilibrium theory is clearly an extension of the neutral mutation-random drift hypothesis of Kimura (1968) and King and Jukes (1969). It stresses negative purifying selection rather than balancing selection or positive directional selection.

In fact, the models used (e.g., King and Ohta, 1975; Ohta and Kimura, 1975) envision alleles as either neutral or deleterious; there is no provision in the models for the possibility of positive directional selection. The quote from Ohta (1974) above includes the phrase, "once the structure and function of a molecule are determined in the course of evolution," but does not deal with the fundamental evolutionary question of how the structure and function of the molecule are determined in the first place. The implication seems to be that once the molecule is perfected by some unspecified evolutionary mechanism, subsequent evolution merely involves protection of the molecule from degenerative change. Given the complexities of physiological and biochemical adaptation, this view of evolution seems very limited and unimaginative. Again it should be noted that the mutation-equilibrium theory was developed in an effort to get a better fit than had been possible with the neutral mutation-random drift theory between the predictions from theory and the primary data available, which are the distribution and frequency of electromorphs in natural populations.

HETEROGENEITY WITHIN ELECTROMORPHS

From the very outset (e.g., Hubby and Throckmorton, 1965; Lewontin and Hubby, 1966), it was realized that the electrophoretic technique was unable to detect all of the genetic variation in a population because a number of amino-acid substitutions may occur in a protein that do not have a detectable effect on net charge. The actual proportion detected was unknown, but Lewontin and Hubby originally estimated that they were detecting about one-half of the amino-acid substitutions. Later, Lewontin (1974) estimated the proportion to be even lower, about one-third, and King and Wilson's estimate (1975) of 0.27 was still lower. Despite the realization that there must be a considerable amount of cryptic genetic variation not revealed by electrophoresis and the fact that the estimates of the fraction detected seemed to get smaller as time passed, the data on electromorph distributions and frequencies both within and between species were used to draw some rather far-reaching conclusions. The discovery of many new alleles through additional screening techniques was rather disconcerting, for it raised questions not only about the validity of the conclusions but even about the validity of the data on which they were based. Throckmorton (1977), for example, headed a section dealing with these discoveries "Twelve years of work down the drain?" Another section title, "The bandwagon era," gives some indication of the enormous amount of work done during these twelve years. However, as Lewontin showed in his 1974 review of the subject, the data on genic polymorphisms revealed by gel electrophoresis did not fit either the neutralist or the selectionist theories comfortably and, in fact, were contradictory to these theories in one way or another. With the discovery of the wealth of additional alleles, the possibility had to be faced that all the previous data were wrong or at least seriously misleading.

Among the first to demonstrate the heterogeneity within electromorphs were Gibson and Miklovich (1971) and Bernstein, Throckmorton, and Hubby (1973). Gibson and Miklovich found evidence in *D. melanogaster* suggesting variation in enzyme activity and thermolabile properties both within and between electrophoretic forms of alcohol dehydrogenase. Bernstein, Throckmorton, and Hubby reexamined the amount of genetic variation at the xanthine dehydrogenase locus in the *D. virilis* group of species by using heat denaturation of the proteins as well as electrophoresis. Where 11 electromorphs had been revealed by electrophoresis alone, the number of detectable differences increased to 32 thermoelectromorphs. Moreover, the degree of genetic similarity among these species, previously thought to be high, was markedly reduced. A comparable study of the octanol dehydrogenase locus in the same group of species (Singh, Hubby, and Throckmorton, 1975) revealed that three different electrophoretic allozymes gave rise to 18 different "alleles" with heat denaturation. Moreover, in three of four species in which the locus was thought to be monomorphic, it turned out to be polymorphic when the added test was used. Therefore, the use of one additional test revealed a considerable amount of hidden genetic variability within electrophoretic allozyme classes. The heat denaturation studies revealed that the electrophoretic techniques underestimated the number of genetic variants or "alleles" and the degree of heterozygosity and polymorphism, but overestimated the similarity of "alleles" within and between populations.

Perhaps the best illustration of the need to reassess the earlier work on electrophoretic variants is the study of *D. pseudoobscura* by Singh, Lewontin, and Felton (1976). Their study of xanthine dehydrogenase used, in sequence, four different electrophoretic conditions and a heat stability test. These five tests revealed a total of 37 "alleles" in 146 genomes tested where previously only 6 "alleles" had been detected under the standard conditions for electrophoresis. A parallel study by Coyne (1976) of xanthine dehydrogenase in the sibling species, *D. persimilis*, revealed 23 "alleles" where only 5 "alleles" had been identified previously. Coyne, however, used only the four electrophoretic criteria of Singh, Lewontin, and Felton and not the heat denaturation test so that, presumably, still more "alleles" might have been discovered in *D. persimilis* if the additional test had been used.

The distribution of *D. pseudoobscura* ranges over much of western North America, from British Columbia to Guatemala and from the Pacific Ocean to Texas. In addition, an isolated population has been found in the South American Andes in the vicinity of Bogotá, Colombia, some 1,500 miles from the range of the rest of the species. In 1969 Prakash, Lewontin, and Hubby reported on a study of genetic variation at 24 loci in *D. pseudoobscura* from central (California), marginal (Colorado, Texas) and isolated (Bogotá) populations. Despite the distance between populations, they reported a remarkable similarity of allelic frequency distributions in these widely separated populations, which they interpreted as evidence for some form of balancing selection in the maintenance of this genic variation. They found the Bogotá population to have only about one-third of the genic variation of the other populations, and this depauperate variation of the Bogotá flies was notable in that the commonest, or the only alleles in Bogotá, with one exception, were also the ones with the highest frequency in the North American populations. They interpreted this to mean that the Bogotá population stemmed from a recent colonization by a small number of individuals—so recent that mutation had not yet had time to build up the number of alleles again. More specifically, Lewontin (1974) postulated that *D. pseudo-*

obscura must have reached the Bogotá area not much before 1960, a very recent invasion indeed.

Prakash (1972) had reported reproductive isolation between the flies from Bogotá and flies from California, Colorado, and Guatemala. In crosses between Bogotá females and males from the other three areas (but not in the reciprocal crosses), sterile F_1 hybrid males were produced. Such a finding suggested very rapid evolution of reproductive isolation if the population in Bogotá had only reached there just before 1960, as Prakash also inferred. Furthermore, it meant that reproductive isolation was evolving in a population that did not appear to differ greatly genetically from the other populations except in having a lower level of genic variability (Prakash, Lewontin, and Hubby, 1969).

On the basis of his genetic analysis of the hybrid sterility in the Bogotá-North American hybrids, Dobzhansky (1974) proposed an ancient origin for the Bogotá population rather than a very recent one, as had been suggested by Prakash, Lewontin, and Hubby (1969) and Prakash (1972). Because of the incipient reproductive isolation and because they found that the Bogotá population differed from the rest of the species at about 20% of 44 loci tested, Ayala and Dobzhansky (1974) named the Bogotá population a separate subspecies, *D. pseudoobscura bogotana,* distinct from the nominate subspecies *D. p. pseudoobscura* found in the rest of the range.

The work of Singh, Lewontin, and Felton (1976) not only revealed many additional alleles at the xanthine dehydrogenase locus when further tests were used, but also forced them to revise their previous conclusions about the origin and composition of the isolated Bogotá population of *D. pseudoobscura.* The Bogotá flies had previously been reported to be monomorphic for the *Xdh* allele that was most common in the rest of the species. Instead, the added tests revealed that the Bogotá population was polymorphic at this locus, and, even more significant, it was polymorphic for some unique alleles found nowhere else in the species. Thus, the added tests brought the electrophoretic data more in line with the evidence from other sources, and the isolated Bogotá population appears to merit its status as a separate subspecies.

In a study of the morphologically very similar sibling species, *D. pseudoobscura* and *D. persimilis,* Prakash (1969) found striking similarities in allelic frequencies at 24 loci. Thus, even though these species are ecologically different and are reproductively isolated, they did not appear to show much differentiation at the molecular level. Lewontin (1974) pointed out that these species showed little or no differentiation in gene frequencies at 88% of the loci tested, clear differentiation at 8%, and some differentiation at 4% of the loci. These findings seemed anomalous, for the two species, *D. pseudoobscura* and *D. persimilis,* appeared to differ at only 25% of their loci whereas two subspecies of a single species, *D. p. pseudoobscura* and *D. p. bogotana,* differed at about 20% of the 44 loci tested.

Coyne's (1976) reanalysis of *D. persimilis,* undertaken in conjunction with the study by Singh, Lewontin, and Felton (1976) of *D. pseudoobscura,* has cast new light on the relationship between these species. With the added tests Coyne showed that instead of being quite similar, the interspecific distribution of alleles became almost completely disjunct. "Alleles" in *D. persimilis,* formerly thought to be homologous to those in *D. pseudoobscura,* turned out to be unique to *D. persimilis.* Therefore, estimates of genetic identity or genetic distance based on a single electrophoretic test may be very wide of the mark.

Several other studies have been directed toward the detection of additional alleles by using discriminating tests other than routine electrophoretic screening. Heat denaturation and sequential electrophoresis with different gel concentrations and pH's have already been mentioned in relation to the work of Bernstein, Throckmorton, and Hubby (1973), Singh, Hubby, and Throckmorton (1975), Coyne (1976), and Singh, Lewontin, and Felton (1976). Johnson (1976b) used gel sieving and Milkman and Koehler (1976) employed isoelectric focusing as still other methods to reveal additional genic variation in *Colias* butterflies and the bacterium *Escherichia coli,* respectively. The papers by Singh, Lewontin, and Felton (1976) and Johnson (1977a) outline still other methods for the detection of additional variants.

Using heat denaturation, Cochrane (1976) increased the number of known alleles at the esterase-6 locus in *D. melanogaster* from four to seven. Cobbs (1976), McDowell and Prakash (1976), and Cobbs and Prakash (1977a) reinvestigated the esterase-5 locus in *D. pseudoobscura, D. persimilis,* and *D. miranda* with a battery of tests and considerably increased the known number of alleles at this locus. In *D. pseudoobscura,* for instance, McDowell and Prakash identified nine alleles where only two had been recognized previously.

The number of known alleles at a locus did not always greatly increase when added tests were used. Coyne and Felton (1977), for example, analyzed the alcohol dehydrogenase-6 (*Adh-6*) locus and the octanol dehydrogenase (*Odh*) locus in *D. pseudoobscura* and *D. persimilis.* The number of *Adh-6* alleles was increased from eight to eighteen, but the number of known alleles only increased from nine to eleven at the *Odh* locus, which is largely monomorphic. Thus, the *Adh-6* locus, like the xanthine dehydrogenase locus (*Xdh*) discussed previously, revealed considerable added heterogeneity with added tests, but the monomorphic *Odh* locus did not. Furthermore, Beckenbach and Prakash (1977) failed to find any additional alleles at the hexokinase-1 (*hex-1*) locus in *D. pseudoobscura* and *D. persimilis,* in stark contrast to the highly polymorphic *Xdh* and *Est-5* loci in these species. Therefore, additional screening tests do not necessarily produce a proportional increase in the number of alleles known at different loci. Each locus must be tested individually with a battery of tests to determine the degree of heterozygosity at that locus.

The mutation-equilibrium theory was framed in terms of the stepwise mutation model or unit-charge model of King and Ohta (1975). The unit-charge model has been questioned by Johnson (1977a, 1977b) because the size, the conformation, and the isoelectric point of a protein as well as its net charge will all influence the protein's mobility in an electric field. Cobbs and Prakash (1977b) seemed to feel that the unit charge model might be a useful first approximation for use in studies of polymorphism at a locus; but they, too, found indications that neutral amino-acid substitutions could be detected by electrophoresis. Johnson also suggested ways to investigate the catalytic properties of enzymes that could reveal additional hidden variability at a locus.

At present, a new era appears to be at hand in which more sophisticated techniques will be used to obtain a better approximation of the actual amount of genic polymorphism at a locus. Although it is not possible to predict just where this information will lead, it seems clear that at most loci the known number of alleles will increase. Moreover, the unit-charge model, with its implications of fixed decreases in fitness with each step-wise mutation away from the central optimum allele, seems overly simple compared with the realities of variation being revealed by the new techniques. However, the discovery of this additional variation in some ways merely compounds the puzzle. The attempts to account for the presence of a limited number of electromorphs generated considerable controversy between the neutralists and the selectionists. Now that the number of alleles at many loci has been demonstrated by added tests to be far greater than was previously known, the problem of accounting for the presence of so much genetic variability may be even more perplexing, but at least the true dimensions of the problem seem finally to be emerging.

CURRENT TRENDS

Some indications of the trends in thought following the discovery of added numbers of alleles with additional tests can be gleaned from some recent papers. Bernstein, Throckmorton, and Hubby (1973), King (1974), and Singh, Hubby, and Throckmorton (1975) all suggested that both selective and random processes are operating simultaneously in most populations and that both are responsible for the observed distributions of allele frequencies in natural populations. It should be noted that this is the position that has been consistently advocated by Sewall Wright since the 1930s. In these recent papers there appears to be a tendency to believe that some fraction of the alleles is undergoing selection while the rest are neutral—in other words, that alleles have fixed s values rather than values that vary with the conditions. The clarification of the relationship between this wealth of variation at the molecular level and fitness is going to require studies both in the laboratory and in the field, very much in the tradition of ecological genetics.

Lewontin, Ginzburg, and Tuljapurkar (1978) finally have concluded that heterosis is an unlikely type of balancing selection to maintain high levels of genic polymorphism. Heterosis has always had considerable appeal to theoreticians apparently because the relative fitness of the heterozygotes and the corresponding homozygotes is rather easy to determine. If heterozygote advantage is found under one set of conditions, it is generally assumed that it will exist under other conditions, which may or may not be true. Nonetheless, the assumption focuses attention on the fitness of the genotypes relative to one another and minimizes the relationship between fitness and the environment. Despite the popularity of heterozygote advantage as the form of balancing selection responsible for genetic polymorphisms, this explanation has always been suspect because of the intolerable genetic loads supposedly generated. Now, at last, it appears that other forms of balancing selection such as frequency dependent selection, density dependent selection, and selection varying over space or time may be investigated as alternative means of maintaining balanced polymorphisms.

In the past, the theories attempting to account for genic polymorphism have generally assumed that the populations are in equilibrium, presumably to simplify the treatment to some extent. However, given the range of fluctuations in the physical and biotic conditions that confront just about any population, it seems unlikely that many populations ever achieve an equilibrium in gene frequencies. This fact was recognized by Li (1978) in his modified version of Kimura and Ohta's neutral mutation hypothesis, for a major difference in his theory is that he did not assume that large natural populations are in a stable mutation-selection equilibrium.

One of the assumptions of the work with electrophoresis was that the loci being tested were a random sample of the genes in the genome. In fact, the loci being tested were structural genes, and, for the most part, structural genes that coded for enzymes rather than for structural proteins. It is clear that structural genes, especially in higher organisms, comprise only a small proportion of the total genome, probably less than 10% of the total DNA; the rest of the DNA appears to be regulatory in function. Since all the loci being tested, both with the standard electrophoretic techniques and more recently with the added discriminatory tests, fall in the category of structural genes, it seems legitimate to ask whether these techniques are in fact testing a random sample of loci for the entire genome.

Wilson (1975), apparently in an effort to reconcile the differences between the neutralists and the selectionists, suggested that "regulatory mutations play the major part in adaptive evolution," and that "evolution at the organismal level may depend primarily on regulatory mutations which alter patterns of gene expression." In other words, mutations in the structural genes may be adaptively neutral, but the adaptations observed at the organismal level are due to mutations in the regulatory DNA. Although it is generally agreed that structural genes code for messenger RNA, which in turn is translated into polypeptides or proteins, the definition of regulatory genes is less clear. If regulatory genes are those that control or modify the activity of other genes, then structural genes and regulatory genes are not mutually exclusive classes, for some genes are both structural and regulatory. An example is the gene coding for an effector polypeptide that regulates the transcription of other genes. Furthermore, this simple scheme does not allow for the genes coding for ribosomal RNA and transfer RNA, which may also have regulatory functions.

Having concluded that adaptive evolution in morphology, physiology, or behavior at the organismal level depends on "regulatory mutations" rather than mutations in the structural genes, King and Wilson (1975) further concluded that molecular evolution and morphological evolution may proceed at independent rates. As proof of this conclusion they offered a comparison of evolution in humans and chimpanzees. They pointed out that at the biochemical level, the proteins and nucleic acids of humans are remarkably similar to those of chimpanzees, more so than is true for most pairs of species within a single genus, whether of vertebrates or invertebrates. However, humans and chimpanzees are classified taxonomically not just in separate genera but in separate families.

King and Wilson (1975) attributed this taxonomic separation to morphologists who perceived the morphological differences between humans and chimpanzees as being quite large. However, this particular pair of species is atypical and, thus, seems a rather poor choice to illustrate a general concept. In the first place, the taxonomy of the superfamily Hominoidea is rather a mess, primarily because of the tendency to place specimens or populations in higher taxonomic categories than would be the case if they were in groups less closely identified with humans (Mayr, 1950). This anthropocentric effect is understandable; nonetheless, it probably inflates the taxonomic differences recognized in the Hominoidea. Furthermore, King and Wilson (1975) apparently believe that a genus or a family in one group of animals is directly comparable to a genus or a family in a different group, but this can hardly be so. All the taxonomic categories except the species are the subjective creations of taxonomists, formed for their convenience in dealing with the groups of species they are studying. Since there are no absolute standards for recognizing a genus or a family, comparisons of genera or families in different taxonomic groups are clearly invalid. As will be seen in the next chapter, species have some degree of objective reality if they are defined in terms of their reproductive behavior in a nondimensional system, but the subjective nature of other taxa makes the comparison of, say, genera of frogs with genera of mammals a highly questionable procedure. Therefore, the discovery that the percent sequence difference in DNA between humans and chimpanzees, which are classified in different families, is 1.1%, compared to 5% in two mouse species in a single genus, *Mus,* 12% in two frog species of the genus *Xenopus,* and 19% in two species of *Drosophila,* simply suggests that, on this evidence, humans and chimps do not really deserve to be placed in separate families, perhaps not even in separate genera, although they are certainly behaving as separate species.

To the tongue-in-cheek suggestion (Merrell, 1975b) that humans and chimpanzees are really quite similar, Cherry, Case, and Wilson (1978) responded with an extensive study that compared the morphological differences between humans and chimpanzees with the morphological differences between different taxa of frogs. Their conclusion was that the morphological differences between humans and chimpanzees, measured with the frog criteria, were greater than those between any of the frog taxa compared, the greatest of which were suborders.

Although they regard their results as demonstrating in an objective way that humans and chimpanzees really are quite different morphologically, they totally ignore the possible effect of differences in mode of locomotion on their measurements. All the frog species are amphibious, adapted to both aquatic and terrestrial life. Moreover, their manner of swimming in the water, a form of breaststroke, and of hopping on land, a saltatorial mode of locomotion, are similar adaptations for comparable modes of life. In short, their morphology reflects similar adaptive compromises that enable them to move about both on land and in the water.

However, the chimpanzees and other primates are primarily arboreal and move through the trees by brachiation. Humans have diverged from their arboreal ancestors and have become aberrant terrestrial primates. Their erect posture and bipedal locomotion represent a major adaptive shift away from their immediate primate relatives and ancestors, which is reflected throughout the human skeleton. Therefore, the differences in the skeletal proportions between humans and chimpanzees are not at all surprising. Perhaps the most obvious conclusion from this study is how fickle a guide morphology can be in systematics.

The same can be said of studies of variation at the molecular level. With the advent of molecular biology, DNA phylogenies, electrophoretic phylogenies, immunological phylogenies, and phylogenies based on the amino-acid sequencing of proteins have all been constructed. Although they provide new, useful, and often enlightening information about the relationships among organisms, they should not be regarded as infallible guides to phylogeny. The use of a single criterion of any kind, whether it be morphology or genetic distances based on molecular data, may be seriously misleading. Apart from erroneous conclusions based on deficient methodology, such as have been revealed as the electrophoretic techniques have been refined, are other possible problems. One is that comparisons of present-day protein or DNA sequences reveal only the existing amino-acid differences or nucleotide substitutions and provide no information about the number of intermediate steps that may have intervened between the ancestral molecule and the molecules now being compared. Because there is no molecular fossil record, these estimates of differences must be regarded as minimum estimates. Another question, seldom raised, is, To what extent are the observed similarities in molecular structure due to similarities in function rather than to common ancestry? In other words, Is convergent molecular evolution possible or likely so that similar molecules are analogous rather than homologous?

Given these possibilities, it is, perhaps, not surprising that anomalies and inconsistencies have appeared in molecular phylogenies. For example, the immunological distance between the Old World monkeys and the apes and humans appears to be greater than that between the Old World and New World monkeys, which is contrary to all the other evidence (Wilson and Prager, 1974). Similarly, a phylogeny based on the amino-acid sequences in cytochrome *c* shows humans and monkeys diverging from the mammalian evolutionary line before the kangaroo, a marsupial mammal (Fitch and Margoliash, 1967). However, all other evidence clearly indicates that the placental mammals, including the primates, are all more closely related to one another than they are to the marsupials.

The truth is, of course, that similar sorts of anomalies may show up in studies at any level from the morphological and behavioral to the biochemical. The most judicious course to take in constructing phylogenies is to use all the available evidence—biochemical, cytological, morphological, and behavioral—plus any other relevant information.

These comments should not be construed as denying King and Wilson's (1975) point that biochemical evolution and morphological evolution may proceed at independent rates. In humans and chimpanzees, the morphological evolution seems to have progressed faster than the biochemical evolution, undoubtedly because of the selection pressures involved in the transition from brachiation to erect bipedal locomotion in humans. They could have strengthened their argument even more by citing the case of the sibling species *D. pseudoobscura* and *D. persimilis,* which are almost indistinguishable morphologically yet have diverged from each another markedly at the biochemical level (Coyne, 1976; Singh, Lewontin, and Felton, 1976; Coyne and Felton, 1977).

It is interesting that the biochemists seem to give greater weight to morphological evidence than to biochemical evidence on the relationship between humans and chimpanzees, for Cherry, Case, and Wilson (1978) interpret their morphological study to support the taxonomists' judgment that humans and chimpanzees belong in separate families. Their own biochemical data show that the genetic distance between humans and chimpanzees is about the same as that between sibling species in a wide variety of organisms, including fruit flies, horseshoe crabs, salamanders, lizards, fishes, bats, and rodents, and, thus, support the notion (Merrell, 1975b) that humans and chimpanzees may not be so different after all. If the comparison of humans and chimpanzees holds any lesson, it is that marked evolutionary changes can occur unaccompanied by any comparable change at the level of the structural genes and proteins. The knowledge that the percent sequence difference in DNA is only 1.1% between humans and chimpanzees tells one very little about the nature of the differences between chimpanzees and humans.

The two central problems in evolution are adaptation and speciation. In the past, biologists have tended to think of adaptation in terms of physiological, morphological, and behavioral traits. The study of adaptation, even under the best of circumstances, is never easy, for it is difficult to measure the effect of particular traits on reproductive fitness. The adaptive significance of alternative physiological, morphological, or behavioral traits can usually be inferred from the nature of the traits; thus, appropriate experimental tests of their effects on fitness can be devised. However, inferences about fitness at the molecular level are usually difficult to make because the functional role of the molecule in the organism is not well understood. In the case of alcohol dehydrogenase in *Drosophila,* the role of the enzyme in the life of the flies has been identified, and the adaptive role of different alleles at this locus has been demonstrated. But the same cannot be said for many of the other enzymes identified by electrophoresis. Although the controversy between the neutralists and the selectionists has not yet been resolved, the differences between the two positions seem to be gradually diminishing. Perhaps the controversy will be settled only when one is able to think about and study adaptation at the molecular level in the same fashion one now thinks of physiological, morphological, and behavioral adaptations. No matter what the outcome, the controversy has stimulated a remarkable amount of research and thought into the nature of the evolutionary process. The research has revealed a hitherto unsuspected wealth of genetic variation in natural populations. The thinking has brought about a reassessment of some of the fundamental tenets of evolutionary thought, which, until about a decade ago, seemed to be hardening into dogma. The other central problem in evolution — speciation — will be addressed in the next chapter. With speciation, too, the dogma of the recent past is being challenged.

CHAPTER 14

The Species Concept

SPECIES DEFINITIONS

The full title of Darwin's major work was *The Origin of Species by Means of Natural Selection or The Preservation of Favoured Races in the Struggle for Life*. In the second chapter, "Variation under nature," he discussed the problems involved in distinguishing between individual differences and varieties, between varieties and subspecies, and between subspecies and species, and then pointed out that no clear lines of demarcation had as yet been drawn among these different forms. He used this as an argument favoring the idea that species are not immutable creations, that these different forms are different stages in the process that ultimately may give rise to distinct species.

He used such terms as variety, breed, race, subspecies, geographical race, species, and even polymorphic in much the same sense that they are used today. Although he wrote about the origin of species, the definition of a species gave him trouble. For example, he wrote, "Nor shall I here discuss the various definitions which have been given of the term species. No one definition has satisfied all naturalists; yet every naturalist knows vaguely what he means when he speaks of a species." Nonetheless, three pages later he referred to "good and true species" as if he did, in fact, have a clear conception of what a species is. This impression was weakened in the next paragraph where he wrote, concerning doubtful cases, "Hence, in determining whether a form should be ranked as a species or a variety, the opinion of naturalists having sound judgment and wide experience seems the only guide to follow. We must, however, in many cases, decide by a majority of naturalists, for few well-marked and well-known varieties can be named which have not been ranked as species by at least some competent judges." This is a most interesting statement. In doubtful cases, the taxonomic status of a group should be determined by an appeal to authority, and since any one authority is apt to be unreliable, the best procedure is to determine taxonomic status by majority vote. Even more remarkable is the fact that the situation has not improved much since Darwin wrote those words more than a century ago. Biologists still have difficulty in formulating a satisfactory, workable definition of a species. That a concept so central to biological thought and so crucial to biological research should still be rife with uncertainties and ambiguities is a commentary on the difficulty of the problem. It is crucial to research because, whatever the nature of the experiment, it is essential that the experimenter identify the species being used if others are to validate the work. At the same time, this difficulty remains as one of the best evidences that species are not fixed, static units, but are dynamic entities in a state of flux. In other words, the difficulties of the taxonomist stem from and are evidence for the evolutionary process.

It is noteworthy that many biologists still subscribe to the idea that a species is what a competent taxonomist considers to be a species. The problem then becomes to classify taxonomists rather than organisms, which may or may not be simpler—for, as Darwin pointed out, even the experts disagree. In any case, an appeal to authority seems alien to the nature of scientific work. Furthermore, it suggests that species are the subjective creations of taxonomists, which is true of other taxonomic categories both lower and higher than the species, but is not altogether true of the species.

At the other end of the spectrum of species definitions is that of Wilhelmi (1940), who states, "'Species' of helminths may be defined tentatively as a group of organisms the lipid-free antigen of which, when diluted 1:4000 or more, yields a positive precipitin test within one hour with a rabbit antiserum produced by injecting 40 mg. of dry-weight, lipid-free antigenic material and withdrawn ten to twelve days after the last of four intravenous injections administered every third day." This definition is quantitative, precise, and objective, but just how meaningful is open to question.

In the past, the origin of species posed no great intellectual challenge. Species were thought to have originated by special creation. If that were the case, then each species must be a distinct entity, and if there were difficulties in distinguishing among them, the fault lay with the biologist rather than with the material being studied. Modern taxonomy dates from the tenth edition (1758) of *Systema Naturae* by Linnaeus, a Swedish botanist. Early in his career, Linnaeus believed in the fixity of species, and his binomial system of nomenclature started from the premise that each species could be described and given a binomial that would distinguish it from all similar groups. Taxonomy had its origins in the attempts to classify local faunas and floras. In a local, essentially nondimensional system, species distinctions are usually rather clear-cut. However, as information begins to accumulate from a wider geographical area, the clear-cut distinctions begin to be blurred by geographical variation. As Linnaeus's fame spread, he became familiar with material from many parts of the world. It was probably this broadened experience that led him, later in life, to modify his belief in the fixity of species, at least to the point of recognizing the possibility of change within a genus.

Not only was special creation widely accepted, but belief in the spontaneous generation of living organisms persisted into the nineteenth century. It is difficult to know to what extent this belief influenced thought about the origin of species, but it seems probable that a belief in spontaneous generation would make the problem of the origin of species seem relatively unimportant. If living organisms could be formed spontaneously at almost any time or place, one would not tend to puzzle very long over the origin of species. It is interesting that Pasteur's experiments, which finally put the quietus on the ancient concept of spontaneous generation, were completed less than a decade after the publication of *The Origin of Species*.

The concept of evolution had a dramatic impact on the rationale used in the classification of organisms. *Taxonomy* has been defined as the theory and practice of naming, describing, and classifying organisms. *Systematics* has been defined as the study of the kinds of organisms, of the similarities and variations within and between groups of organisms, and of the relationships among them. Although systematics and taxonomy are often used as synonyms, the distinction between them is worth retaining—The rationale in taxonomy is classification, in systematics, phylogeny. Thus Darwin's concept gave impetus to the development of modern systematics, which is devoted not just to classification, but to a classification that reflects the evolutionary relationships among organisms.

In the past taxonomists relied primarily on morphological similarities and differences among organisms to reach taxonomic decisions. It is not surprising, therefore, that morphological species definitions have been suggested. It is surprising, however, that the ultimate in morphological definitions was given by A. H. Sturtevant, a geneticist as well as a *Drosophila* taxonomist, who wrote in 1942, "Distinct species must be separable on the basis of ordinary preserved material. This is in order to make it possible for a museum man to apply a name to his material." Morphological criteria are the only ones available to paleontologists and are still used in alpha taxonomy—that is, in the initial descriptive phase of the study of poorly known living groups.

For years the *type concept* dominated taxonomic thought. This postulates that all members of a taxon conform to a single type, and thus tends to exaggerate the constancy and discreteness of the taxon and minimize the variability present. The type concept, of course, was based on morphology, and when individuals different from the type were discovered, there was a tendency to tack a name on them, thus creating a new species. For example, in 1922 Weed described two new species of frogs, *Rana burnsi* and *R. kandiyohi*, because they looked so different from other frogs in the U.S. Upper Midwest where they occurred. Only decades later were the unspotted Burnsi type and the mottled Kandiyohi type shown to be due merely to two different, unlinked, dominant genes in the familiar and common leopard frog, *R. pipiens*. Hence, *R. burnsi* and *R. kandiyohi* were not valid species at all.

This tendency to confer species status on anything out of the ordinary and the prestige for a taxonomist in the discovery, description, and naming of a new species have probably led to the creation of a number of invalid species. In birds, for instance, as the groups have become better known, the number of species has been reduced from approximately 25,000 to 8,000. The trend toward reducing the number of species is healthy, for it usually helps to bring some degree of order out of the chaos of variation and at least simplifies the number of names that must be mastered. However, the *lumpers* are not necessarily any more infallible than the *splitters*, for cases are known in *Drosophila*, for example, where 28 distinct species were recognized in what was originally thought to be one "good" species, *D. repleta*.

The change, more than any other, that separates modern from classical taxonomy is the transition from the type concept to a population approach in taxonomic work. Rather than thinking in terms of an idealized type representative of an entire species, taxonomists now realize that they are dealing with variable populations and must, if they are to grasp the essential characteristics of these populations, study samples from them to determine means and variances for the traits of interest.

Despite this shift in approach, taxonomists still designate type specimens in their work. However, the function of the type has changed. Repeatedly in the history of taxonomy, the material on which the original description of a species was based is later discovered to contain several species. Modern taxonomists select a single specimen from the individuals in the original sample and designate it as the type specimen or *holotype*. Then if this material is later found to contain several species, the type specimen can be reexamined and the name given by the original author will be retained for this individual and the species to which it belongs. Thus the type specimen functions as a name bearer and is not really typical of anything. It is useful in averting problems in nomenclature, but has no descriptive significance. The uselessness of the type specimen for descriptive purposes is easily demonstrated. Suppose you were asked to designate a type specimen for *Homo sapiens*. Would you choose a child or an adult, a male or a female, an Eskimo, an Amerindian, a Caucasian, a Mongoloid, a Negro? No matter what you chose, it would give no real indication of the diversity that is classified under the label *Homo sapiens*.

As pointed out above, morphological criteria are still used in some taxonomic work. However, another major change in modern taxonomy is the shift from morphological species definitions to what are known as *biological* species definitions. They are called biological because they rest, not on the degree of morphological similarity or difference, but on the biological criterion of reproductive isolation.

A sampling of biological species definitions will provide some insight into their nature. Mayr (1942), for example, wrote that "Species are groups of actually or potentially interbreeding natural populations, which are reproductively isolated from other such groups." In 1970 he gave a slightly, but significantly, different version. "Species are groups of interbreeding natural populations that are reproductively isolated from other such groups." Dobzhansky, one of the most influential in the development of the biological species concept, wrote (1951):

> Species are formed when a once actually or potentially interbreeding array of Mendelian populations becomes segregated in two or more reproductively isolated arrays. Species are, accordingly, groups of populations the gene exchange between which is limited or prevented in nature by one,

or by a combination of several, reproductive isolating mechanisms. In short, a species is the most inclusive Mendelian population.

The common denominator in these definitions is the emphasis on reproductive isolation as the criterion for distinguishing between species.

The biological species definitions, in one sense, are negative. Groups of organisms that do not interbreed with one another are to be considered as separate species. A more positive definition is suggested by the idea that a species is the most inclusive Mendelian population. Members of the same species have a common gene pool; they are bound together by bonds of mating and parentage. The gene pool concept implies that the species has continuity, that individuals are merely the temporary custodians of the genes from the gene pool of the species. In sexually reproducing species, each individual has a unique genotype drawn from the gene pool of the species, and returned to it when, and if, the individual reproduces. Individuals may come and go, but the gene pool lingers on.

Another implication of the quote from Dobzhansky is that species are a stage in the evolutionary process—that they are dynamic rather than static entities. In essence, the groups of populations will have a common gene pool and remain members of the same species until reproductive barriers become established that isolate some of the breeding populations from others. Once that point is reached, each reproductively isolated group of populations will pursue its own independent evolutionary course. At the point of fixation of the discontinuity between groups, they achieve the status of separate species and before that, they are races of the same species.

Consider what is meant by reproductive isolation. If, when members of two different species mate, the offspring are not viable, then clearly the species are reproductively isolated. If the offspring are viable, but sterile, again the species are isolated reproductively. Some have interpreted the biological species definition to mean that fertility is the criterion for determining species status. If the matings produce fertile hybrids, the groups must belong to the same species; if the hybrids are sterile, the parents belong to different species. However, the biological species definition is more subtle than this.

According to the definition, the question is not whether they can interbreed and produce fertile hybrid offspring, but whether, in fact, they do interbreed in nature. Even more fundamental is the question of whether the exchange of genes between the two groups is sufficient to lead to a breakdown in the isolation of the gene pools of the two species. Thus reproductive isolation does not have to be complete; an occasional hybrid may be formed in nature even between "good" species so long as the hybridization does not lead to a merger of the gene pools of the species. Ultimately, the question boils down to the relationship between selection and gene flow. If the gene-frequency changes stemming from gene flow are greater than those due to the selective pressures maintaining the individuality of each species, the species will not retain their separate identities, but will fuse. If the changes due to gene flow are less than those due to selection, the species will remain separate and distinct. Therefore, the mere knowledge that members of two groups can mate and produce fertile F_1, F_2, and backcross progeny is not proof that they belong to the same species. The true test is their behavior under natural conditions. If the genetic exchange in nature is minimal, the groups will retain their separate identities and will pursue independent evolutionary paths, and, hence, must be regarded as separate species despite their interfertility. This reliance on the behavior of the organisms themselves to determine whether they are members of the same or different species distinguishes the species from all the other taxonomic categories. All the others, both subspecific and supraspecific, are based on subjective judgments of taxonomists. The biological species, based on the behavior of the organisms themselves, is an objective natural unit, in contrast to subspecies, genera, families, orders, and so on.

Although the biological species definition is regarded as a major advance over previous types of definitions, it is not without drawbacks. The most obvious is that it can be applied only to sexually reproducing, cross-fertilizing species. It is not applicable to organisms that reproduce only asexually or to hermaphrodites with compulsory self-fertilization, for in such cases, each organism is reproductively isolated from all others. Although a variety of types of asexual reproduction exist (fission, parthenogenesis, apomixis, etc.), in many cases, both asexual and sexual reproduction occur in the same species. Sometimes there is a regular alternation of sexual and asexual generations, but in other cases, sexual reproduction leading to genetic recombination may occur sporadically in species that ordinarily reproduce asexually. In all these cases, the genetic composition of the populations resembles that of sexually reproducing species more than that of organisms that reproduce only asexually, and therefore the biological species definition is applicable. White (1978) estimated that only 1 in 1,000 known animal species has an exclusively asexual mode of reproduction. A similar estimate is not available for plants. However, a greater variety of asexual modes of reproduction is known in plants than in animals, and vast agamic complexes have developed in certain plant genera as the result of a combination of apomixis, hybridization, and polyploidy. Only a limited number of such complexes is known; furthermore, the members of a complex are often facultative apomicts so that the whole complex is still tied together by sexual reproduction. Thus, despite the widespread incidence of various forms of asexual reproduction in plants, asexuality is generally facultative rather than obligatory, and asexuality is secondarily derived from sexual forms, which can often still be identified. However, both Stebbins (1950) and Grant (1971) emphasized that neither apomicts, nor even facultative apomicts, have ever been able to evolve a new genus or even a new subgenus of plants. The same is apparently true in animals (White, 1978), where, with one possible exception, there is no large genus or higher taxonomic group in which all members reproduce asexually. Therefore, in the long run, asexuality appears to lead to an evolutionary dead end. It can be concluded that the biological species definition is applicable to the great majority of animal species, and Stebbins (1963) estimates it to be applicable to 70 to 80% of the species of higher plants.

Even though the biological species definition does not apply to uniparental organisms, species are still recognized in such groups. In these cases, what is recognized are not "biological species" but "taxonomic species" (Grant 1963, 1971) — that is, groups of organisms similar enough to be lumped together as a single species in practical, routine taxonomic work. The closest approach to the biological species concept possible with uniparental organisms is the "evolutionary species" (Simpson, 1961; Grant, 1971). Here, the biological basis for grouping individuals into species is the genotypic similarity among related individuals, which stems from a common ancestry. The use of electrophoresis to determine genetic similarities and genetic distances should provide a much better measure of relationships among asexually reproducing forms than has previously been available.

Another drawback to the biological species definition is its lack of any morphological criterion for separating species. For paleontologists or for taxonomists working with preserved specimens, the biological definition is useless. Of necessity, they must work with morphological traits. Paleontologists have no other choice; taxonomists, working with living species, may eventually be able to study reproductive isolation in their material, but usually their preliminary work is done with dead organisms, which can hardly be tested for reproductive behavior.

With the biological species definition, it is possible for two morphologically indistinguishable groups to be called separate species if they are reproductively isolated from one another. Probably the most dramatic confrontation between the biological and the morphological species definitions came when Dobzhansky and Epling (1944) named what had previously been known as *Race B* of *D. pseudoobscura* as a new species, *D. persimilis*, even though at the time no simple, reliable morphological differences between the two groups were known. However, it was known that the two groups rarely if ever hybridized in nature and artificial crosses produced sterile hybrid males and partially sterile females so that they clearly were reproductively isolated. Since they behaved as "good" species and were pursuing independent evolutionary paths, they were called separate species even though they were morphologically indistinguishable. Subsequently, Rizki (1951) found slight differences in the male genitalia by which the species of individual males could be determined, but females of the two species still cannot be identified by inspection. Additional cases of *sibling species* of *Drosophila* have since been identified (e.g., *D. melanogaster* and *D. simulans*, and *D. willistoni* and *D. equinoxialis*). The latter pair are still not separable morphologically even though they have diverged to the point where they will no longer cross in the lab.

Conversely, it is also possible for two morphologically quite distinct groups to be completely interfertile. If such groups are normally allopatric, for example, the plane trees or sycamores, *Platanus occidentalis*, of the eastern United States and *P. orientalis* of the eastern Mediterranean region, they are usually regarded as separate species even though the artificial hybrids between them are vigorous and highly fertile (Stebbins, 1950). This separation is probably reasonable because, apart from their geographical isolation, their ecological requirements are thought to be so different that they would rarely, if ever, exchange genes under natural conditions. More difficult is the case of *Catalpa ovata* of China and *C. bignonioides* of the eastern United States, which are completely interfertile and have similar ecological requirements (Stebbins, 1950). However, they remain geographically isolated and are morphologically quite distinct so that their status as separate species is unchallenged. Where the ranges of such species overlap, the problem becomes more complex. Among the oaks, for example, hybridization has often been reported. The blackjack oak, *Quercus marilandica*, and the scrub oak, *Q. ilicifolia*, of the black oak subgenus retain their distinctive morphological and ecological characteristics throughout much of the zone of overlap, but extensive hybridization, both F_1 and backcross hybrids, was found in the New Jersey Pine Barrens (Stebbins, 1950). Among the white oaks, widespread intergradation between the swamp white oak, *Q. bicolor*, and the bur oak, *Q. macrocarpa*, of the central United States has been reported (Stebbins, 1950). Although they are sympatric over most of their ranges, have similar ecological requirements, and hybridize extensively, they have retained their identities. In the junipers, the eastern red cedar, *Juniperus virginiana*, intergrades geographically with three other members of the genus: *J. horizontalis* to the north, *J. scopulorum* to the west, and *J. barbadense* to the south (Anderson, 1953). In this case, hybridization and intergradation are so extensive that these groups resemble geographic races of a single species rather than several separate species. In these examples drawn from plants, all the species differ morphologically, yet are interfertile. However, under the biological species definition, their taxonomic status is determined not by their degree of morphological resemblance or their relative fertility, but by whether or not they retain their separate identities under natural conditions.

This statement brings us to still another difficulty with the biological species definition: It is nondimensional. The presence or absence of reproductive isolation can be established only if members of both groups occur together in breeding condition at the same time and place. Mayr's earlier definition quoted above (1942) referred to "actually or potentially interbreeding" natural populations as did Dobzhansky (1951), but Mayr's later definition (1970) referred only to "groups of interbreeding natural populations." The reason for the change, of course, is that any decisions about the potential for interbreeding involve subjective judgments by biologists. If the biological species definition is to maintain its claim to being objective because it depends on the behavior of the organisms themselves rather than on the subjective judgment of taxonomists, its has to be purged of the subjective element introduced by the phrase *potentially interbreeding*. However, to do that is to limit the utility of the definition to sympatric populations whose breeding behavior can be observed under natural conditions. Such a limitation considerably restricts the application of the biological definition. One cannot, for example, be altogether certain that the robins, *Turdus migratorius*, of Minnesota would interbreed with the robins of Massachusetts. Their songs are slightly different, but one has no way of knowing whether this would affect their interbreeding because they do not normally come in contact with one another. Similarly, one cannot be altogether certain that the robins living in Minnesota today would interbreed with the robins living in Minnesota at the time of the Civil War. In both cases one infers that they would interbreed, but these are inferences rather than a valid application of the biological definition. Viewed in this light, the biological definition is not universally applicable. In a nondimensional system, the definition works. If the concept is expanded in space or time, difficulties start to develop that are inherent in the evolutionary process itself. Species are not static entities, but may change both in space and through time.

The fundamental question is: What does one wish to accomplish with the classification of organisms? Does one want a species definition of the sort suggested by Sturtevant that would permit the pigeonholing of specimens, or does one want one that will express, as nearly as possible, the natural biological relationships among the various groups? For most biologists, the latter is probably preferable. The biological species definition, despite its shortcomings, comes closer to that goal than any other. It should be added that, where such comparisons are possible, the "morphological species" and the "biological species" usually coincide because, apparently, morphology is a reflection of the physiology and functioning of the organisms. When populations have changed sufficiently to become reproductively isolated, the genetic changes leading to this divergence nearly always result in morphological changes as well.

By now it should be realized that no species definition will be completely satisfactory. The complications introduced by asexuality, hybridization, introgression, polyploidy, geographic variation, and various combinations of these factors all conspire to erode the concept of species as discrete entities, unique and distinct from one another. However, the difficulties are inherent in the material and stem from the evolutionary process itself. If the goal is to describe what is happening in natural populations, one must accept the fact that not all of them will fall into clear-cut species. This fact stems from the way evolution occurs. The first major process is what might be termed a *transformation in time*, in which a single lineage evolves into a series or succession of species through time, species A evolving into B, which in turn evolves into C, and so on. This type of evolution is also known as *phyletic evolution* or, more recently, *anagenesis*. In reality, of course, in a single lineage, there will be no sharp discontinuity between species A and species B. Unless some saltational form of evolution occurs, species A will undergo gradual evolutionary changes until eventually it becomes recognizable as species B. Any discontinuity between A and B ordinarily is due to a gap in the fossil record.

A second major evolutionary process may be thought of as a *multiplication in space*. In this case, a single ancestral species splits into two or more descendant contemporaneous species. This process is known as *speciation* in the limited sense of the word, or more recently, *cladogenesis*. The mechanism of phyletic evolution seems relatively straightforward. An ancestral species population, subjected to the combined effects of mutation pressure, selection pressure, and, possibly, random genetic drift, is apt to undergo evolutionary change through time unless very strong stabilizing selection acts to maintain existing gene frequencies. The mechanism of speciation, however, is not so simple. The fundamental problem in speciation is to explain how a freely interbreeding ancestral species population becomes separated into two or more reproductively isolated groups. The problem is two-fold: First, What causes divergence among populations within the same species? Second, How does reproductive isolation between such populations become established?

ISOLATING MECHANISMS

The orthodox explanation for speciation, which has been elevated almost to the level of dogma in the works of Mayr (1963, 1970), is that speciation is allopatric — that is, the same processes that give rise to geographic races may, if carried on long enough, give rise to separate reproductively isolated species. Populations of the same species that are physically separate from one another will pursue independent evolutionary paths. Even though initially the gene pools of the populations may be very similar, the physical and biological conditions in any two areas are unlikely to be exactly alike. Therefore, the selection pressure in physically or geographically isolated populations will be different, and the adaptive responses of the populations will lead to genetic divergence. In addition, both mutation and random genetic drift lead to divergence in finite populations so that all three forces tend to promote genetic divergence among isolated populations.

In the consideration of geographical races we saw that when previously isolated races of a species again come in contact with one another, gene flow between them occurs in a zone of secondary intergradation. However, in some such cases, gene flow is restricted;

hybridization no longer occurs freely. The populations that previously were unable to interbreed because they were physically isolated from one another and never had the opportunity to mate now no longer interbreed even when the opportunity becomes available. They are reproductively isolated by some mechanism other than mere physical separation. A number of such *isolating mechanisms* have been identified. Geographical isolation serves to isolate different populations, but it is not an isolating mechanism. Isolating mechanisms are under genetic control, and serve to reduce or prevent interbreeding between sympatric members of different populations.

First, the various kinds of reproductive isolating mechanisms will be described. A number of similar classifications of genetic isolating mechanisms have been proposed. The following classification outlines the major ways that populations are isolated from one another.

I. *Spatial (geographical) isolation.*
II. *Reproductive (genetic) isolating mechanisms.*
   A. *Premating and prezygotic isolating mechanisms* prevent crossing and the formation of hybrid zygotes.
      1. *Habitat (ecological) isolation.* The populations occupy the same general area, but potential mates do not meet because of differences in habitat requirements.
      2. *Temporal (seasonal) isolation.* The time of mating or flowering of the populations does not coincide.
      3. *Behavioral (ethological or sexual) isolation.* Potential mates meet but do not mate because courtship behavior patterns do not mesh.
   B. *Postmating but prezygotic isolating mechanisms* prevent fertilization and zygote formation even when mating is attempted.
      1. *Structural (mechanical or morphological) isolation.* Copulation in animals or pollen transfer in plants is prevented or restricted by differences in the structure of the genitalia or the flower parts.
      2. *Gametic incompatibility.*
         a. With *external* fertilization, gametes are not attracted to one another.
         b. With *internal* fertilization, gametes are weak or inviable in the female reproductive tract or style of alien species.

C. *Postmating and postzygotic isolating mechanisms* reduce the viability or fertility of hybrid zygotes.
   1. *Hybrid inviability or weakness.* F_1 hybrid zygotes are inviable or have reduced viability.
   2. *Hybrid sterility.* F_1 hybrids are viable but sterile.
      a. *Developmental.* F_1 hybrids are sterile because the gonads develop abnormally or meiosis itself is abnormal.
      b. *Segregational.* F_1 hybrids are sterile because chromosomal segregation is abnormal, the gametes containing abnormal, unbalanced combinations of chromosomes or chromosomal segments.
   3. *Hybrid breakdown.* F_1 hybrids are normal, vigorous, and fertile, but the F_2 or backcross hybrids have reduced viability or fertility.

A great deal has been written about reproductive isolating mechanisms, but we shall discuss only a few examples. Note that premating isolating mechanisms limit interspecific or interpopulation crosses and thus prevent the wastage of gametes on the production of inviable, sterile, or poorly adapted hybrids. Postmating mechanisms do not avert this wastage, hence seem a less efficient way to maintain reproductive isolation.

A simple example of *habitat isolation* exists in two species of spiderworts or *Tradescantia* (Anderson and Hubricht, 1938). *T. canaliculata* and *T. subaspera* occur sympatrically in the Ozark region of the central United States where *T. canaliculata* grows on rocky slopes in full sun and *T. subaspera* grows on rich soil in deep shade. Their habitat requirements are so different that the two species do not ordinarily interbreed even though their habitats may be so close that *T. canaliculata* grows at the top and *T. subaspera* at the base of the same cliff. Where the habitats are less sharply separated, and intermediate conditions prevail, some backcross hybrids have been found occupying the intermediate ecological zone. However, as is so often the case, more than one isolating mechanism helps to maintain the reproductive isolation between two species. In this case, the species are isolated not only by ecological isolation but also by a difference in flowering seasons and by partial hybrid sterility.

An interesting case of habitat isolation in animals has been described in the deer mouse, *Peromyscus maniculatus*, a widespread, complex species ranging from the Arctic Circle in Canada to southern Mexico. As many as 65 subspecies have been described that fall into two main groups, one occupying woodlands and the other, grasslands. In the Great Lakes region, two such subspecies coexist without interbreeding. The short-tailed, small-eared, pale prairie deer mouse (*P. m. bairdi*) lives in open grass-covered areas. The long-tailed, long-eared, dark woodland deer mouse (*P. m. gracilis*) inhabits forested areas. Even though immediately adjacent to one another (the populations are *parapatric*), the habitat preferences are so strong that the two subspecies retain their distinctive features and do not interbreed. Thus, in Michigan they are behaving like separate species. If it were not for connecting populations of this species to the south, west, and north that are intermediate in morphology and habits to these two subspecies, they would probably be elevated to full species status.

Temporal isolation may be a very effective isolating mechanism. In humans, reproduction is continuous; mating may occur and babies are born at any time throughout the year. In most species of plants and animals, however, the flowering season or the mating season is limited to certain periods of the year. If related species of plants flower at different seasons or related species of animals come into breeding condition at different times, the reproductive isolation between them may be essentially complete even though otherwise they are completely interfertile.

A number of cases of seasonal isolation have been reported. The Monterey pine, *Pinus radiata*, grows along the coast in central and southern California; the Bishop pine, *P. muricata*, lives along the coast of northern California. The ranges of these two species overlap in the vicinity of Monterey Bay, yet even though they are sympatric and wind-pollinated, they seldom hybridize because the Monterey pine, adapted to mild winters, normally sheds its pollen in early February, whereas the Bishop pine does not shed pollen until April (Stebbins, 1950). There are ecological differences as well, for the Bishop pine grows on exposed ridge crests, and the Monterey pine grows in the better soils surrounding the ridges. Occasional hybrids are found, but they constitute less than 1% of the stand.

Another species, the knobcone pine, *P. attenuata*, is found primarily in the middle and inner coast ranges and the Sierra Nevada mountains of California, but it, too, is sympatric with the Monterey pine in the Monterey Bay area. The species are ecologically isolated, the Monterey pine growing on the seaward side of the ridges and the knobcone pine on the landward side of the ridge crests; small numbers of hybrids, again less than 1% of the population, have been found in areas where the species are contiguous or parapatric. Here, too, seasonal isolation must be a significant isolating mechanism, for the knobcone pine, like the Bishop pine, does not shed its pollen until April, much later than the Monterey pine. The hybrids between the Monterey and knobcone pines are known to be vigorous and fertile; those between the Monterey and Bishop pines seem somewhat less so than the parent species. Curiously, Stebbins did not comment on whether or not hybrids had been found between the knobcone and Bishop pines, both of which shed their pollen in April in the Monterey Bay area. However, it appears that seasonal isolation serves as a major barrier to gene flow between the Monterey pine and the other two species.

The fact that hybrids exist despite the two-month difference in the usual time of pollen production suggests that seasonal isolation is not complete in these species. The probable reason is that year-to-year variations in the weather cause the pollination periods to overlap in some years so that hybridization can occur. Thus these species appear to retain their identities in the Monterey Bay area through a combination of seasonal isolation and habitat isolation.

That seasonal isolation alone might be an unreliable isolating mechanism is suggested by an experience in northern Minnesota. Apparently because of its long, harsh, winters, Minnesota has a depauperate amphibian fauna (Breckenridge, 1944). Among the species living in northern Minnesota are the wood frog, *Rana sylvatica*, the leopard frog, *R. pipiens*, the American toad, *Bufo americanus*, the spring peeper, *Hyla crucifer*, and the chorus frog, *Pseudacris triseriata*. The chorus frog starts breeding in early spring and the spring peeper somewhat later, their breeding seasons usually extending over a period of several weeks from late March into May. In contrast, the wood frog is an explosive breeder that completes mating and egg laying in a few days in very early spring. The wood frog is ordinarily the first amphibian to breed in Minnesota; the leopard frog breeds shortly thereafter and its season is over in about a week. Their breeding seasons normally end in late March or early April, depending on the weather. In contrast, the American toad does not start to breed until May. Thus, in a normal year, *R. sylvatica*, *R. pipiens*, and *B. americanus* have separate breeding seasons. Of course, there are those who would claim that in Minnesota, where summer has been defined as two weeks of lousy skating, there is no such thing as a normal year. A case in point is 1960, for after a late, cold spring, the temperature in northern Minnesota zoomed into the 80s (°F) on May 14. The transition was so sudden and the weather so warm that you could hear plants growing in the woods as they pushed their way up through the dry leaf litter. All five species whose breeding seasons are ordinarily quite different were found calling and breeding in the same pond on an island in Lake Vermilion. The unusual weather conditions had led to the telescoping of otherwise rather distinct breeding seasons into the same time period. In this case, there was no evidence of a breakdown in reproductive isolation, but the situation is different in the case of *B. americanus* and *B. fowleri*.

The American toad and Fowler's toad are widely sympatric in the eastern United States, but have retained their identity as separate species even though hybrids between them are fully viable and fertile (Volpe, 1952; Cory and Manion, 1955). These two species are ordinarily seasonally isolated, for in a given locality, *B. americanus* will breed earlier than *B. fowleri* because its temperature threshold for breeding is about 60°F and the threshhold for *B. fowleri* is slightly less than 70°F. (Cory and Manion, 1955). Cory and Manion also reported habitat isolation, for they collected typical *B. americanus* only in completely wooded areas and typical *B. fowleri* only in open fields. Despite the seasonal isolation (clearly shown in Cory and Manion) hybridization has apparently occurred frequently in natural populations (Volpe, 1952; Cory and Manion, 1955). Evidently, the clearing of the wooded areas of the eastern United States has so disrupted the habitats of these species that they now come in contact with one another and hybridize, the hybrid populations occupying the ecologically disturbed areas. This example again suggests that seasonal isolation alone is not an effective isolating mechanism.

It is possible that diurnal differences in the time of mating or flowering might give rise to temporal isolation. For example, most *Drosophila* species will mate in total darkness, but *D. subobscura* and *D. auraria* will not (Spieth, 1952), and the sibling species *D. pseudoobscura* and *D. persimilis* differ in their preferred hour for mating (Spieth, 1958). However, these differences apparently play little if any role in reproductive isolation, and other similar cases of diurnal differences in animals are not known to lead to reproductive isolation. On the other hand two species of *Oenothera* in the western American desert are apparently isolated by diurnal differences in flowering time (Raven, 1962). *O. brevipes* and *O. clavaeformis* grow and flower together and both are pollinated by solitary bees, yet they rarely hybridize. The flowers of *O. brevipes* open before dawn and are visited by bees that fly early in the morning. *O. clavaeformis* blossoms open late in the afternoon and are visited by bees that fly at dusk. The rarity of such examples in the literature suggests that diurnal differences in breeding behavior seldom act as effective reproductive isolating mechanisms or else that biologists have not given much attention to this possibility.

At the other extreme on the time scale from diurnal differences in reproductive behavior are the periodical cicadas, *Magicicada*. They are remarkable in having the longest life cycles known for insects. They spend either 13 or 17 years underground as nymphs sucking the juices from the roots of deciduous trees in the eastern United States. They are called *periodical* because in any one population the development of virtually all individuals is synchronized so that they are all the same age. The mature nymphs finally emerge from the ground, molt to become adults, mate, lay eggs, and die, all within the same few weeks. Members of not just one species but three different species of *Magicicada* have their emergence synchronized in this way so that in a given locality enormous numbers of cicadas emerge every 13 or every 17 years. Furthermore, each of the three species with a 17-year cycle has a sibling species with a 13-year cycle that is otherwise indistinguishable from it. The three species with the 17-year cycles have a more northerly distribution; the three with the 13-year cycles a more southerly distribution. While in any one locality the mass emergence of cicadas occurs only once every 13 or 17 years, in different regions different "broods" of periodical cicadas are out of synchrony and emerge in different years. A number of different broods have been identified and mapped. Since the synchrony is so strong, it is clear that the temporal isolation between members of the same species in different broods must be virtually complete. This is the most unusual case of temporal isolation yet reported. (For more on this fascinating group of species and the evolutionary questions they pose, see Alexander and Moore, 1962, and Lloyd and Dybas, 1966a, 1966b.)

Behavioral or ethological isolation is found primarily in animals. The most complex behavior is usually associated in some way with reproduction—for example, in courtship patterns, nest building, territorial behavior, care of the young, and so on.

Courtship usually involves a complex sequence of events. When a male and female encounter one another, one of them, usually the male, initiates the courtship with some action that serves as a signal and a stimulus to the female. If the signal is appropriate and the female is receptive, she will respond in some way that serves as a signal and a stimulus to the male. Having received a favorable or appropriate response, the male will continue to court the female, and together they will complete the courtship ritual with each action and reaction serving as both stimulus and response. If the male's signal is inappropriate or the female is not receptive, she will not respond appropriately to the male's initiative and may try to break off the courtship.

Courtship seems to have several functions. It establishes that the individuals are of the appropriate sex and species to mate and to produce viable, fertile offspring, and in some species it may also serve to establish individual mating preferences for particular partners. In addition, courtship arouses readiness to mate and synchronizes the mating behavior of the male and female. If courtship begins between males and females of different species or, as sometimes happens, between two males of the same species, it is usually broken off quickly because the stimulus-response patterns of the individuals do not mesh. The courtship beween male and female may involve visual, auditory, tactile, or chemical stimuli. The visual and auditory cues act at a distance and so may certain pheromones, for the molecules, even at extremely low concentrations, may be detected at considerable distance. In other cases, direct contact between male and female is essential during courtship so that tactile stimuli or taste or its equivalent are a necessary part of the ritual preliminary to mating.

Visual cues are clearly important to birds and fish where brightly colored males with distinctive species-specific colors or color patterns court the females with complex behavioral rituals and displays that emphasize the distinctive coloration. The flash code of fireflies is another instance where visual signals are of primary significance in bringing males and females of the same species together for mating (Carlson and Copeland, 1978). The pattern of male flashes and female responses is species-specific and serves as an isolating mechanism. Even more interesting is the aggressive mimicry of the females of some species. These *femmes fatales* mimic the photic responses of females of other species and thus lure unsuspecting males seeking only a mate to their doom, for the females pounce on them and devour them.

Auditory signals are part of the courtship repertoire of groups such as frogs and toads, birds, and some insects. Stridulation by male orthopterans (grasshoppers, crickets, etc.) is species-specific, for the females seek out only the stridulating males of their own species (Alexander, 1960; Walker, 1974). Similar behavior has been observed in the anura (Mecham, 1961; Blair, 1964; Littlejohn, 1969), although in this case, both males and females are attracted to a calling male. As a result, choruses of conspecific calling males assemble in the breeding ponds (Merrell, 1977). The romantic notion celebrated by poets of the male skylark or bobolink singing in the throes of ecstasy to his ladylove has been replaced by a more pragmatic view of the function of song in birds. There is no doubt that singing reaches its peak during the breeding season, but its function appears to be more complex than a simple outpouring of affection. It is, for example, a means by which a male announces his presence on a breeding territory, thus serving as a warning to other males as well as a signal to females, which are attracted to the calls of conspecific males (Carpenter, 1958). Here, too, the songs are species-specific, and in birds, as well as in insects and frogs and toads, the calls appear to serve as isolating mechanisms.

The *pheromones* are probably the most remarkable type of chemical stimulus involved in mating yet to be identified. Pheromones have been defined as biologically active substances secreted externally by an individual that causes a specific reaction in another individual of the same species. Two types of pheromones have been distinguished: *releaser* pheromones, which produce a more or less immediate and reversible change in the behavior of the recipient, and *primer* pheromones, which initiate a chain of physiological or developmental events in the recipient without necessarily triggering any immediate behavioral response. Releaser pheromones have been found to serve a variety of functions—foraging, alarm, care of the young, aggregation, dispersal, sex attractant, food exchange, oviposition, territorial marking, "aphrodisiac," and so on. Most of the work on pheromones has been done with insects and to a lesser extent with mammals, but pheromones are apparently ubiquitous in the animal kingdom with the possible exception of birds.

Sex pheromones have a role in reproduction and sexual isolation. Attractants emitted by female moths have been detected by males more than a mile away. The concentration of the pheromone is so low that it appears that males may be able to detect single molecules of the substance. Sexual dimorphism in the antennae of such species as the giant silkworm moth and the gypsy moth is quite striking. The antennae of the males are much broader and more finely subdivided than those of the females and are adapted to detect minute quantities of the pheromone. It has been shown (Wilson, 1963; Shorey, 1973) that a female moth can advertise her presence over a wide area with a minimum expenditure of energy. When a male moth detects the pheromone, he responds by flying upwind in the general direction of the female. Only after he gets fairly close to the females is there enough of a concentration gradient for him to move along the gradient to the female. The pheromone in this case actually causes a hierarchy of male responses. First, the male is aroused from a resting state. Then he starts to fly, first upwind and then along the concentration gradient. The final response is an attempt at copulation. The strength of the female sex pheromone is indicated by the fact that males are attracted to a female in a gauze cage even though they cannot see her, but will ignore a female clearly visible in a sealed glass cage from which no pheromone can escape. Even though the male copulatory response can be used as a bioassay for sex pheromones in the absence of all other female-produced stimuli, normal mating behavior in these species probably involves other close-range stimuli—for example, during the orientation of the male to the female when he attempts copulation.

Once the existence of pheromones was recognized, efforts were made to isolate them and to determine their chemical structure. A number of such molecules have now been characterized chemically, a considerable technical feat for molecules produced in such small quantities. In some cases, a single type of molecule may elicit a complete response; in others, a mixture of several components may be present. Pheromone molecules are relatively small, molecular weights ranging from about 100 to 300. They are small enough to be volatile but large enough to have a narrow specificity, for the pheromones in closely related species have been shown to be species-specific, males responding only to the sex attractant from females of their own species. Thus, sex pheromones also serve as a very efficient reproductive isolating mechanism, and as many as 30 species of a single moth genus may coexist without interbreeding owing in large measure to the species-specific nature of their sex attractants.

Pheromones are also involved in a rare male mating advantage with different karyotypes of *D. pseudoobscura* (Ehrman and Probber, 1978). The experimental system consisted of a double chamber, the upper separated from the lower by cheesecloth. When equal numbers of males and females of two different karyotypes were confined in the upper chamber, and large numbers of males of one of the karyotypes were confined in the lower chamber, males of the other karyotype became rare in the two-chambered system. Under these circumstances, mating in the upper chamber was non-random. Even though both kinds of males were present in equal frequency in the upper chamber, the males different in karyotype from the males in the lower chamber mated significantly more often. Further analysis showed that the rare male mating advantage was dependent on airborne olfactory cues.

This sort of frequency-dependent mating raises some intriguing questions. The pheromones permit the females not only to distinguish between males having different inversion karyotypes but also to tell which is common and which is rare. Moreover, the female is then more receptive to the rare male type, whether like herself or different. Hence, it is not just a matter of positive or negative assortative mating.

It seems probable that tactile stimuli are also important in courtship and mating in a number of animal species although not a great deal has been reported on them. One probable reason is that the sensory apparatus in other species is so different from our own that it is difficult to know whether the stimuli involved are olfactory, gustatory, or tactile. In any case they should not be overlooked as a possible isolating mechanism.

In the study of sexual isolation, various types of mating tests have been used. In the so-called "male-choice" tests, one type of male is placed with two different types of females. In "female-choice" experiments, one type of female is placed with two different types of males. In "multiple-choice" tests, both types of males and females are placed in the mating chamber so that both sexes have a "choice" of partners. With single-pair matings the frequency of matings between different male and female types is used to estimate the degree of sexual isolation. Usually only one type of test is used to measure sexual isolation. However, use of different types can sometimes be quite informative. For example, the sibling species *D. pseudoobscura* and *D. persimilis* are known to be reproductively isolated even though morphologically they are virtually indistinguishable. Three types of mating tests were used to study sexual isolation beween these species. The results, Table 14-1 (Merrell, 1954), show that the degree of sexual isolation is much lower with "male-choice" experiments than with "female-choice" or "multiple-choice" experiments, in which sexual isolation was nearly complete. Thus, it appears that the ability of the females to discriminate between males of their own species and males of a different species when given a "choice" of both male types is much greater than the ability of the males when given a similar "choice" between females. In fact, it is probable that the males mate at random (Streisinger, 1948), and the nonrandomness in the "male-choice" experiments was due to the reluctance of the females to mate with alien males rather than any "choice" on the part of the males.

A problem in the study of sexual isolation is that it is due to nonrandom mating, but not all nonrandom mating is due to sexual

Table 14 - 1. Sexual Isolation Between *Drosophila pseudoobscura* and *Drosophila persimilis* as Measured by Different Types of Mating Tests.

Type of Test	Males	Females	Type of Mating				Isolation Estimate
			Homogamic		Heterogamic		
			n	%	n	%	
"Male choice"	per	per pse	45	73.3	46	39.1	
	pse	per pse	41	90.2	42	33.3	.457
"Female choice"	per pse	per	46	65.2		4.3	
	per pse	pse	50	84.0		6.0	.069
"Multiple choice"	per	per	49	71.4		0.0	.058
	pse	pse	51	66.7		7.8	

Source: Reprinted from D. J. Merrell, Sexual isolation between *Drosophila persimilis* and *Drosophila pseudoobscura. American Naturalist* 1954, 88:93-99.

isolation. Sexual isolation is the result of positive assortative mating, the mating of like with like or an excess of homogamic over heterogamic matings. However, nonrandom mating may also result from some form of selective mating in which one type of male or female mates more often in proportion to its numbers than other types. Assortative mating and selective mating are similar in being trait specific, but differ in that assortative mating does not lead to changes in gene frequency, but selective mating does. Selective mating may occur, for example, if some types of males or females are more successful in mating than others owing to "greater vigor" or "sex drive" or to a lower threshold of receptivity in the females or a quicker readiness to mate in the males.

Therefore, if nonrandom mating is observed in a particular study, it may be due to sexual isolation, or selective mating, or both. Since mating is the result of the interactions between a male and a female, it is dependent on the characteristics of the individual partners and also on the nature of their interactions. Thus the degree of sexual isolation can be estimated only if the confounding effects of other forms of nonrandom mating can first be removed from the data. A number of indices of sexual isolation have been proposed, but none has been entirely satisfactory in disentangling the effects of various types of nonrandom mating (Stalker, 1942; Levene, 1949;

Merrell, 1950, 1960; Malogolowkin-Cohen et al., 1965; Ehrman and Petit, 1968).

It seems probable that in animals ethological or behavioral isolation is one of the first isolating mechanisms to become established. Many sympatric species are known to produce fully viable and fertile F_1, F_2, and backcross progeny in captivity or in the laboratory, but natural hybrids are extremely rare. The mallard and the pintail, for example, occupy the same breeding grounds over much of the Northern Hemisphere, yet even though these freshwater ducks are completely interfertile, natural hybrids have a frequency of only about 1 in 100,000. They breed and nest near the same ponds and sloughs but not with one another; courtship between males and females of these species is exceedingly rare.

It might be thought that ethological isolation would be confined to animals. However, because some insects show species-specific flower-visiting behavior, a second-order type of behavioral isolation is found in some groups of plants (Grant, 1949, 1950, 1971; Levin and Kerster, 1967). Bees, hawkmoths, and to a lesser extent, some other insects and hummingbirds are able to recognize a plant species by the distinctive color, form, or odor of its flowers. Once they learn how to gather nectar and pollen successfully from one species of flower, they tend to feed preferentially on that species and ignore others. This flower constancy by the insects results in most cross-pollinations being intraspecific, and the different plant species are reproductively isolated from one another by the behavior of the pollinators.

In some instances mating is attempted but fertilization does not occur and zygotes are not formed. Two major types of postmating but prezygotic isolating mechanisms have been recognized. One type, *structural* (or *mechanical* or *morphological*) *isolation*, is generally cited as an isolating mechanism in animals, but the evidence for it is rather sketchy. The idea originally was developed because the distinctive differences in the genitalia of closely related species of insects were thought to serve as a lock-and-key mechanism preventing interspecific mating; this was subsequently shown to be untrue (review in Mayr, 1963). In most groups of animals structural isolation is not a significant isolating mechanism.

However, structural differences in the flowers of related plant species have been shown to prevent interspecific cross-pollination (review in Grant, 1963). This type of structural isolation has been found primarily in plant families with complex floral mechanisms such as the Orchidaceae (orchid), Fabaceae (pea), Lamiaceae (mint),

Scrophulariaceae (figwort), and Asclepiadaceae (milkweed). A good example of structural isolation is found in two species of louse-wort growing sympatrically in the Sierra Nevada mountains of California. Both *Pedicularis groenlandica* and *P. attollens* are polli-nated primarily by the same species of bumblebees, but do not hybrid-ize because the shape of the flowers causes the bee's head to trans-port the pollen from the anthers of one *P. attollens* flower to the stigma of another, whereas the bee's abdominal venter is the vehicle for pollination in *P. groenlandica* (Sprague, 1962). A number of similar cases have been described in which the species are sympatric, interfertile, and bloom at the same time, but fail to hybridize in nature because of structural isolation. In the case of the two louse-worts, ethological isolation is also a factor, for individual bees were observed in mixed colonies of *Pedicularis* to gather food exclu-sively from one species even though they had to bypass the flowers of the other in their travels.

The other type of postmating but prezygotic isolating mechanism is *gametic incompatibility*. In species with external fertilization, even if the eggs and sperm of two different closely related species are shed at the same time, fertilization may be rare or absent because of a lack of attraction between the sperm and egg. For example, eggs of the sea urchin, *Strongylocentrotus franciscanus*, were nearly all fertilized when placed in sea water with spermatozoa of the same species. However, less than 2% of such eggs were fertilized by the same concentration of sperm from a different species, *S. purpuratus*. The reciprocal experiment gave similar, but less extreme results (Lillie, 1921). A mechanism comparable to an immune reaction has been shown to help bring the eggs and sperm of the same species together in some cases. The specificity of this reaction promotes intraspecific fertilization and inhibits interspecific fertilization and thus is an effective isolating mechanism.

In species with internal fertilization, even though interspecific copulation or pollination may occur, fertilization may not occur because of incompatibility between the sperm, or pollen, and the reproductive tract of the alien female. In many species of *Drosophila*, an "insemination reaction" has been observed following copulation (Patterson and Stone, 1952), which is marked by a rapid secretion of fluid into the vagina and enlargement of the vagina to three or four times its normal size. With homogamic matings, for instance between males and females of *D. buzzatii*, the fluid mass is soon expelled and the vagina returns to normal size in a few hours. In heterogamic matings, however, for example between female *D.*

buzzatii and males of the closely related *D. arizonensis*, the reaction mass becomes hard and is retained in the vagina for long periods of time. Its presence interferes with normal fertilization of the eggs and also damages the female reproductive tract to the point where normal ovulation, fertilization, and oviposition of the eggs are virtually impossible. Thus, even though ethological isolation may be weak, the insemination reaction is a very effective barrier to hybridization. This phenomenon has been demonstrated in many species of *Drosophila*, but to what extent a similar type of reaction is present in other groups of animals is unknown.

Gametic incompatibility in flowering plants has also been reported. If alien pollen lands on a stigma, it may fail to germinate, as in the case of *Datura stramonium* pollen on a *D. meteloides* stigma (Avery, Satina, and Rietsema, 1959). Even if the pollen germinates, the pollen tube may burst as it grows down the foreign style, as Smith and Clarkson (1956) showed with *Iris tenuis* pollen growing in *I. tenax* styles. Even if it does not burst, the pollen tube may grow so slowly that the alien pollen cannot compete successfully in fertilization with pollen from flowers of the same species as Latimer (1958) demonstrated with *Gilia splendens* and *G. australis*. Finally, even if the alien pollen tube reaches the ovule, the sperm nuclei sometimes fail to fertilize the egg and endosperm nuclei in the ovule to form the embryo and the endosperm respectively. The failure in this case resembles the gametic incompatibility in species with external fertilization.

If mating is successful and fertilization occurs, the hybrid may fail to develop into a normal adult. *Hybrid inviability* may be manifested at almost any point in the course of development. The more remote the relationship between the parents, the sooner development is apt to cease. Hybridization is possible, for example, between sheep and goats, which belong to different genera (*Ovis* x *Capra*), but even though the early embryos appear normal, they soon die. Although, in some cases, the hybrid embryo seems to run down like a clock, and development may cease at almost any point, in others, all the hybrid embryos seem to be blocked at about the same stage of development—often at gastrulation, or much later, for example, at the time of pupation or of emergence from the pupa case in insects. This type of blockage seems to come most often at these critical stages.

The exact causes of hybrid inviability are often unknown, but in some cases clues are available. For instance, when two species of flax were crossed (*Linum perenne* ♀ x *L. austriacum* ♂), the hybrid seeds failed to germinate. However, if the embryos were freed from the seed coat, a maternal tissue, and cultured in a nutrient solution, germination occurred and vigorous fertile hybrid plants developed (Laibach, 1925).

The classic example of *hybrid sterility* is the mule, the product of an intrageneric cross beween the horse, *Equus caballus*, and the ass, *Equus asinus*. The mule is not only viable but vigorous, for it is renowned for its physical endurance and surefootedness as well as its disposition, and it is often cited as an example of hybrid vigor or heterosis. Despite this vigor, mules are sterile, and this serves as a barrier to gene exchange.

There is abundant literature on hybrid sterility for both animals and plants (Stebbins, 1950, 1958; Dobzhansky, 1951, 1970; White, 1954, 1973; Grant, 1963, 1971). The causes of hybrid sterility are diverse and often complex. In essence, even though the hybrid genome may support the development of a normal or even a vigorous hybrid individual, it may not support normal differentiation and development of the gonads or the gametes. Again, as in hybrid inviability, the point at which differentiation is blocked differs in different hybrids. In some cases, the gonads remain small and undifferentiated so that no gametogenesis can occur. In others, the gonads differentiate and spermatogonial mitosis seems normal, but the spermatocytes degenerate in the first meiotic prophase. Sometimes spindle formation is abnormal, or if the meiotic spindle is normal, the chromosomes fail to pair or else pairing is incomplete. Even when meiosis seems normal and bivalents are formed, the hybrids still may be sterile, either because of abnormal meiotic behavior subsequent to first meiotic prophase or abnormalities in gametogenesis itself. In other words, gonadogenesis, meiosis, and gametogenesis are all under genetic control, and disharmonies between the components of the hybrid genome may lead to a breakdown in normal differentiation and development at almost any stage.

Efforts to categorize the causes of hybrid sterility have led to the distinction between genic and chromosomal sterility (Dobzhansky, 1970), or diplontic and haplontic sterility (Stebbins, 1958), or segregational and developmental sterility (Stebbins, 1971b). *Chromosomal sterility* results when the chromosome complements from the parental species are structurally so dissimilar that synapsis is abnormal and abnormal, unbalanced combinations of genes are segregated to the gametes. *Genic sterility* is due to the unfavorable or disharmonious interactions of the gene complements from two different species in the cells of the hybrids. As in the case of hybrid inviability, the effects of this genic disharmony may be expressed at various stages of development in different hybrids. *Haplontic* or *gametic sterility* acts on the haploid stage, the gametes or gametophytes; *diplontic sterility* affects the diploid tissues. Genic sterility typically affects the diploid cells and hence is diplontic. However, at times, unfavorable combinations of genes may segregate to the gametes and produce haplontic genic sterility. This type of genic sterility may be very difficult to distinguish from chromosomal sterility so that the practical value of the distinction between genic and chromosomal sterility is diminished.

Stebbins (1971b), accordingly, has suggested that the more appropriate terms are *developmental sterility* (rather than genic sterility) and *segregational sterility* (rather than chromosomal sterility). Developmental sterility is the result of genetic disharmonies expressed in the somatic tissues of the F_1 hybrid and leads to abnormal development of the gonads and the reproductive system. Segregational sterility results from differences in the chromosomal patterns in the hybrid genome, which lead to abnormal segregation at meiosis and hence to the formation of abnormal gametes. Developmental sterility is far more common in animals than in plants, apparently because differentiation is more complex in animals. Developmental sterility also comes into play earlier in development than segregational sterility. Operationally, it is possible to distinguish between segregational and developmental hybrid sterility because fertility can be restored in hybrids with segregational sterility by doubling the chromosome number but not in hybrids with developmental sterility. In the allopolyploids thus created each chromosome now has a homologue, pairing at meiosis is normal, and fertility is restored. The unfavorable genetic interactions in the hybrid genome responsible for developmental sterility, on the other hand, persist in polyploids. Since segregational hybrid sterility limits fertility in plant hybrids more often than in animal hybrids, the high frequency of fertile allopolyploids in plants is not surprising.

Although the distinction between developmental and segregational sterility may be useful, some cases of sterility may involve both. Furthermore, in some cases, the sterility may be cytoplasmic or else stem from maternal effects. In other cases, the hybrid sterility may result from abnormal sexual differentiation that gives rise to intersexes. A well-studied case is the analysis of intersexes produced by crosses of "strong" and "weak" races of the gypsy moth (Goldschmidt, 1934).

One of the few generalizations that have emerged from the wealth of literature on hybrid sterility and hybrid inviability is Haldane's rule. Haldane observed that when one sex is absent, rare, or sterile in the F_1 progeny from an interspecific cross in animals, that sex is the heterogametic sex. The male is heterogametic (XY) in mammals and most other groups; the female is heterogametic (ZW) in birds, butterflies, and moths. Thus, for example, the cross between *D. pseudoobscura* and *D. persimilis* produces relatively fertile F_1 hybrid females, but the males are sterile. The reverse is true in lepidopteran hybrids where the females may be missing or sterile. However, there are also a number of exceptions to Haldane's rule. Where it holds true, it may be due to an unfavorable interaction between the Y chromosome (or W) of one species and the X chromosome, autosomes, or cytoplasm of the other, or else to the imbalance in the genome of the heterogametic sex as compared to that of the homogametic sex. The latter has a sex chromosome (X or Z) from each parent as well as a complete set of autosomes so that the hybrid genome is balanced ($X_1 X_2 A_1 A_2$ or $Z_1 Z_2 A_1 A_2$). The heterogametic sex ($X_1 Y_2 A_1 A_2$ or $Z_1 W_2 A_1 A_2$), lacking a complete haploid genome because of the presence of the Y or W chromosome, apparently suffers from unfavorable interactions stemming from this imbalance. The exact nature of the unfavorable interactions in hybrid inviability or hybrid sterility for the most part remain obscure at the molecular level.

A final type of postzygotic isolating mechanism is *hybrid breakdown*. In this case, the F_1 hybrids are vigorous and fertile, but the F_2 or backcross individuals have reduced viability or fertility. For example, Stephens (1950) reported in detail on hybrid breakdown from species crosses in cotton, *Gossypium*, in which many of the F_2 plants died or were of poor viability even though the F_1 hybrids were vigorous and fertile. In such cases, the F_1 has two complete haploid genomes that apparently are able jointly to support normal development and fertility, but the segregation and recombination, both genic and chromosomal, involved in the formation of the F_2 give rise to unbalanced, disharmonious genomes that cannot support normal development.

THE ORIGIN OF REPRODUCTIVE ISOLATION

Although we have discussed each of the different types of isolating mechanisms separately, the point has emerged several times that different species are isolated by more than one type of isolating mechanism. In fact, it seems safe to say that several different isolating mechanisms are normally responsible for the reproductive isolation between related species (Grant, 1963; Dobzhansky, 1970; Stebbins, 1971b). Although any one isolating mechanism may produce only partial reproductive isolation, the combined effects of several may completely cut off gene flow between closely related species. Since interbreeding may be prevented by a combination of several premating and postmating isolating mechanisms as diverse as ecological, temporal, behavioral, and structural isolation and hybrid incompatibility, inviability, sterility, and breakdown, they obviously must be essentially independent of one another. Furthermore, reproductive isolating mechanisms, as pointed out previously, are inherited, and thus must be independently inherited. In addition, when the genetic basis of a single well-developed isolating mechanism such as a difference in habitat preference, or breeding season, or courtship pattern, etc., is analyzed, segregation for these differences among the progeny of species hybrids does not fit a simple Mendelian ratio as it would if the trait difference were controlled by one or just a few loci. Instead, the F_2 consists of a range of intermediate types with a frequency distribution typical of multiple factor inheritance. This type of result is commonly found both with premating and postmating isolating mechanisms such as hybrid inviability and sterility. Therefore, if single isolating mechanisms are controlled by a number of loci, and reproductive isolation between related species is due to the combined effects of several independent mechanisms, then reproductive isolation must be due to the accumulation of genetic differences at a number of different loci and also, of course, may be due to chromosomal differences as well. Hence, one must conclude that speciation does not ordinarily occur at a single step but by the accumulation of numerous genetic differences.

The discussion in previous chapters has already brought out the fact that populations can diverge from one another both morphologically and in their adaptation to different habitats without necessarily becoming reproductively isolated. Therefore, it seems safe to assume that the genetic basis for reproductive isolation is separate from the genetic basis for adaptive radiation or for phyletic evolution, for that matter, and it is appropriate to inquire into the origin of reproductive isolation. Two different theories have been proposed. One is that reproductive isolation develops as an incidental by-product of genetic divergence. The other, that reproductive isolation is a product of natural selection — that is, if selection acts against species hybrids because they are less fit than the parental species, the ultimate result will be selection, not just against the hybrids, but against hybridization itself. Before studying these theories in more detail, consider the basic requirements for reproductive isolation.

The development of reproductive isolation is the crucial step in speciation. From an initial species population having a common gene pool, the population evolves into two reproductively isolated populations, say X and Y, in which interbreeding occurs freely among members of X or among members of Y, but is limited or nonexistent between X and Y individuals. Analysis has shown (e.g., Dobzhansky, 1970) that a single mutation would not suffice to establish reproductive isolation in sexual, outbreeding forms. The evolution of reproductive isolation involves building different systems of complementary genes in each population; the minimum number of loci that can be involved is two. The reasoning is as follows: Suppose the original population has the genetic composition *aabb*. A single mutation cannot very well give rise to reproductive isolation in a sexual, outcrossing population because the original individual, say *Aabb,* would lack mates, being reproductively isolated from the rest of the *aabb* population. Therefore, *aabb, Aabb,* and *AAbb* individuals presumably must be able to interbreed freely. However, if the original population separates into two populations, X and Y, and *A* becomes established in X as *AAbb,* and *B* in a similar fashion in Y as *aaBB,* then the basis for reproductive isolation has been established. Even though *aabb, Aabb,* and *AAbb* individuals can interbreed freely and *aabb, aaBb,* and *aaBB* individuals can interbreed freely—and thus there was no reproductive barrier along the path from the initial aabb population to X *(AAbb)* or to Y *(aaBB)*—if there is an unfavorable interaction between *A* and *B,* then the cross between *AAbb* and *aaBB* would be difficult if not impossible. The interaction causing reproductive isolation could be premating leading to habitat, temporal, or behavioral isolation, or postmating if, for example, the *AaBb* hybrids were formed but were inviable or infertile. This model is, of course, simplistic and is, furthermore, based on an allopatric model of speciation. The evidence suggests that many more loci than two are involved in reproductive isolation, but the model helps to clarify the minimal requirements for the establishment of reproductive isolation.

These two theories of the origin of isolating mechanisms are not conflicting but complementary. Furthermore, both are based on the implicit assumption of allopatric speciation. The idea that isolating mechanisms arose as an incidental by-product of genetic divergence had existed for some time before Muller (1940, 1942) formulated it in genetic terms. The idea, essentially, is that, as isolated populations of a species diverge owing to the combined effects of selection, mutation, and drift, and give rise to entities variously labeled as ecotypes, ecological races, geographic races, and the like, the genetic changes involved may, quite incidentally, give rise to reproductive isolation. If, for example, differences in habitat preference, or breeding season, or sexual behavior evolve in these isolated populations, they may be sufficient to serve as isolating mechanisms if the populations again become sympatric. Many such differences may result from different selection pressures in the diverging populations as they adapt to different environmental conditions. Although the origin of premating isolating mechanisms in this way may seem fairly obvious, the origin of postmating mechanisms such as gametic incompatibility, or hybrid inviability, or sterility may not seem quite so straightforward. However, the gene pool of a given population is an integrated genetic system. As two populations diverge in isolation, they accumulate numerous genetic differences. Eventually, the point is reached at which the genetic systems are so different that they will no longer support normal growth and differentiation when combined in a hybrid.

The second theory, that reproductive isolation may result from natural selection, has been traced back to A. R. Wallace (1889), but its modern treatment stems primarily from Dobzhansky (1940, 1970). This theory presupposes that different populations of a species have been geographically isolated from one another and have undergone some degree of genetic divergence during this period, which may, or may not, lead to the development of various kinds of isolating mechanisms. If these divergent populations again come in contact—that is, become secondarily sympatric—any isolating mechanisms or incipient isolating mechanisms that have developed during the period of isolation may be reinforced by the direct action of natural selection. For example, if interbreeding between members of the two populations produces hybrids that are poorly adapted to the existing environmental conditions, or are partially sterile, or have low viability, natural selection will act against these hybrid individuals. However, selection eliminates not only the hybrids, but the genes of the parents that hybridized. Any individual that hybridizes with a member of the other group is wasting its gametes on the production of poorly adapted progeny and is, thus, less fit than individuals that mate with members of their own group. Therefore, any genes in either population that lead to homogamic rather than heterogamic matings will be favored by natural selection and should increase in frequency in the populations. Experimental demonstrations of the validity of this hypothesis were carried out by Koopman (1950) and Kessler (1966). Using the sibling species *D. pseudoobscura* and *D. persimilis*, Koopman showed that sexual isolation between them could be enhanced, whereas Kessler later showed that artificial selection could weaken as well as strengthen the degree of sexual isolation between them.

If natural selection plays a role in the establishment of isolating mechanisms in wild populations, then reproductive isolation between two species should be more complete in areas where they are sympatric than where they are allopatric. This postulate has been tested in a number of cases and supporting evidence has sometimes been found (Grant, 1963; Dobzhansky, 1970). Although both premating and postmating isolating mechanisms could originate as a by-product of genetic divergence, it has been argued that premating isolating mechanisms are more apt to be enhanced by natural selection since they actually prevent the wastage of gametes whereas postmating mechanisms do not.

Obviously, more must be learned about the origin of isolating mechanisms. We need to know more about the genetics of reproductive isolation, about the level of isolation in sympatric as compared to allopatric populations, and about the role of natural selection in the origin of reproductive isolation in nature.

Character displacement (Brown and Wilson, 1956), is also supposed to result from selection in areas of geographical overlap. In other words, similar traits in related species are expected to be less alike in areas of sympatry than in areas of allopatry. However, in some instances (e.g., Moynihan, 1968), the species are more alike in the zones of overlap than elsewhere. This greater resemblance may be due to similar responses to similar selection pressures, leading to convergence, or in some cases it may be due to gene flow or introgression in the zone of overlap. The strengthening of isolating mechanisms by natural selection can be regarded as a form of character displacement, but it is difficult to generalize about the origin of reproductive isolation without considerably more information from a number of different groups of organisms.

CHALLENGES TO THE BIOLOGICAL SPECIES CONCEPT

Although most biologists regard the biological species definition, despite its shortcomings, as the most meaningful species definition available, not all biologists concur. Ehrlich and Holm (1962, 1963), for example, doubt the existence of species, regarding them as artifacts of the procedures of taxonomy. They challenge the idea that organisms occur as well-defined clusters of "good" species reproductively isolated from one another. They argue that the proportion of "good" species in both higher plants and higher animals is far less than it is generally thought to be. Their reasoning seems to stem from two sources—one, the practical difficulties inherent in applying a definition based on reproductive isolation; the other, their belief that gene flow within a species is so restricted that populations of the same species do not have a common gene pool but are reproductively isolated (Ehrlich and Raven, 1969). If this is the case, the species is not the evolutionary unit; it is merely a grouping of organisms showing "phenetic" resemblance. In their view, the Mendelian population (in the restricted sense) is the significant evolutionary unit.

In the first place, Ehrlich and Holm seem to have overstated the proportion of poorly defined species in animals and plants. Stebbins (1963) estimated that 70 to 80% of higher plant species conform to the biological species definition, and the proportion is undoubtedly greater in higher animals where the complications due to hybridization and polyploidy are much less common than in plants. We have already dealt at some length with the practical difficulties in using the biological species definition, but have also pointed out that in a nondimensional system, it works. Thus, contrary to Ehrlich and Holm, the biological world does not consist primarily of poorly defined groups that intergrade morphologically, but rather of well-defined species, for the most part, reproductively isolated from one another.

In support of their argument that gene flow between populations of the same species is so insignificant that gene flow has no bearing on their evolution and cannot serve as a cohesive force holding plant or animal species together, Ehrlich and Raven (1969) cited evidence that indicates a lack of gene flow in a number of plant and animal species. The most obvious criticism of these examples is that they all represent short-term studies. The lack of gene flow between populations of *Euphydryas* butterflies (Ehrlich 1961, 1965) in the course of a few years' study is no assurance that these populations will retain their identities over time spans of a century or more. Environmental fluctuations over a longer period seem likely to disrupt the stability of the population structure, leading to extinction of some populations, and displacement, expansion, or dispersal in others. Consequently, even though gene flow may be sporadic, and thus difficult to detect in studies such as Ehrlich's (1961, 1965) with butterflies or Twitty's (1959) with California newts, so little gene flow is required to prevent genetic divergence that the mere fact that it is not detected in short-term studies is no proof that it never occurs. The frequently cited study of Selander (1970) indicating a lack of gene flow between populations of the house mouse, *Mus musculus*, in a single large barn is a case in point. Although the mice were apparently not exchanging genes at the time, obviously they could only have taken up residence in the barn after it was built, hence must have migrated from elsewhere. When the barn collapses or is destroyed, the mice presumably will be displaced. Thus, in the long term view, the existing population structure of the barn must be regarded as ephemeral. It is noteworthy that even though Ehrlich and Raven (1969) argued that gene flow in nature is of little significance, Camin and Ehrlich (1958) attributed the frequency and distribution of banded and unbanded water snakes, *Natrix sipedon*, on islands in Lake Erie to the combined effects of selection favoring the unbanded type on the islands and constant migration of the banded type from the mainland to the islands. In this case, the phenotypes were so different that the effects of gene flow were readily apparent. In other cases, gene flow may not be so obvious, but electrophoresis has provided a new approach through which inferences about gene flow may become possible. Electrophoresis permits the analysis of a number of different loci, and, if gene flow is occurring, its effects should be apparent at all these loci, which is not true of other evolutionary factors.

Since up to one-half of all species of flowering plants are thought to be allopolyploids—that is, the product of hybridization between species followed by chromosome doubling—obviously, if we define hybridization as a form of gene flow, which it is, then gene flow has played a very significant role in the evolution of plants. In addition, introgressive hybridization has been reported in many cases in plants (Anderson, 1949; Grant, 1971) and represents gene flow from one species to another. Introgression is often difficult to detect without special statistical or other techniques. Interspecific hybridization is undoubtedly less common than intraspecific hybridization—that is, gene flow in the usual sense—so that it is difficult to escape the conclusion that gene flow must be occurring between populations of the same species (including subspecies and ecotypes as well as less well-differentiated populations) and may play an equally important, though more subtle, role than introgression or allopolyploidy in the evolution of higher plants. Although ordinary gene flow may be more common than introgression, it may be even more difficult to detect.

Mayr (1963) discounted the importance of hybridization in the evolution of animals, and this certainly seems true if one compares the frequency of polyploidy or of introgression in animals with their incidence in plants. Even though cases of polyploidy and of introgression have been reported in animals (Mayr, 1963; White, 1978), they are far less common in animals than in plants. We have already discussed the reasons why reproductive isolation may be more complete between animal species than plant species. One is that behavioral isolation is absent in plants; another is the greater developmental complexity of animals, which may reduce the fitness of hybrids in a variety of ways; a third is the chromosome imbalance stemming from segregation of the sex chromosomes, especially in polyploids, which may lead to the development of partly or completely sterile intersexes. It is noteworthy that in recent decades a number of unisexual vertebrate populations have been reported. Some are diploid, some are triploid; some are gynogenetic, others parthenogenetic. Of particular interest is Uzzell's (1970) statement that there is reason to believe that all these unisexual vertebrates, which include fishes, amphibia, and reptiles, have a hybrid origin and are permanent hybrids. This finding suggests that interspecific animal hybrids may sometimes be formed in nature but can have little impact unless the difficulties enumerated above can somehow be circumvented. Uzzell added that there was no evidence of antiquity for any of these unisexual populations and that they appear to have limited evolutionary potential. However, these cases do establish that, unlikely though it may seem, species crosses do sometimes occur in animals, and if crosses occur at this level, it seems even more probable that gene flow will occur between populations of the same species. The question, however, is, How much gene flow occurs, and does it have any evolutionary significance? Ehrlich and Raven (1969) have argued that very little gene flow occurs at any level, hence that gene flow has minimal evolutionary significance. Mayr (1963) accepts the importance of intraspecific gene flow in animals but believes that interspecific hybridization has been unimportant in animal evolution. Grant (1971) has reviewed the extensive role of hybridization in the evolution of plants. Stebbins (1950, 1959, 1963, 1971b) has not only dealt with the importance of hybridization and gene flow in plant evolution, but has argued for their importance in the evolution of animals as well, particularly in closely related populations between which reproductive barriers are not well developed.

The differences recited here are not over the facts, but over their interpretation. Population genetics has dealt with gene flow as a recurrent phenomenon. Therefore, if one attempts to estimate a migration coefficient over the course of a few generations and obtains an estimate of zero, there is a natural tendency to regard gene flow as unimportant. However, if one views hybridization or gene flow as a rare or even unique event that may generate exceptional new genotypes for natural selection to work on, its possible importance may seem much greater. Stebbins seems to take the latter position in contrast to the other authors. At the present time, it seems safe to say that hybridization and gene flow have played a significant role in plant evolution and have undoubtedly played a more significant role in animal evolution than that envisioned by Ehrlich and Raven. Much more remains to be learned about gene flow in nature, but an attack on the biological species concept, because gene flow is thought to be insufficient to serve as a cohesive force within the species, hardly seems warranted even in the light of present knowledge.

Sokal and Crovello (1970) mounted a somewhat different attack on the biological species concept. In essence, they believe that the only workable species definition is a phenetic species definition stemming from the delineation of operational taxonomic units by the methods of numerical taxonomy (Sokal and Sneath, 1963; Sneath and Sokal, 1973). Although phenetic traits are often considered to be equivalent to morphological traits and much of the work in numerical taxonomy has dealt with morphology, Sokal and Crovello specifically state that they conceive of phenetics in a much broader sense to include not just morphology, but physiological, biochemical, behavioral, and ecological traits as well. From their analysis of the biological species concept, they conclude that it is not operational and thus not useful for practical taxonomy, that it is neither necessary nor useful for evolutionary taxonomy, and that it lacks heuristic value in generating hypotheses about evolution. They also developed a flow chart, based on their conception of the biological species definition that consists of seven steps necessary to delimit a biological species, most of which, they argue, are either largely or entirely phenetic, even in theory.

However, it is clear that they regard the difficulties in applying the biological species concept to actual populations as so great that their flow chart is actually based on a morphological species concept and not on the biological species concept at all. This is obvious from the first step in the flow chart, which is "Assemble phenetically similar individuals." The whole point of the biological definition is that species status is determined by the breeding behavior of individuals with respect to one another in nature. Their first step violates the very essence of the biological definition. Although they consider the possibility of observing organisms breeding in nature to determine their species status, they dismiss it as so unlikely as to be impractical. However, difficult or not, the only way to test the biological species concept is to study the breeding behavior and reproduction of individuals in their natural habitats. Such studies require a more intimate knowledge of natural populations than is usual in taxonomic work. However, if the ultimate goal is an understanding of the relationships among organisms under natural conditions, this sort of information will have to be sought eventually anyway. To paraphrase Wordsworth (1798), we should seek to study nature, not dead specimens. Although the numerical taxonomists stress the practical advantages and operational nature of their phenetic approach, one cannot help but wonder if an equal amount of time spent studying the species in the field might ultimately be more informative about species relationships than the same amount of time spent measuring a multitude of characters in the laboratory.

To bring this discussion to a practical level, consider the case of *Rana burnsi*. As mentioned earlier, these unspotted frogs were described as a new species by Weed in 1922. Not until 1942 did Moore demonstrate by experimental crosses in the laboratory that the unspotted Burnsi frogs differed from the spotted leopard frogs, *R. pipiens*, at only a single locus. The unspotted phenotype was due to the action of a simple dominant allele, and crosses between "*R. burnsi*" and *R. pipiens* produced fully viable and fertile offspring segregating in the expected Mendelian ratios. From these results, Moore drew the eminently reasonable conclusion that *R. burnsi* did not deserve the status of a species or subspecies but should be reduced to synonymy with *R. pipiens*. After dealing with the improbability of observing breeding behavior and reproduction in nature to test the biological species concept, Sokal and Crovello (1970) pointed out the inherent limitations and difficulties with crossing experiments as tests of the concept. In essence, they argued that crossing experiments are also impractical either because the specimens of interest so often are already dead or because the results of breeding tests in the laboratory may be difficult to relate to the breeding behavior and reproduction of the organisms in the field. Only after dismissing field observations and laboratory crosses as too impractical did Sokal and Crovello reach step one in their flow chart, "Assemble phenetically similar individuals." However, Moore, by a single series of crosses, was able to shed new light on a question that morphological and other approaches had been unable to resolve for twenty years, and drew the reasonable inference that there was no such species as *R. burnsi*. Therefore, despite the supposed difficulties, the experimental crosses produced much better and more direct evidence about the relationship between *R. burnsi* and *R. pipiens* than could the more elaborate procedures used by numerical taxonomists to delineate phenetic species.

Nevertheless, it should be realized that even though crossing experiments can be very informative and Moore's conclusion was entirely reasonable, it was not a direct test of the species status of these two forms. If the strikingly different Burnsi and pipiens types failed to interbreed in nature, they would, under the biological species definition, have to be regarded as separate species despite the simple Mendelian difference between them. Therefore, it should be added that unspotted Burnsi frogs have, on occasion, been captured in amplexus with spotted "wild-type" frogs in natural breeding populations of *R. pipiens* and that the tadpoles in these ponds have metamorphosed into both Burnsi and spotted young frogs so that the matings were successful. On this, the ultimate test of the biological species concept, the two types clearly belong to the same species. Thus, in this case at least, it is possible to apply the biological species concept to reach a taxonomic decision. Equally important or perhaps even more important, the study of these populations in the field provided information unavailable with the phenetic approach. This approach may be useful, or even necessary, for a first approximation in the classification of organisms, but most biologists are interested in more than just the classification or phylogeny of the species they are studying and ultimately will have to study them in their natural habitats, difficult though it may sometimes be.

Like Ehrlich and Raven (1969), Sokal and Crovello (1970) regard the local breeding population rather than the species as the most useful unit for evolutionary study, reflecting their belief in the unimportance of gene flow in natural populations. Here, too, the Burnsi story provides food for thought. The Burnsi allele is widely and rather uniformly distributed over an area of some 100,000 square miles in Minnesota and parts of adjacent states (Merrell, 1970) at frequencies ranging from less than 1% to about 5%. This area is greater than all of New England or Great Britain, yet is only a small portion of the range of *R. pipiens*. The Burnsi distribution cuts across the ecotones in this region involving the deciduous and coniferous forests and the prairie, but is confined almost exclusively to the recently glaciated area from which the Wisconsin ice retreated only some 10,000 years ago. Despite the long, harsh winters, the environment in the range of the Burnsi polymorphism is extremely favorable for *R. pipiens*, and this area has long been a primary source of supply for frog dealers. If one speculates about how the unspotted Burnsi allele became established over so wide an area, one must also attempt to explain why it has not become established in the rest of the range of *R. pipiens* where it is either absent or extremely rare. Several otherwise reasonable explanations fail when the latter question is asked. Given the short period of time since the retreat of the glaciers and the high vagility of *R. pipiens* (Merrell, 1977), the most plausible explanation for the present distribution and frequency of Burnsi is that it originated in the population that reoccupied the recently glaciated area and spread rapidly to its present distribution as the ice retreated. One can go a step further and speculate that all the Burnsi alleles in this region are identical by descent from a single ancestral allele. Although, at first glance, this last assumption may seem highly improbable, one must remember that if mutation pressure were responsible for the frequency of Burnsi in the Minnesota region, then one must explain why mutation pressure does not produce similar frequencies of Burnsi in the rest of the range of *R. pipiens*. Similarly, if selection were maintaining the Burnsi allele in these populations, and there is some evidence that Burnsi is associated with the ability to survive the winter or other types of stress (Merrell and Rodell, 1968; Dapkus, 1976), other populations in other parts of the range are subject to equally severe winters or other types of stress, but they lack Burnsi.

With estimates of population size, generation length, and the mutation rate, an estimate of the total number of mutations to Burnsi over the past 10,000 years in the Minnesota region can be made. Although this number is large, the probability of loss for each mutation is so high that even if they have a slight selective advantage, nearly all will be lost by chance, as discussed earlier in Chapter 4. These assumptions are so simplistic that they must not be taken too seriously; but if one assumes 100 frogs per square mile over 100,000 square miles and two years per generation, the total number of frogs in the area over the past 10,000 years was 5×10^{10}. If the mutation rate to Burnsi is 1×10^{-6}, the expected number of Burnsi mutations among these 50 billion frogs is 100,000. However, even if the Burnsi allele has a 1% selective advantage, after 127 generations about 97% or 97,000 of these mutations would have been lost by chance. The remaining 3,000 mutations would be distributed among 50 billion frogs over 5,000 generations, or less than 1 per generation—that is, less, than 1 per 10 million frogs. Given these sorts of figures plus the possibility of a small initial population or of population bottlenecks, the likelihood that the Burnsi alleles in the Minnesota area are identical by descent seems much less implausible. Furthermore, if this is true, the Burnsi allele can serve as a marker for gene flow over this large area. Even if these speculations remain no more than that, the breeding behavior and migratory behavior of *R. pipiens* (Merrell, 1977) are so different from the behavior of Twitty's newts (Twitty, 1966), for example, that gene flow must be significant in *R. pipiens* populations. Local breeding populations can be identified more easily in leopard frogs than in most species because each breeding pond contains a separate deme and its progeny. However, the subsequent migrations to the summer feeding range, to the overwintering sites, and to the breeding ponds in the following years destroy the identity of the demes. Homing to a given breeding pond cannot be a significant factor in this species because different ponds are used for breeding in different years, often because the temporary ponds used for breeding one year may be dried up the next. Therefore, whatever the evolutionary unit may be in *R. pipiens*, it clearly is larger than the local breeding population.

The challenges to the biological species concept seem to center around two problems: First is the belief that observation of the breeding behavior and reproduction of individuals in natural populations is so difficult as to be impractical. Second is the belief that what is usually regarded as a species is actually not an evolutionary unit bound together by bonds of mating and parentage because gene flow between different populations of the same species is so low that they do not have a common gene pool. These criticisms do not invalidate the biological species concept. Instead, they point up the need for more information based on studies of natural populations in the field. Only through observations of breeding behavior and reproduction under natural conditions can decisions be reached as to whether individuals belong to the same or to different species. Only by studies of the actual amounts of gene flow between natural populations of the same species can one determine whether or not they have a common gene pool. Of the four major evolutionary factors, natural selection and mutation have received by far the most attention and study by population biologists. Although gene flow (e.g., Emlen, 1978) or random genetic drift (e.g., Jaenike, 1978) are often suggested as explanations in the interpretation of field data, actual attempts to estimate effective population sizes and the magnitude of random genetic drift or the amount of gene flow in natural populations are rare. Admittedly, such data are difficult to obtain, but they are essential if one is ever to resolve the questions about the nature and origin of species. The approach of ecological genetics, which involves a judicious combination of field and laboratory studies, seems best suited to provide the answers.

CHAPTER 15

The Origin of Species

ALLOPATRIC SPECIATION

The origin of species is a complex process. For the most part, thus far we have adhered to the conventional concept of the origin of species, that of geographic or allopatric speciation, whose most vigorous advocate has been Mayr (1963, 1970). The following statement (Mayr, 1970) reflects a widespread attitude about the origin of species.

> That geographic speciation is the almost exclusive mode of speciation among animals, and most likely the prevailing mode even in plants, is now quite generally accepted. [Further on he wrote] The theory of geographic speciation . . . states that in sexually reproducing animals a new species develops when a population that is geographically isolated from the other populations of its parental species acquires during this period of isolation characters that promote or guarantee reproductive isolation after the external barriers break down.

According to this concept, spatial isolation or physical separation of breeding populations of the same species is a necessary prerequisite to speciation. It is assumed that speciation is a gradual process, involving the accumulation of genetic differences at a number of gene loci, and that reproductive isolating mechanisms arise as an incidental by-product of the genetic divergence in these isolated populations.

Most models of allopatric speciation envision isolated populations more or less equal in size. Mayr (1942, 1954), however, suggested the "founder effect" or "founder principle" as a form of allopatric speciation involving very small, isolated, peripheral populations; more recently Carson (1968, 1971, 1973, 1975) has suggested that species may originate in Hawaiian *Drosophila* from a population as small as a single gravid female. Both concepts involve a combination of genetic drift and natural selection in these small populations. Carson also invoked population flushes and crashes and open and closed systems of genetic variability to account for speciation in these populations. However, these concepts are reminiscent of the "shifting balance" theory developed by Wright (e.g., 1978) and are special cases of geographic or allopatric speciation.

In recent years a number of questions have arisen about the universality of the model of geographic or allopatric speciation outlined above. One such question is whether speciation is always allopatric or whether sympatric speciation may also occur. Another question is whether evolutionary divergence is always gradual or whether some type of saltational change or quantum evolution may give rise to new species.

The possibility of saltational evolution is usually given short shrift. Nonetheless, about one-half of all species of higher plants are polyploid, and virtually all of these are allopolyploids having a hybrid origin. Because the origin of allopolyploidy through hybridization followed by chromosome doubling is a saltational process that has given rise to many plant species, it seems unwarranted to dismiss saltational evolution out of hand as unimportant. Moreover, as the initial hybrid must have been formed sympatrically with its parental diploids, half the species of higher plants have arisen through sympatric speciation.

Cases are also known in which the formation of new species has been accompanied by a major reorganization of the genome. If these karyotypic differences serve as reproductive isolating mechanisms between the original and the derived species, as is often the case, this, too, represents a form of saltational evolution and could permit sympatric speciation to occur.

Even in animals, it is doubtful that geographic speciation is as prevalent as Mayr suggests. In the first place, every free-living animal species is parasitized by at least one species and usually by several parasitic species. Therefore, there must be more parasitic than free-living species in the animal kingdom. In addition, in insects especially, there are several groups of related species, each of which is monophagous on a different host plant. Speciation in parasitic or monophagous species will generally involve a shift in host preference, and such shifts seem much more apt to occur sympatrically than speciation in free-living forms. Although knowledge of evolution in such groups is limited at present, it seems premature, given the number of species involved, to conclude that geographic speciation is "the almost exclusive mode of speciation among animals."

PATTERNS OF EVOLUTION

Speciation, in the narrow sense, is a multiplication in the total number of species through the splitting or branching of a single ancestral line into two or more contemporaneous species. Also known as cladogenesis, speciation is often thought of as a multiplication of species in space or adaptive radiation.

Phyletic evolution or anagenesis is the transformation through time of a single evolutionary line from species A into species B. If the evolution from A to B has involved the gradual accumulation of numerous genetic differences, it will be impossible to draw a sharp line of demarcation separating A from B. Since the fossil record is usually incomplete, the separations in cases such as these ordinarily coincide with gaps in the fossil record. However, if some form of discontinuous evolutionary process is involved, then a sharp separation between species A and species B is possible. For example, if the polyploids replace their diploid ancestors or if the new karyotypes supplant the old, in essence, saltational phyletic evolution will have occurred.

A third possible pattern of evolution is *reticulate evolution*, the basis for which is hybridization. We discussed the importance of allopolyploidy and introgressive hybridization in plant evolution in a previous chapter; clearly, a reticulate pattern is important in plants. We also discussed there the possible importance of hybridization between more closely related forms of both plants and animals, which would be more difficult to detect.

Thus species may originate by the transformation of a species through time, by the multiplication of species in space, or by reticulate evolution, in which hybridization unites previously separate groups into a single evolutionary line.

THE EVOLUTION OF GENETIC SYSTEMS

In the development of evolutionary theory, the genetic system usually assumed is a diploid species with separate sexes and homosequential chromosomes. Although other types of genetic systems sometimes seem to be regarded as aberrations and the diploid species with separate sexes the norm, there is such an array of other possible genetic systems that it seems advisable to outline these possibilities because together they constitute a significant proportion of all species. Just as not all species' origins conform to the gradual allopatric model, not all genetic systems conform to the diploid type with separate sexes.

To help clarify some of the complexities and bring a degree of order to what otherwise may seem a chaos of possibilities, we shall pose a series of questions, almost a key to the possible types of genetic systems.

1. Is reproduction sexual or asexual or both?
2. If sexual, are the sexes separate or are the individuals hermaphroditic?
3. If the sexes are separate, what is the mechanism of sex determination?
4. If the individuals are hermaphroditic, does self-fertilization or cross-fertilization occur?
5. In sexual reproduction, there is an alternation of haploid and diploid phases in the life cycle, and four fundamental processes occur in various sequences: meiosis, mitosis, gamete formation, and fertilization. What is the sequence of these events in the life cycle of the species?
6. Has the karyotype remained unchanged or homosequential, or have there been changes in the karyotype?
7. If karyotypic changes exist, are they examples of chromosomal polymorphism or do different karyotypes characterize different populations?
8. Do the karyotypic differences observed serve as reproductive isolating mechanisms or not?
9. Are all of the species diploid or are some polyploid?
10. If polyploid, are they auto- or allopolyploid?
11. If reproduction is asexual, do the individuals have a hybrid origin or are they descended from a single ancestral line?

Asexual reproduction can take place in a variety of ways, recently summarized by Grant (1971) for plants and White (1973) for animals; however, it is not necessary to pursue this subject in detail. One thing common to the various types of asexual reproduction is that new combinations of genes are not being generated.

It should be obvious that, if one were to apply the above key to existing species, they would key out into quite an array of different genetic systems. In view of this array, it is valid to ask if there are any underlying trends or patterns that can account for this diversity of genetic systems.

One trend found in the evolution leading to both higher plants and higher animals is the tendency toward the prolongation or dominance of the diploid phase. Although in the life cycles of lower forms of life, meiosis may follow immediately after fertilization so that the rest of the cycle is haploid, or else both haploid and diploid phases occupy significant portions of the life cycle, in higher animals meiosis occurs immediately before gamete formation and fertilization so that the gametes are the only haploid cells in the cycle. In higher plants, the haploid gametophyte has been reduced to a handful of nuclei. The usual explanation for this trend is that diploids can carry a considerable amount of unexpressed variability in the heterozygous condition. The release of this variability through recombination each generation permits the population to respond to new selection pressures if they arise while at the same time remaining well-adapted to the existing conditions. In addition, diploidy permits gene interactions not open to haploids such as interallelic, epistatic, and heterotic interactions. Thus diploidy seems better suited to provide complex developmental control as well as a means for the conservation and gradual release of genetic variability.

The prokaryotes—the viruses, bacteria, and blue-green algae—lack a nucleus, and the hereditary material is in the form of a naked nucleic acid molecule. In eukaryotes, the DNA is intimately associated with basic proteins to form complex structures, the chromosomes, which are enclosed within the nucleus. The organization of the genetic material into chromosomes and the regularity in the distribution of the chromosomes ensured by mitosis and meiosis are the basis for the genetic system in eukaryotes. One way to view the variations in the genetic system is to study their effect on recombination, which appears to be their most significant effect.

A number of mechanisms are known to increase genetic recombination. Sexual reproduction itself, through meiosis and fertilization, provides for a regular reshuffling of the genetic material each generation. If the sexes are separate, cross-fertilization is mandatory, but even in hermaphrodites, mechanisms that reduce or prevent selfing are common, e.g., self-sterility alleles in plants. Recombination will also be enhanced by a large number of small chromosomes, a high chiasma frequency, or an absence of chromosomal rearrangements such as inversions and translocations.

On the other hand, recombination may be limited in several ways. Asexual reproduction, of course, is the most effective; but inbreeding and the ultimate in inbreeding, self-fertilization, will also reduce recombination. The amount of recombination will be minimized if the genes are organized into a small number of large chromosomes, especially if the chiasma frequency is also reduced. The ultimate in this direction is the absence of crossing over in *Drosophila* males. Heterozygosity for chromosomal rearrangements also reduces recombination, for the blocks of genes within an inversion, for example, tend to be transmitted intact.

In general, the changes in the genetic system that restrict recombination have occurred in species in which immediate fitness and a high reproductive rate are at a premium. Genetic recombination is limited in three significant ways: asexual reproduction, an increase in self-fertilization, or a reduction in chromosome number or in chiasma formation. These mechanisms tend to be mutually exclusive so that if self-fertilization is found in a group, for example, the other two will not occur to any significant extent. This finding suggests that the adaptive significance of the different genetic systems is related to their effect on recombination, a point that Carson (1975 and earlier) has been making for some time. However, the retreat from the cross-fertilizing, diploid condition, though it may confer an immediate adaptive advantage, appears to do so at the expense of long-range adaptability, for such groups generally appear to be evolutionary dead ends.

In a number of cases, the genetic system seems designed to maintain a permanent hybrid state. We have already seen that it is highly probable that all the asexually reproducing fish, amphibia, and reptiles have a hybrid origin. Similarly, virtually all of the naturally occurring polyploids in plants are allopolyploids having a hybrid origin, which is perpetuated and stabilized by chromosome doubling to form the polyploid. Permanent translocation heterozygotes such as those in the plant genus *Oenothera* represent still another mechanism for the maintenance of hybridity. The cases of inversion polymorphism (e.g., in *Drosophila*) or translocation polymorphism (e.g., in *Paeonia*) represent still another way that structural and, of course, genetic hybridity can be maintained. The most reasonable explanation for the maintenance of hybridity in these cases is that the modified genetic systems ensure permanent heterosis. However, even though these mechanisms confer a short-term adaptive and evolutionary advantage, in the long run they may be maladaptive because they also tend to prevent or inhibit genetic recombination and thus prevent major adaptive shifts from occurring.

SALTATIONAL EVOLUTION

We have already established that saltational evolution occurs in the formation of allopolyploid plant species. Now let us consider the evidence for other possible forms of saltational evolution.

One of the first to challenge the idea that the origin of species involves the gradual accumulation of numerous genetic differences, or micromutations as he called them, was Goldschmidt (1940). He distinguished between microevolution, evolution within the species giving rise to geographic races, ecotypes, and the like, and macroevolution, giving rise to species, genera, and the higher taxonomic categories. The former was mediated by micromutations, the latter by systemic mutations or macromutations. To Goldschmidt, a systemic mutation consisted of a change in the intrachromosomal pattern and was distinct from gene mutations. The concept apparently grew out of his work on position effect in which phenotypic effects result from new new chromosomal patterns or arrangements of existing genes. These systemic mutations could give rise to "hopeful monsters" that could occupy a new environmental niche and thus, give rise to a new species in a single step. Goldschmmidt's idea has never won acceptance (Grant, 1963; Mayr, 1963). The main objections seem to center around the low viability and fertility of the postulated "hopeful monsters," their lack of suitable mates, and their inability to become established as independent, reproductively isolated populations. Goldschmidt developed his theory before the era of modern molecular genetics so that it is difficult to reformulate it in modern terms. Even more serious, he was not primarily a population geneticist, hence did not address himself to the question of just how a systemic mutation, if and when it occurred, could give rise to a reproductively isolated, distinct species in a single step.

More recently, however, new theories of speciation centered on chromosomal rearrangements have been advanced (Lewis, 1966, 1973; White, 1969, 1978). Although White calls his theory "the stasipatric model of chomosomal speciation," which was advanced to explain the enormous number of closely related species of animals with visibly different karyotypes, and Lewis calls his theory "speciation resulting from saltational reorganization of the chromosomes," which was proposed to explain speciation in annual species of flowering plants, the two theories are very similar.

Both Lewis and White recognize that gradual speciation does occur in which the chromosome arrangements remain essentially unchanged or homosequential, but gradual ecogeographic differentiation into ecotypes or subspecies takes place. If reproductive isolation develops, the process may lead to the origin of new species. In other words, they accept geographic or allopatric speciation as a reality in both plants and animals, but argue that evolution involving chromosomal rearrangements is a second, quite distinct and fairly common mode of speciation. Furthermore, Lewis, in particular, believes that this type of speciation is saltational.

Chromosomal rearrangements based on breakage and reunion seem to be of three types. The first and most common type is so deleterious that it will be quickly eliminated by natural selection. The few survivors, estimated by White to be about 1 in 10^4 or 1 in 10^5, give rise to the other two types. One leads to balanced chromosomal polymorphism, either because of heterosis in the chromosome heterozygotes, or frequency-dependent selection, or some similar mechanism. The third and final type, estimated to be as rare as 1 in 10^6 or 1 in 10^7, has an adverse effect on the fitness of the heterozygotes, but does not adversely affect the fitness of the new homozygous type. In some cases, the new homozygote may even be more fit than the original homozygote.

Although it appears that the chromosomal rearrangements that lead to balanced chromosomal polymorphism are usually not the same as those that lead to chromosomal differences between closely related species—that it, to chromosomal speciation (White, 1978)—more evidence on the relationship, if any, between chromosomal polymorphism and chromosomal speciation seems desirable.

It should be added that the rearrangements sometimes lead to "Robertsonian" changes in chromosome number either through fusion, leading to a decrease in the chromosome number, or dissociation, leading to an increase in chromosome number. In this way, the chromosome number in closely related species may come to differ, but the number of chromosome arms, the *nombre fondamental* (*N.F.*) of Matthey, may remain the same.

Chromosomal speciation is dependent, then, on the occurrence of rare, or unique, chromosomal changes that reduce the fitness of the heterozygotes for the old and new arrangements, but produce new homozygotes as fit or fitter than the old type homozygote. Alternatively, the new homozygotes in some way are protected from competition with the original type. A major difficulty for Goldschmidt's theory of systematic mutations was an inadequate explanation of how they might become established. A similar difficulty exists for Lewis's and White's theories of chromosomal speciation. At its origin, the new arrangement will be rare, and thus will be present in the heterozygous condition, which is postulated to have reduced fecundity or viability. Therefore, either it will be eliminated by selection, or some means must exist by which homozygotes for the new arrangement can be produced immediately.

Lewis (1966) postulated that saltational speciation is initiated by one or more individuals heterozygous for structural rearrangements that reduce fertility. If these heterozygotes are, to some extent, isolated from the parental population and retain some degree of fertility, they can produce homozygotes of either type. If of the original type, nothing will have changed. However, if the homozygotes are of the new chromosome type and their chromosomal differences form a reproductive barrier with the parental homozygotes, then the population with the new arrangement can maintain its identity in the presence of the parental species even if it consists of less fit individuals. Therefore, Lewis envisioned, at the outset, some degree of spatial isolation, however slight, to permit the new type to become established. He thought this was most apt to occur at the ecological limits of the species where dispersal might carry an individual to an unoccupied site, or "catastrophic selection" might have eliminated virtually all the population. Furthermore, the species Lewis studied differed by multiple chromosome breaks rather than a single rearrangement. Therefore, in keeping with his saltational model, he argued that multiple chromosome breakage resulted from enforced inbreeding in these small marginal populations of normally outcrossed individuals.

White (1978), however, believed that the chromosomal repatterning could be more gradual, involving one rearrangement at a time, but he, too, believed that each rearrangement was unique and thus had to spread from its origin in a single individual to its present distribution. He thought the rearrangements could originate anywhere within the species range, not just in peripheral populations. He also believed that the rearrangements involved in chromosomal speciation differed from those responsible for chromosomal polymorphism and that they produced adaptively inferior heterozygotes. He envisioned their spread as being facilitated by genetic drift in small isolated demes that might occur anywhere within the species range or possibly by meiotic drive in the heterozygotes.

Both Lewis and White considered the chromosomal rearrangements to play a primary role in initiating speciation in the groups they studied. The fact that so many closely related, similar species show marked karyotypic differences is a strong argument in favor of their interpretation. Genic and adaptive divergence then is thought to be secondary and subsequent to the establishment of the new species, which are isolated by postmating isolating mechanisms from the outset. Lewis and White also stress that chromosomal speciation seems to occur primarily in species of low vagility. The weakest part of their argument seems to center on the origin and establishment of the chromosomal rearrangements that give rise to new species, but now that the possibility of this type of speciation is recognized, better understanding of this mechanism should be forthcoming.

SYMPATRIC SPECIATION

White, in *Modes of Speciation* (1978), considered seven models of speciation based on the relationship between the speciation process and the geographical distribution of the populations involved. To these he added asexual speciation and speciation due to polyploidy. Among the geographic models considered were allopatric speciation, clinal speciation, area-effect speciation, stasipatric speciation, and sympatric speciation. One type not included by White, but often used by others, is parapatric speciation (e.g., Bush, 1975). This tendency to emphasize the geographical component of the speciation process tends to overshadow the underlying genetic mechanisms in speciation, which, as has been seen, may be quite diverse. Furthermore, the words themselves may lead to confusion. Speciation is a process involving populations of a species. Although allopatric, sympatric, and parapatric are often used in reference to populations, we saw in Chapter 11 that the words can be defined more rigorously only in terms of individuals. Viewed in this light, only two of the words emerge unscathed: sympatric and allopatric. Parapatric, for example, is used to refer to contiguous populations that hybridize along their common boundary. However, the individuals in these populations that actually hybridize are sympatric by definition; individuals that do not encounter members of the other population and, hence, cannot mate with them are allopatric. Therefore, the so-called parapatric populations are composed of sympatric and allopatric individuals, a somewhat confusing situation.

Similar criticisms can be directed to the concepts of clinal speciation and area-effect speciation. Here, too, the important factor is whether the individuals are sympatric or allopatric, and, consequently, how much gene flow is actually occurring. Stasipatric was coined by White and his associates to characterize their model of chromosomal speciation. Although the word seems to resemble some of the others we have been discussing, in this case the genetic mechanism is more important than the geographical aspect.

We have already seen that sympatric speciation can occur through allopolyploidy. It is also conceivable, though both Lewis and White suggested that spatial isolation was an essential part of their theories of chromosomal speciation, that chromosomal speciation could lead to the sympatric origin of species if the chromosomal rearrangements lead directly to reproductive isolation. This possibility seems reasonable enough to require one to keep an open mind on the subject until more data are available.

Advocates of sympatric speciation have also argued that disruptive or diversifying selection, if it is strong enough, might cause sympatric speciation. Attempts to demonstrate genetic divergence and the development of reproductive isolation through disruptive selection were reported to be successful by Thoday and his co-workers (Thoday and Boam, 1959; Thoday and Gibson, 1962, 1970, 1971; Thoday, 1965, 1972). Unfortunately, however, repeated efforts to duplicate these results have led to failure (Scharloo, 1964, 1971; Scharloo, den Boer, and Hoogmoed, 1967; Scharloo, Hoogmoed, and ter Kuile, 1967; Chabora, 1968; Barker and Cummins, 1969; Robertson, 1970). Therefore, these experiments can hardly be regarded as providing very strong support for the sympatric origin of species through disruptive selection.

The best evidence for the effectiveness of disruptive selection comes from plants (reviewed by Antonovics, Bradshaw, and Turner, 1971, and Bradshaw, 1972). Plants growing on soil contaminated by heavy metals such as copper, lead, or zinc from mine tailings have been shown to be tolerant to the heavy metal whereas plants of the same species growing on immediately adjacent uncontaminated soil cannot tolerate the heavy metal and die if transplanted into toxic soil. These differences have been shown to be genetic and therefore, must have evolved, in some cases, in a very short time. One of the more spectacular cases involved a galvanized (zinc-coated) fence installed thirty years previously beneath which zinc-tolerant plants had evolved in the zinc-contaminated soil, but the plants a few centimeters away on either side of the fence on uncontaminated soil were not tolerant (Antonovics, 1971). It should be added that similar results were obtained where human disturbance of the environment had not been a factor, for example, the study by Aston and Bradshaw (1966) on the grass, *Agrostis stolonifera*, in protected and unprotected maritime habitats. In the various cases reported, the environmental discontinuities are sharp, especially those involving the contaminated and uncontaminated soils. The boundaries between the tolerant and intolerant plant populations are also sharp and match the environmental discontinuities closely. In some instances selection has been going on much longer than thirty years; conversely, experimental tests have shown an immediate response to selection for metal tolerance in *Agrostis tenuis* in a single generation (McNeilly and Bradshaw, 1968; Wu, Bradshaw, and Thurman, 1975).

In some cases, genetically determined partial barriers to gene flow have become established. For instance, mine populations of both *Agrostis tenuis* and *Anthoxanthum odoratum* (sweet vernal grass) flower about a week earlier than the closely adjacent, non-tolerant pasture populations (McNeilly and Antonovics, 1968). Furthermore, self-fertility has become more frequent in small metal-tolerant populations of these same two species than in the normal populations (Antonovics, 1968). However, even though disruptive selection has led to the genetic divergence of populations immediately adjacent to one another, the populations have not become reproductively isolated from one another to the point where they would be considered separate species by any means. The sharpness of the boundary between the tolerant and nontolerant populations might seem to indicate a lack of gene flow between these populations in keeping with the position taken by Ehrlich and Holm (1962), Ehrlich and Raven (1969), and most recently by Levin (1979). However, the distances involved are so small that even with the estimates for gene flow potential used by these authors (e.g., Levin and Kerster, 1974), it is clear that gene flow is entirely possible. Furthermore, there is evidence that gene flow is actually occurring between these populations (Aston and Bradshaw, 1966; McNeilly, 1968; McNeilly and Bradshaw, 1968; Antonovics and Bradshaw, 1970; Bradshaw, 1971, 1972). The sharp boundaries between the populations are due, not to reproductive isolation between them or to a lack of gene flow between them, but to strong selection pressures maintaining the distinctiveness of these populations in spite of the gene flow between them. Even though detectable gene flow is occurring, the selection pressures are so large (Antonovics, 1971) that the populations maintain their separate identities. We discussed the relationship between the selection coefficient and the migration coefficient in Chapter 11, and these examples illustrate the case in which selection outweighs the effects of migration. Furthermore, since in some cases the beginnings of reproductive isolation have been detected (McNeilly and Antonovics, 1968; Antonovics, 1968), it requires little stretch of the imagination to see that a continuation of this process over a long period could lead to sympatric speciation. These populations have been called parapatric, but as we saw earlier, individual members of these populations are interbreeding. Therefore, they must be sympatric, and the word parapatric is ambiguous.

A final word about dispersal and gene flow seems appropriate. Levin and Kerster (1974) and Levin (1979) stress that gene flow in plants is a very short-range phenomenon, seldom covering more than about 100 meters. This sort of value is difficult to reconcile with the rapidity with which oceanic islands of recent volcanic origin have been populated by plants or recently glaciated areas have been reoccupied by plants. For example, the Des Moines Lobe of the Wisconsin glaciation retreated from the upper midwestern region of the United States some 10,000 years ago. If plants were able to disperse only some 100 meters per year, they would only now be approaching Winnipeg, Canada, which seems patently absurd. Levin (1979) apparently makes a distinction between gene flow and colonization, but this seems somewhat artificial. The phenomenon is the same in either case; the dispersal of propagules is unlikely to be influenced by the presence or absence of populations of the same species in adjacent or remote areas. The outcome, however, may be quite different, depending on whether unoccupied areas open to colonization are available or the propagules must become established in the face of competition with existing, well-established populations of the same species. Once again, we are confronted by the contrast between short-term, short-range recurrent phenomena such as those reported by Levin and Kerster (1974) and the possibility of rare, or even unique, events of considerable importance to the evolution of the species with no easy way to evaluate their relative importance.

Sympatric speciation can occur if reproductive isolation is established in a single step as it is with allopolyploidy or with chromosomal speciation of the type envisioned by Lewis and White. Furthermore, extremely strong disruptive selection has led to genetic divergence and to the beginnings of reproductive isolation between populations of metal tolerant and intolerant plants despite continued gene flow. If this process were prolonged, it, too, should lead to sympatric speciation. In this case, the divergence is related to dramatic microhabitat differences, with the metal-contaminated soil lethal to plants from uncontaminated soil with a few rare exceptions. The heritability of metal tolerance is high, and it seems likely that relatively few genes are involved.

Most of the other suggested cases of sympatric speciation involve habitat diversity or temporal diversity or both. Speciation arising from temporal reproductive barriers between sympatric populations has been called *allochronic speciation*. An analysis by Alexander and Bigelow (1960) of what was originally thought to be a single species of field cricket widely distributed in northeastern North America revealed, instead, two closely related sympatric species, one diapausing in the egg stage, the other in the nymphal stage. Clear-cut morphological differences were absent, although most females from the same locality could be separated by ovipositor length. The females of the egg-over-winterer, *Acheta pennsylvanicus*, had longer ovipositors than females of the nymph-over-winterer, *A. veletis*. Furthermore, only *A. pennsylvanicus* occurs in Nova Scotia, which was interpreted to mean that *pennsylvanicus* eggs are better able to survive harsh winters than *veletis* nymphs. The calls of the two species were indistinguishable, but despite all their similarities, laboratory crosses between the two species failed to produce viable eggs. In nature, the breeding populations of the two species are seasonally isolated, for *A. veletis*, which overwinter as nymphs, live and breed as adults for four to eight weeks in late spring and early summer while *A. pennsylvanicus*, the egg-over-winterers, emerge and breed during a similar span in late summer and early fall. Alexander and Bigelow interpreted their observations as a case of sympatric, allochronic speciation brought about by the selective elimination of all but two widely separated overwintering stages, the eggs and the nymphs, in the ancestral population. This selective pressure then gave rise to two distinct, seasonally and reproductively isolated populations of adults. This interpretation was strongly criticized by Mayr (1963, 1970), who believed that geographical or allopatric speciation provided a more reasonable explanation of the observations. However, Alexander gave a comprehensive defense of allochronic, sympatric speciation in these species in a 1968 paper.

The entire situation was complicated by the report of Lim, Vickery, and Kevan (1973), who found that the two species have very different karyotypes, indicating extensive karyotypic reorganization. They reported on samples from only one locality so that as yet the extent of karyotypic variability within each species is unknown. As seen earlier, seasonal or temporal isolation is probably one of the less reliable isolating mechanisms so that the report of Lim et al. would seem to increase the likelihood of sympatric speciation in this instance.

Mayr's advocacy of allopatric or geographic speciation, which has won many adherents, is apparently based on his understanding of the nature of the genetic differences between species, illustrated by the following quotation (1970): "Most species differences, however, seem to be controlled by a large number of genetic factors with small individual effects. The genetic basis of isolating mechanisms, in particular, seems to consist largely of such genes." Although this may be true in many of the species that have been studied, it must be remembered that this sort of information is available for only a limited, and not particularly representative, group of species of animals and plants. As more detailed information for a wider variety of species becomes available, this perception of the nature of species differences may change, as it did in the case of *A. veletis* and *A. pennsylvanicus*. At this point, it may be noted that White (1978) estimated that somewhere between 90 and 99% of related species of animals and plants show karyotypic differences, or, conversely, that at an upper limit, 10% or as a more likely figure, only 1% of related species are homosequential in their karyotypes. If these estimates are anywhere near correct, and if the karyotypic changes are in any way associated with speciation, possibilities it no longer seems safe to ignore, then many present assumptions about the process of speciation will need reexamination.

The assumption that related species differ in a large number of genes, each of small effect, is not true in two species of green lacewings, *Chrysopa carnea* and *C. downesi* (Tauber, Tauber and Nechols, 1977; Tauber and Tauber, 1977a, 1977b). In this case, F_1 and F_2 species hybrids were completely viable and fertile in laboratory crosses, but the two sympatric species were reproductively isolated in nature by a combination of seasonal isolation and habitat isolation. The habitat isolation is associated with segregation at a single locus. The pale green *C. carnea* (*gg*) occurs primarily in grassy meadows during the warm months but in autumn moves from the grassy habitat onto the dead leaves of deciduous trees and changes color from pale green to reddish-brown so that they continue to be camouflaged in their overwintering site. The dark green *C. downesi* (*GG*) is restricted to conifers throughout the year and, unlike *C. carnea*, does not change color. The *G* allele is semidominant, for *Gg* individuals are intermediate in color.

The crosses also revealed that single allelic substitutions at two unlinked autosomal loci produce differential responses to photoperiod, which are responsible for the seasonal isolation between the two species. The pale green *C. carnea* is multivoltine $(D_1 D_1 D_2 D_2)$, breeding first in late winter before the vernal equinox, the rest of the breeding season extending from late spring to the end of summer. At that time, the adults enter reproductive diapause and move to the overwintering sites on dead deciduous leaves. The dark green *C. downesi* is univoltine $(d_1 d_1 d_2 d_2)$, breeding only in early spring just after the vernal equinox, and on emergence in June enters a reproductive diapause that persists through the summer, autumn, and winter until the following spring. Thus the breeding seasons of these two sympatric species do not overlap because of differences at two loci, and their habitat preferences are controlled by one locus so that just three major loci are responsible for the reproductive isolation between these species.

Tauber and Tauber (1977b) interpreted their findings to indicate that *C. downesi* is derived from a *C. carnea*-like ancestor. They postulated that the initial step involved the establishment of a stable polymorphism at the *G* locus, with the homozygous dark green *GG* individuals able to occupy the coniferous habitat. This habitat is unfavorable to light green *gg* individuals, which are conspicuous on the dark green background though cryptically colored in their usual grassy habitat. As the *Gg* heterozygotes are intermediate in color, they are at a competitive disadvantage to the homozygotes in either habitat. Consequently, the population was subjected to disruptive selection, with each homozygous type favored in a different habitat and the heterozygote favored in neither.

With each morph favored in a different habitat, mating undoubtedly was no longer random, for homogamic matings between similar morphs in the same habitat would be more likely to occur than matings between light green and dark green individuals. Furthermore, each morph would now be subject to selection pressures to enhance its fitness in its own habitat. In conifers the most favorable conditions for the predaceous larvae and the adults of *C. downesi* occur in early spring so that selection favored a univoltine, early-spring reproductive cycle. In the original grassy habitat, it was postulated that a high reproductive rate and multivoltine life cycle continued to be selectively advantageous, and thus allochronic reproduction became established as a result of disruptive selection in diverse habitats. When reproduction became asynchronous, interbreeding between the two types essentially ceased, and *C. downesi* became established as a species separate from *C. carnea*. After the two populations became reproductively isolated, selection is presumed to have continued to improve the adaptation of each species to its respective habitat, for the Taubers have found additional genetically controlled differences between the species.

Therefore, in this case, the genetic differences between two closely related sympatric species were not numerous, each with a small effect, but rather few in number and with major effects. The Taubers' interpretation of the speciation process in these lacewings seems generally plausible. However, one assumption seems unnecessary. They assumed that the initial step was the establishment of a stable polymorphism for the color morphs in the two niches. This assumption was also made in the theoretical treatments of sympatric speciation by Maynard Smith (1966) and by Dickinson and Antonovics (1973), although Maynard Smith pointed out that the conditions to be satisfied for a stable polymorphism were severe. However, if the disruptive selection was strong in the lacewings and the coniferous niche was open, the polymorphism may have been transient rather than stable. There are enough examples of rapid evolutionary change related, for example, to the development of pesticide resistance to suggest that stable polymorphism is not a necessary prerequisite to genetic divergence.

This speciation in lacewings is of particular interest because it appears to be a case of sympatric speciation due to disruptive selection in polyphagous, non-host-specific animals. Bush (1975) wrote that "Sympatric speciation appears to be limited to special kinds of animals, namely, phytophagous and zoophagous parasites and parasitoids." The lacewings do not fall in that category, thus this case opens up the possibility of sympatric speciation in groups other than the monophagous or parasitic species. Furthermore, unlike the field crickets where major karyotypic reorganization and hybrid sterility were found, in the lacewings the F_1 and F_2 hybrids were viable and fertile, the allelic differences between the species were few in number, and, as yet at any rate, there have been no reports of karyotypic differences. Thus, sympatric speciation may be more general in scope than has been suspected, even by some of its advocates such as Bush (1975).

Speciation of the sort envisioned by Bush can be illustrated by his work on the true fruit flies of the family Tephritidae, an extensively studied group of considerable economic importance because it contains a number of pests of cultivated fruit. The *Rhagoletis pomonella* group consists of four sibling species, each quite similar morphologically, but isolated reproductively and differing biologically in that each species infests fruits of a different plant family (*R. pomonella*, Rosaceae; *R. mendax*, Ericaceae; *R. cornivora*, Cornaceae; *R. zephyria*, Caprifoliaceae). *R. pomonella*, itself, originally infested only hawthorns, *Crataegus*, throughout the eastern United States (Bush, 1969). Although cultivated apples had been inroduced into North America some 200 years previously, only in the 1860s was a "host race" of *R. pomonella* discovered on introduced apples in the Hudson River Valley. It spread rapidly over most of the northeastern United States and lower Canada, reaching Winnipeg by 1916, but did not invade the southeastern states where still only the hawthorn "host race" is found.

At present, sympatric populations of these two host races of *R. pomonella* differ slightly in size, the number of postorbital bristles, and ovipositor length, but differ significantly in their time of emergence. Individuals of the apple "race" emerge from the pupa from about June 15 to the end of August, with a peak about July 25. The hawthorn "race" emerges from about August 5 to October 15 with a peak about September 12. In each case the time of emergence coincides with the time when their respective host fruits are in a suitable condition for oviposition—that is, about a month before the fruits ripen. This means, however, that the apple race emerges and begins oviposition about four to five weeks before the hawthorn race, and the populations are almost completely allochronically isolated because the life span of *R. pomonella* is estimated at about 20 to 30 days in nature (Bush, 1974). Native crabapples are not infested by *R. pomonella* so that the apple race is thought to have originated directly from the hawthorn race.

The apple race, in turn, is thought to have been the source of a new and quite localized race of *R. pomonella* just discovered in the 1960s living on cultivated cherries grown on the Door Peninsula in Wisconsin. The fruiting season of the cherries is even earlier than that of the apples so that early emerging adults of the apple race would be more likely to make the transition to cherries than the much later emerging adults of the hawthorn race.

Furthermore, in *Rhagoletis*, courtship and mating take place on the fruit of the host plant where the female oviposits. Therefore, once a shift to a new host has occurred, the insects attracted to the new host will be reproductively isolated from those drawn to the original host. This isolation, plus the allochronic emergence of adults with different hosts, produces a strong barrier to genetic exchange between the two populations. Rather than host races, they appear to behave as separate species, and the circumstances of their origin suggest that they have originated sympatrically, a possibility that even Mayr (1970) now recognizes.

The sequence of events for sympatric speciation in *Rhagoletis* was envisioned by Bush (1974) as follows: First, the shift to a new host plant would occur where both old and new host plants occur together. Second, the fruiting times of the two host plants should overlap to some extent. Under these circumstances, repeated host shifts might be "attempted" until individuals with the right combination of genes for host selection and survival on the new host make the transition. The following assumptions were also made:

1. Diapause and emergence times are genetically controlled.
2. Host recognition basically is controlled by alleles at a single locus, which influences orientation to and selection of a host plant through chemical cues. For example, $H_1 H_1$ individuals might be attracted only to the original host and $H_2 H_2$ only to the new host. $H_1 H_2$ is presumed to be attracted to both hosts but might, because of conditioning or polygenic modifiers, be attracted to one or more of a range of hosts.
3. Survival on the host would depend on alleles at another locus, with $S_1 S_1$ individuals surviving on the original host, $S_2 S_2$ individuals on the new host, and $S_1 S_2$ individuals on either.
4. Courtship and mating occur on the host plant so that homogamy is the rule.

Thus Bush assumed that allelic differences at two loci were sufficient to mediate a host plant shift. If allochronic isolation were also involved, at least three loci would be necessary.

This model may seem simplistic, and ambiguous in the assumptions about the heterozygotes. However, some evidence in support of these assumptions is available. First are the results of Tauber and Tauber (1977b), discussed earlier, who found three loci responsible for the reproductive isolation between two sympatric species of green lacewings. More evidence was provided by Huettel and Bush (1972) in a study of two very closely related sibling species of essentially monophagous gall-forming flies of the genus *Procecidochares* (Tephritidae). One species, *P. australis*, infests the camphor weed (*Heterotheca* spp. of the tribe Asterae, family Asteraceae); the other, *P. actitis*, infests the camphor daisy (*Macroanthera phyllocephala*, Asterae, Asteraceae). Crosses have established that a single locus controls host recognition, for flies homozygous for one allele will select only one type of host plant, which means that they can distinguish between the two host plants and, of course, other members of the Asteraceae as well. The F_1 hybrid larvae can be reared on either host plant, but the F_1 adults oviposit nearly all their eggs on the host species on which they were reared, which indicates some sort of conditioning. Backcross hybrids, however, revealed that about half the B_1 females laid eggs on the host plant on which they were not reared. Therefore, there must be a separate genetic component controling the tendency to oviposit. Furthermore, forced oviposition experiments showed that *P. australis* larvae die on *M. phyllocephala* and *P. actitis* larvae cannot survive on *Heterotheca*. Therefore, adaptation to one host plant or the other involves genetic changes at loci controling three separate processes: larval survival, host recognition, and oviposition "induction." Moreover, relatively few loci appear to be involved.

Another similar carefully studied case involving a number of species of diprionid sawflies was reported by Knerer and Atwood (1973). Some 85 species have been recognized, but probably many more remain to be identified. Among them is the *Neodiprion abietis* complex in which three *N. abietis* "strains" found in Nova Scotia showed a rigid host selection: the females of each strain would oviposit only in their preferred host—balsam fir, white spruce, or black spruce. If deprived of their normal host, the females would die with their full egg complement rather than lay on an alternate host. This rigid host selection was apparently controlled at a single locus. Since many conifers contain toxic chemicals, larvae often could not survive when transferred to an alien host. Detoxification of such chemicals and, hence, larval survival also seem to be under genetic control. In addition, temporal isolation between strains has appeared owing to differences in time of larval hatching and adult emergence. Thus differences in host preference, larval survival, and developmental rates all serve to keep sympatric "strains" such as the three "strains" or "races" of *N. abietis* in Nova Scotia reproductively isolated from one another. Although morphologically they are extremely similar, they are behaving as separate species. Yet the differences between them seem to be under relatively simple genetic control.

It is not essential to review in detail all possible cases of sympatric speciation, in part because the subject has been reviewed recently by Bush (1975) and by White (1978), and also because so few of the possible cases of sympatric speciation have been studied thoroughly enough for a decision to be made about their validity. It is worth noting, however, that two of the recent textbooks on evolution (Dobzhansky et al., 1977; Futuyma, 1979) barely mention sympatric speciation.

A number of other possible modes of sympatric speciation have been suggested. In one group, changes in behavior are thought to lead to reproductive isolation. For example, if pheromones are involved in mating, a simple change in the pheromone molecule could lead to isolation. Similarly, changes in the courtship pattern could produce isolation if disruptive selection is operative. In plants with polymorphic flowers, flower constancy by the insects pollinating these different flower types has been suggested as a means by which sympatric isolation could be achieved. The genetic control in all of these cases could be rather simple, but all could lead to sympatric isolation. The internal and external animal parasites and parasitoids constitute an extremely large group of species in which sympatric speciation has been postulated, and in many cases a sympatric shift to a new host often seems more plausible than allopatric speciation for the origin of some of these species. Although the clusters of closely related endemic species in freshwater lakes have been assumed to originate by allopatric speciation (Mayr, 1963), the arguments are not always convincing. Further study of these remarkable species swarms in such lakes as Baikal in Russia, Malawi and Tanganyika in Africa, Ochrid on the Albanian-Yugoslavian border, and Lanao in the Philippines may lead to a reinterpretation of the data (review in White, 1978). White also called attention to endemic species groups on extremely small oceanic islands of volcanic origin as likely examples of sympatric speciation.

Allopatric speciation involves the geographic isolation of populations and their gradual divergence genetically as they adapt to their environments. Reproductive isolation arises as an incidental by-product of this genetic divergence although it may be reinforced by natural selection if the populations later become sympatric.

Sympatric speciation appears to involve the origin of some form of reproductive isolation as the initial, crucial step. The reproductive isolation may result from chromosomal change (allopolyploidy, chromosomal rearrangements), change to a new host as in *Rhagoletis* and *Chrysopa*, or perhaps from extremely strong disruptive selection as in the metal-tolerant plants. In all of these cases, the genetic differences responsible for the reproductive isolation are initially relatively few. A major problem in this transition is for the newly isolated sympatric population to reproduce itself. If the individuals are self-fertile hermaphrodites, or if an asexual generation intervenes to permit the buildup of numbers, this should be no particular problem. In species with separate sexes, if the reproductive rate is high enough, several individuals with the same new favorable combination of genes may be produced by a single set of parents. In species with low reproductive rates, however, lack of a mate for the new type could be an insuperable barrier to sympatric speciation. Once reproductive isolation is established between the sympatric populations, they will each adapt to their particular environmental circumstances and thus will gradually diverge from one another genetically. Therefore, the process of sympatric speciation is almost the reverse of allopatric speciation, for in sympatric speciation the initial step is the achievement of reproductive isolation followed by gradual adaptive genetic divergence. Two factors make sympatric speciation seem more reasonable than it was formerly thought to be. One is a greater appreciation of the opportunistic nature of evolution as demonstrated by studies of introduced species, pesticide resistance, and the like. The other is the extremely high selection pressures that have been observed in nature. To these should be added the fact that more instances are constantly being reported in which sympatric speciation is the most reasonable interpretation of the data.

This brief review of the ways that species may originate indicates that the patterns of evolution and the genetic systems that mediate evolutionary change can be quite diverse. In a way this is unfortunate. No longer is it safe to assume that one mode of speciation, allopatric speciation, is the norm and all others are aberrations from it. It means that speciation in each group must be studied on its own terms and in greater detail than ever before. For example, it means that, if White is anywhere near correct in his estimate that 90% or more of all closely related species differ in karyotype, karyotypic analysis should become a routine part of any phylogenetic study. Similarly, data ranging from the ecological and the ethological level to the biochemical and genetical level will be desirable or necessary to delineate the nature of species differences. This is unfortunate in the sense that so much more work will be required to produce an adequate study, but it is fortunate in the sense that once the work is done, the understanding of the process of evolution in the group will be so much more complete.

CHAPTER 16

Competition

DEFINITIONS

Thus far, we have considered only the population dynamics of a single species. However, now we shall consider the nature of interactions between two different species.

The population size of a species is affected by its own innate characteristics, its r value, for example, and by the carrying capacity of the environment—that is, by the environmental conditions that determine K. In addition, the size of a population may be influenced in positive or negative ways by the presence of another species, the possibilities of which are outlined in Table 16-1. Interactions 7 through 9 are positive. Interactions 5 and 6 are favorable to one population, the predator or the parasite, and unfavorable to the other, the prey or the host. In the discussion of population fluctuations and cycles, it was seen how the abundance or scarcity of the prey species may affect the population size of the predator. The table brings out the similarity between predator-prey interactions and host-parasite interactions. Interactions 2 through 4 are negative in that the presence of another species either directly or indirectly affects population growth adversely. Where both species are affected adversely, the phenomenon is usually called *competition*. From the table, it is clear that competition differs from predation and from parasitism.

The word *competition* is another widely used term in biology that has been variously defined. Birch (1957) wrote a useful paper on the

Table 16 - 1. Types of Species Interactions

Type of Interaction	Effect on Species		Nature of Interaction
	1	2	
1. Neutralism	0*	0	None
2. Direct competition	-	-	Each species inhibits the other directly.
3. Indirect competition for resources	-	-	Indirect inhibition if resource is in short supply.
4. Amensalism	-	0	Species 1 inhibited, 2 unaffected.
5. Parasitism	+	-	Species 1, the parasite, usually smaller than 2, the host
6. Predation	+	-	Species 1, the predator, usually larger than 2, the prey.
7. Commensalism	+	0	Species 1, the commensal, benefitted; 2 unaffected.
8. Protocooperation	+	+	Favorable to both but not obligatory.
9. Mutualism	+	+	Favorable to both and obligatory.

Source: Fundamentals of Ecology, Third Edition, by Eugene P. Odum. Copyright © 1971 by W. B. Saunders Company. Copyright 1953 and 1959 by W. B. Saunders Company. Reprinted by permission of Holt, Rinehart and Winston.

0 = No significant interaction.

+ = Favorable interaction; positive term added to growth equation.

$^-$ = Negative interaction; negative term added to growth equation.

meanings of competition. The definition he favored was, "Competition occurs when a number of animals (of the same or of different species) utilize common resources the supply which is short; or if the resources are not in short supply, competition occurs when the animals seeking that resource nevertheless harm one or other in the process." Andrewartha (1961) feels that the definition of competition should be restricted to just the first half of Birch's definition, that is, that competition occurs only when the common resources are in short supply.

Elton and Miller (1954), on the other hand, gave the following definition: "Inter-specific competition, in the more limited and correct use of the notion, refers to the situations in which one species affects the population of another by a process of *interference,* i.e., by reducing the reproductive efficiency or increasing the mortality of its competitor. Or both species may be acting in such a way on each other." In this case, the emphasis is on interference; there is no mention of common resources, which both Birch and Andrewartha stress. Moreover, Elton and Miller refer only to interspecific competition and apparently exclude intraspecific competition from their definition.

Birch cites both etymology and Darwin in support of his definition, and points out that competition is derived from *com* and *petere,* which mean "together seek." However, the etymology carried only this far leaves out the idea of a contest, which is implicit in Birch's definition and is found among the meanings of *competere.* His definition also refers only to animals, and thus seems to suggest that competition does not occur among plants. He also excludes as incorrect the use of the word to refer to competition between different genotypes of the same species, a usage common among geneticists and evolutionists. He states that competition is separate from and not synonymous with natural selection and that Darwin did not regard *competition* as equivalent to the *struggle for existence.*

The following is DeBach's (1966) definition:

> "Competition"—the attempted or actual utilization by two or more organisms of common resources or requisites involves competition, even if one does not directly harm or interfere with (in the sense of "bother") the other in the process. An apparent abundance of food or other requisites does not preclude the occurrence of competition. . . . It is my contention, and a major point of difference with most others, that competition and competitive displacement can occur even when the supply of food (or requisites) is abundant in relation to the animals' immediate needs.

Miller (1967) proposed a modified version of a definition suggested by Clements and Shelford in 1939: "Biological competition is the active demand by two or more individuals of the same species population (intraspecies competition) or members of two or more species at the same trophic level (interspecies competition) for a common resource or requirement that is actually or potentially limiting."

Emlen (1973) offers the following as the "only workable, noncircular definition" of interspecific competition that retains the spirit of Birch's definition: "(interspecific) competition occurs when two or more species experience depressed fitness (r_o or K) attributable to their mutual presence in an area." However, earlier, following a general discussion of the nature of competition he states, "Intraspecific competition is the process through which natural selection acts on gene frequencies." In this, he clearly is at odds with Birch. Ayala (1970b) also seems to follow Birch's definition, but he, too, broadens it to include some aspects of natural selection.

Attempts have been made to distinguish between the various forms that competition may take by the use of such expressions as "exploitation competition" and "interference competition" (Park, 1954) or "scramble competition" and "contest competition" (Nicholson, 1954) or "active competition" and "passive competition" (Emlen, 1973). To some degree these phrases are overlapping and their meaning is also dependent on the concept of competition held so that they have not proved particularly useful.

Ricklefs (1973) states, "Competition is perhaps the most elusive and controversial of all ecological phenomena. Competition occurs when the use of a resource—food, water, space, light, and so on—by one individual reduces the availability of that resource to another individual, whether of the same species (intraspecific competition) or of a different species (interspecific competition)." Wilson and Bossert (1971), on the other hand, wrote: "Competition is defined by ecologists as the active demand by two or more organisms for a common vital resource," as if all ecologists were agreed on a definition of competition, which, as the above quotations have made clear, is far from true.

To summarize the differences that have emerged, some definitions apply only to animals, others to all organisms—plants as well as animals; some definitions refer only to interspecific competition, others to both intraspecific and interspecific competition; for competition to occur, resources must be in short supply in some definitions but not in others; sometimes (e.g., Elton and Miller, 1954) the definition is so broad that it does not exclude predator-prey relations, but in others (e.g., Miller, 1967) the same trophic level is specified. Given these differences of opinion, it may be hazardous (not to mention presumptuous) to attempt to reach some workable definition of competition.

Since Darwin was the first to use "competition" in the modern biological sense, it is of interest to study his (1872) use of the word, especially in the third chapter of *The Origin of Species*, "Struggle for Existence." A careful reading leads to the conclusion that he regards competition as one aspect of the struggle for existence, rather than as separate from it as Birch suggests. He states (p.77) for example, "We will now discuss in a little more detail the struggle for existence," and a sentence later, "The elder DeCandolle and Lyell have largely and philosophically shown that all organic beings are exposed to severe competition. In regard to plants, no one has treated this subject with more spirit and ability than W. Herbert, Dean of Manchester. . . . Nothing is easier than to admit in words the truth of the universal struggle for life." There are two points to note here. First, he appears to regard competition as a part of the struggle for existence. Second, he believes that competition involves plants as well as animals.

In the next paragraph, headed "The term, Struggle for Existence, used in a larger sense," he writes,

> I should premise that I use this term in a large and metaphorical sense including dependence of one being on another, and including (which is more important) not only the life of the individual, but success in leaving progeny. Two canine animals, in a time of dearth, may be truly said to struggle with each other which shall get food and live. But a plant on the edge of a desert is said to struggle for life against the drought, though more properly it should be said to be dependent on the moisture. A plant which annually produces a thousand seeds, of which only one of an average comes to maturity, may be more truly said to struggle with the plants of the same and other kinds which already clothe the ground. The mistletoe is dependent on the apple and a few other trees, but can only in a far-fetched sense be said to struggle with these trees, for, if too many of these parasites grow on the same tree, it languishes and dies. But several seedling mistletoes, growing close together on the same branch, may more truly be said to struggle with each other. As the mistletoe is disseminated by birds, its existence depends on them; and it may metaphorically be said to struggle with other fruit-bearing plants, in tempting the birds to devour and thus disseminate its seeds. In these several senses, which pass into each other, I use for convenience' sake the general term of Struggle for Existence.

In the next paragraph the following sentence occurs: "Hence, as more individuals are produced than can possibly survive, there must in every case be a struggle for existence, either one individual with another of the same species, or with the individuals of distinct species, or with the physical conditions of life."

These quotations make it still clearer that he regarded competition between individuals of the same or closely related species of plants and animals as a major factor in the struggle for existence, although he does not use the word competition in these passages. Later, however (p. 84), he does use the word in this context.

> The action of climate seems at first sight to be quite independent of the struggle for existence; but in so far as climate chiefly acts in reducing food, it brings on the most severe struggle between the individuals, whether of the same or of distinct species, which subsist on the same kind of food. . . . Each species, even where it most abounds, is constantly suffering enormous destruction at some period of its life, from enemies or from competitors for the same place and food; and if these enemies or competitors be in the least degree favored by any slight change of climate, they will increase in numbers; and as each area is already fully stocked with inhabitants, the other species must decrease. . . .

That climate acts in the main part indirectly by favouring other species, we clearly see in the prodigious number of plants which in our gardens can perfectly well endure our climate, but which never become naturalized, for they cannot compete with our native plants nor resist destruction by our native animals.

With respect to the severity of competition the following was written (p. 92):

But the struggle will almost invariably be more severe between the individuals of the same species, for they frequent the same districts, require the same food, and are exposed to the same dangers. In the case of varieties of the same species, the struggle will generally be almost equally severe and we sometimes see the contest soon decided: for instance, if several varieties of wheat be sown together, and the mixed seed be resown, some of the varieties which best suit the soil or climate, or are naturally the most fertile, will beat the others and so yield more seed, and will consequently in a few years supplant the other varieties. . . . As the species of the same genus usually have, though by no means invariably, much similarity in habits and constitution, and always in structure, the struggle will generally be more severe between them, if they come into competition with each other, than between species of distinct genera. . . . We can dimly see why the competition should be most severe between allied forms, which fill nearly the same place in the economy of nature; but probably in no one case could we precisely say why one species has been victorious over another in the great battle of life.

Darwin also used competition in relation to sexual selection (p. 108). "This form of selection [sexual selection] depends, not on a struggle for existence in relation to other organic beings or to external conditions, but on a struggle between the individuals of one sex, generally the males, for the possession of the other sex. The result is not death to the unsuccessful competitor, but few or no offspring."

In the use of the word competition in biology, Darwin clearly has priority. Even though he did not formally define competition, a fair and reasonable reading of his writings, as exemplified by the quotations above from *The Origin of Species*, leads to an understanding of his use of the word, and it is clear that his concept is different from that of Birch and other ecologists who have attempted to provide rigorous definitions. From the quotations above the following points emerge:

1. Darwin considered competition to occur among plants as well as animals. Thus Birch's definition, which refers only to animals, is too narrow.

2. Birch's emphasis on the etymology of competition as meaning "together seek" is inappropriate because plants, in a literal sense, do not seek anything together. The more important etymological significance of competition is the idea of a contest between contending organisms.

3. The contest may involve such resources as space or food; it may also involve competition among males for females, which are not a resource in the usual sense of the word, and may or may not be in short supply. In other words, competition involves interference by one contestant with the other in a way that reduces its reproductive efficiency or increases its mortality, as Elton and Miller suggested. Thus the emphasis properly belongs on the contest and not on the resources.

4. Although Darwin did not regard competition as synonymous with the struggle for existence, he considered it an important component of the struggle for existence. In his reference to destruction from enemies or from competitors, he seemed to draw a distinction between competitors and what we today would call predators and parasites.

5. Darwin clearly regarded competition as both intraspecific and interspecific. In fact, he stated that the greater the similarity between the contending organisms, the more severe the competition, and in the example quoted about wheat (and in other examples in the same paragraph not included in the quotation), he explicitly considered the contest between individuals of different varieties (or genotypes) grown together as a form of competition.

6. Therefore competition, in Darwin's view, did constitute a form of natural selection. Furthermore, male competition, together with female choice, formed the mechanism of sexual selection. (Notable, too, is the fact that he recognized the importance of leaving progeny, or reproductive fitness as it is known today, in the "Struggle for Existence," although he is often cited as stressing survival, as in "Survival of the Fittest," rather than reproductive fitness as of primary importance.)

Therefore, in the Darwinian sense, *competition can be defined as a contest between individuals of the same species or of ecologically similar species such that one or both individuals or groups are adversely affected in ways that will ultimately be reflected in reduced survival or reproduction.* Ecologists generally seem to use competition in reference to situations in which both contestants are adversely affected; geneticists and evolutionists use it for situations in which a winner and a loser can be identified. However, this apparent difference is more a matter of emphasis than it is real. In one case, the emphasis is on the unfavorable interactions, but the presumed outcome, as will be seen, is the extinction of one of the competing groups. In the other case, the emphasis is on the outcome, but during the period of competition, the presence of losing competitors can be regarded as having an adverse effect on the ultimate winners. In the above definition, the nature of the "contest" is unspecified because, in competing populations, the effect of competition is measured by the degree of survival and reproduction in the populations when together as compared to when they are separate. To determine the actual nature of the competition leading to the observed adverse effects usually requires further study. With this definition it is difficult to avoid the conclusion that competition is a form of natural selection. Even in interspecific competition, if one species population is eliminated, one group of genotypes has been favored over another group. A final point is that the definition above has been written to exclude predation and parasitism, which are phenomena distinct from competition.

This definition has the advantages of reflecting Darwin's original intentions, excluding predation and parasitism, and indicating an operational measure of competition through the comparison of birth and death rates when populations are in competition and when they are not. Competition will be used as defined above for the remaining discussion.

COMPETITIVE DISPLACEMENT

In the literature, the concept of competition is inextricably interwoven with the concept of *competitive displacement,* which is also known as *competitive exclusion, Gause's principle,* the *Lotka-Volterra law,* and *Grinnell's axiom.*

Opinions of the significance of competitive displacement vary widely. Hutchinson and Deevey (1949), for example, said, "The generalization. . . that two species with the same niche requirements cannot form mixed steady-state populations in the same region has become one of the chief foundations of modern ecology," and they also called it "perhaps the most important theoretical development in general ecology." Hardin (1960) wrote, "In the history of ecology . . . we stand at the threshold of a renaissance of understanding, a renaissance made possible by the explicit acceptance of the competitive exclusion principle." A rather remarkable statement since his formulation of the "exclusion principle" was, "Complete competitors cannot coexist"—a statement so constructed that "every one of the four words is ambiguous." Equally remarkable was his statement, "The 'truth' of the principle is and can be established only by theory, not being subject to proof or disproof by facts." A rather unusual scientific theory!

Contrast these statements with the following from Cole (1960): "It is contended that there is little justification and no necessity for believing in the competitive exclusion principle as usually formulated. There is danger that a trite maxim like this may lead to the neglect of important evidence," and by MacArthur (1968): "The statement 'two species cannot co-exist if their niches are identical' became true but trivial (since no two individuals, let alone two species, have identical niches)." On the one hand, there is a theory that is true, trite, and trivial, and on the other, a coming renaissance based on the most important theoretical development in general ecology.

As with competition, the competitive displacement theory has a fairly long history and has been variously stated. The quotations from Darwin, cited earlier, make clear that he regarded competition between species belonging to the same genus as likely to lead to the exclusion of one species or the other. However, Grinnell (1904) seems to have been one of the first to state the idea explicitly:

> Every animal tends to increase at a geometric ratio, and is checked only by limit of food supply. It is only by adaptations to different sorts of food, or modes of food getting, that more than one species can occupy the same locality. Two species of approximately the same food habits are not likely to remain long enough evenly balanced in numbers in the same region. One will crowd out the other.

Haldane (1924a), Volterra (1926a), and Lotka (1932) gave a mathematical treatment to the concept of competitive exclusion, to which we shall return later, and concluded that two species cannot coexist indefinitely if they use a common resource in limited supply. Nicholson (1933) wrote, "For the steady state (in the coexistence of two or more species) to exist, each species must possess some advantage over all other species with respect to some one, or group, of the control factors to which it is subject." Gause's name became associated with the concept of competitive exclusion because he wrote a much-cited book (1934) about his experimental investigations of the subject.

In 1945 Crombie wrote, "It is easy to demonstrate theoretically that two species with identical ecological niches cannot survive together in the same environment unless density-independent factors keep the population low enough to eliminate interspecific competition." Slobodkin (1961) wrote, "no two species can indefinitely continue to occupy the same ecological niche," and also a corollary, "If two species persist in a particular region, it can be taken as axiomatic that some ecological distinction must exist between them and that their ecological niches, in the restricted sense, do not coincide." DeBach (1966), in a review of the subject of competitive displacement and coexistence, stated, "Different species having identical ecological niches (that is, ecological homologues) cannot coexist for long in the same habitat." More recently Wilson and Bossert (1971) were terse and emphatic, "No two species that are ecologically identical can long coexist." Ricklefs (1973) gave two rather different versions: "The principle of competitive exclusion states that two species with similar ecological requirements cannot coexist in the same environment." Also "No two species can coexist indefinitely on a single limiting resource." A final quote from Emlen (1973): "It is difficult to accept as a basic law a statement that says that competition between two species will inevitably lead to the demise of one provided the two species aren't too different."

These statements are by no means an exhaustive review of all the formulations of the competitive exclusion principle, yet they suffice to bring out several points. First, like the definitions of competition, some make reference to limiting resources, but others do not mention resources at all. Furthermore, some definitions state that exclusion occurs if the ecological niches are "identical"; others only require that they be "similar."

EXPERIMENTS

Some years ago (Merrell, 1951) experiments were run on interspecific competition between two species of *Drosophila, D. melanogaster* and *D. funebris.* Mixed populations were run in units consisting of two half-pint food bottles connected by a rubber hose. The bottles were replaced alternately with a new bottle containing fresh cornmeal-agar food. Thus both species lived in a limited space with a limited food supply on which both were dependent. Three outcomes were observed: (1) both species coexisted for nearly two years until the experiment was terminated; (2) *melanogaster* survived and *funebris* was eliminated; (3) *funebris* survived and *melanogaster* was eliminated. These results were not due to chance, for it was shown that *melanogaster* thrived best on the fresh food and *funebris* did best on the food as it aged and putrefied. If the food bottles were replaced too frequently, *melanogaster* would win out; if they were replaced too infrequently, *funebris* would win; if the replacement was timed properly, both species were able to persist indefinitely.

Therefore, two species living together in a very small space and dependent on a common, limited food supply were able to coexist. This finding does not contradict the Lotka-Volterra dictum if it is phrased that two types with identical needs and habits cannot survive in the same place if they compete for limited resources. The reason it does not is that while they were living in the same place and dependent on the same limited food supply, they did not have identical needs and habits, as additional experiments showed. These experiments demonstrated that the larvae of *melanogaster* developed better on the fresh, yeasted food-medium, but that *funebris* larvae developed better on the old, putrid medium. Thus, even though both species coexisted in a small, limited system, they were exploiting the common resources in the system in sufficiently different ways to permit their coexistence. This is not to say that there was no competition within the system, for there was. Even though the presence of an alien species was shown not to affect egg-laying frequency by the females, larval competition was demonstrated. The percentage of *funebris* larvae able to pupate was significantly reduced in the presence of *melanogaster* larvae although the pupation rate of the *melanogaster* larvae was unaffected in these experiments. (If run on older food, this might have been reversed.) Not only the percentage pupating, but the rate of development and pupation of *funebris* larvae were shown to be slowed in the presence of *melanogaster* larvae as compared to the rate in their absence. Therefore, there must have been competition in the population units even though it did not lead to competitive exclusion.

Lest these results seem to be rather artificial because the experiments were run in the laboratory, a parallel exists in nature in the study by Wagner (1944) of the closely related species *D. mulleri* and *D. aldrichi*. The females of both species lay their eggs in the ripe fruit of the cactus *Opuntia lindheimeri,* and the larvae of each species thus appear to be thrown into direct competition with one another in this small closed ecosystem, and hence, should not be able to coexist. However, Wagner, by an analysis of the intestinal contents of the larvae, showed that they tended to feed on somewhat different species of yeast and bacteria in the fermenting pulp, and thus their ecological niches were sufficiently different to permit them to coexist.

Ross (1957) studied six closely related species of the *lawsoni* complex of the leafhopper genus *Erythroneura,* all of which feed on the eastern sycamore, *Platanus occidentalis.* "All six breed only on the sycamore and are the only leafhoppers to do so. All appear to have identical habits, the generations of the various species maturing synchronously in each locality, hibernating together, and feeding in the same manner, often side by side on the same leaf." He added later that they "appear to occupy the same niche (leaves of one species of plant) simultaneously throughout the same area." Furthermore, he reported that "interspecific competition did not occur between these organisms during the period of this study." From his observations he argued that "several species may occupy the same ecological niche at the same time." However, it is clear in the paper that his niche concept is that of a spatial niche or microhabitat, the leaves of a single tree, rather than the broader niche concept discussed by Hutchinson and others. However, he did find ecological differences among the species: *E. lawsoni,* for instance, seemed better able to live on trees in open, dry, wind-swept situations. Therefore, even though these species coexisted in very close proximity to one another, seemingly in violation of competitive exclusion, this is only true if their ecological niche is equated with their micohabitat on the leaves of the sycamore. They may well, like the examples in *Drosophila,* be utilizing this microhabitat in subtly different ways, which permit their coexistence, perhaps even without competition.

Coexistence was not always the outcome in experimental populations. In many cases, competitive exclusion appeared to occur. For instance, a number of experiments were run on interspecific competition between two species of flour beetles, *Tribolium castaneum* (CS) and *T. confusum* (CF), (Park, 1954). At high temperatures and high humidity, CS invariably won the competition, but at low temperatures and humidity, CF always won. At an intermediate temperature (29°C) and relative humidity (70%), the outcome of the competition was unpredictable. Sometimes one species won, sometimes the other. These results provided a widely cited example of indeterminism in biology, with the outcome thought to be largely decided by stochastic, that is, random events. The process that governed the outcome was considered comparable to tossing a coin. The usual belief was that, under these conditions, the competitive abilities of the two species were so evenly matched that lack of control over minor differences in the environmental conditions between different replicates was sufficient to account for the difference in outcome. Another suggestion was that the unpredictability was the result of complex ecological interactions.

The work of Lerner and Ho (1961) and Lerner and Dempster (1962), however, showed that "When genetic heterogeneity of the founding population is largely eliminated, so is most of the 'indeterminacy'," and that "Within-species variation was demonstrably of over-riding importance" (1962). Thus, the explanation for the indeterminacy lay, not in the variation of ecological conditions, but in genotypic variation. When the genetic variability was reduced by a proper specification of genotypes, the outcome, even at the intermediate environmental conditions, became predictable. In starting his populations, Park had used just two pairs of parents of each species. When Lerner and his associates broadened the genetic base of the founding populations by using 10 pairs of parents from each species, *T. castaneum* won 20 out of 20 times. Furthermore, by inbreeding and then using the different inbred lines as competitors, they demonstrated genetic differences in competitive ability among the different lines of the two species, and again showed that, when the genotypes were specified as well as the conditions, the outcome of the competition was not indeterminate, but predictable. Thus, studies of natural and experimental populations gave diverse results, some seeming to support competitive exclusion and others to refute it.

LOTKA-VOLTERRA EQUATIONS

Since competitive exclusion has drawn so much attention, it may be worth while to develop the mathematical formulations in order to study the concept in that form.

The equations are based on the logistic equation for population growth in a single species, given earlier, which takes the form,

$$\frac{dN_1}{dt} = r_1 N_1 \left(\frac{K_1 - N_1}{K_1}\right) = r_1 N_1 \left(1 - \frac{N_1}{K_1}\right) \tag{16.1}$$

where

$\dfrac{dN_1}{dt}$ = the instantaneous rate of change of population size of species 1,

r_1 = the instantaneous rate of population growth,

N_1 = the size of the population, and

K_1 = a constant, the upper limit to population size, often known as the carrying capacity of the environment.

When $N_1 = K_1$, $dN_1/dt = 0$, the population has reached an equilibrium in size, and no further growth will occur.

In equation 16.1, the rN_1 term is an expression of the unlimited growth potential of the species population in the absence of competition from other individuals of the same species. The other term, $(-N_1/K_1)$, indicates the inhibitory effect on population growth from individuals of the same species. The damping effect on population growth of a single individual of species 1 on its own population growth is then equal to $(1/K_1)$.

If a second species is introduced into the system in competition with species 1, another term can be added to the equation to indicate the inhibitory effect of individuals from species 2 on the growth of the population of species 1, as follows:

$$\frac{dN_1}{dt} = r_1 N_1 \left(1 - \frac{N_1}{K_1} - \frac{aN_2}{K_1} \right) \tag{16.2}$$

where a = the competition coefficient of species 2 with respect to species 1.

The competition coefficient is introduced because it is thought unlikely that an individual of species 2 will have exactly the same effect on the population growth of species 1 as an individual of species 1 itself. The inhibitory effect of one individual of species 2 on the population growth of species 1 then is equal to a/K_1.

Conversely, the growth of a population of species 2 in the presence of species 1 can be written:

$$\frac{dN_2}{dt} = r_2 N_2 \left(1 - \frac{N_2}{K_2} - \frac{\beta N_1}{K_2} \right) \tag{16.3}$$

where β = the competition coefficient of species 1 with respect to species 2.

In a two species system with competition, the growth rate of each population becomes a function of the number of individuals of both species. If $N_2 = 0$, then equation 16.2 reduces to the logistic growth equation of species 1; similarly, if $N_1 = 0$, then equation 16.3 reduces to the logistic equation for species 2.

The two-species system described by equations (16.2) and (16.3) will be at equilibrium when $dN_1/dt = 0$, and $dN_2/dt = 0$. This will be true if r_1, r_2, N_1, or N_2 equals zero, of course, but the more interesting equilibria exist when the terms within the parentheses equal zero — that is, when

$$\left(1 - \frac{N_1}{K_1} - \frac{aN_2}{K_1} = 0 \right)$$

and

$$\left(1 - \frac{N_2}{K_2} - \frac{\beta N_1}{K_2} = 0 \right)$$

or

$$K_1 - N_1 - aN_2 = 0$$

$$K_2 - N_2 - \beta N_1 = 0.$$

Rewritten, these equations become

$$N_1 = K_1 - aN_2$$
$$N_2 = K_2 - \beta N_1$$

which are algebraic representations of straight lines. Therefore, it is possible to graph the possible outcomes of competition and thus to visualize the process on a plane with the rectangular coordinates N_1 and N_2. For the equation $N_1 = K_1 - aN_2$, first N_1 and then N_2 can be set equal to zero, and in this way the two values that intercept the coordinates can be obtained. Thus, when $N_2 = 0$, the intercept on N_1 is K_1. When $N_1 = 0$, the intercept on N_2 is K_1/a. The straight line that joins these two intercepts includes all the points at which $dN_1/dt = 0$. (Figure 16-1). Similarly, for N_2 the intercepts are K_2

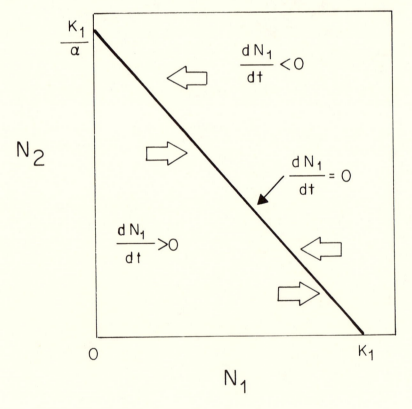

Figure 16.1. The zero-growth curve or isocline of species 1 in competition with species 2. Horizontal arrows indicate direction of change in N_1.

and K_2/β (Figure 16-2), and the straight line joining these intercepts includes all the points at which $dN_2/dt = 0$.

Below the lines, $dN_1/dt = 0$, and $dN_2/dt = 0$, the population size of the species will increase, as shown by the arrows; above the lines the populations will decrease. In other words, when $N_1 < K_1 - aN_2$, then N_1 will increase. When $N_1 > K_1 - aN_2$, then N_1 will decrease. Similarly, when $N_2 < K_2 - \beta N_1$, then N_2 will increase; when $N_2 > K_2 - \beta N_2$, then N_2 will decrease.

It is conceivable, but highly improbable, that $K_1 = K_2$, and $a = \beta$. Such a situation would mean that the carrying capacity of the environment would be the same for both species and that an individual of one species would be exactly equivalent to one individual of the other species—in other words, the competition coefficients would equal unity. In this case, the two species would be identical in all respects and occupy the same ecological niche.

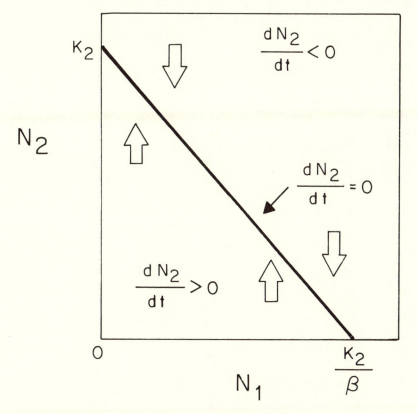

Figure 16-2. The zero-growth curve or isocline of species 2 in competition with species 1. Vertical arrows indicate direction of change in N_2.

It is theoretically possible, at least, for two species to be different and, nonetheless, to occupy the same ecological niche. All that is required for niche identity is that the competition coefficients for the two species be reciprocals of one another. For instance, suppose two species occupy the same environment and that K_1 = 200 for species 1 and K_2 = 100 for species 2. If they use the environment in exactly the same way, then an individual of species 2 requires twice the ecological niche space of an individual of species 1 — that is, a = 2. By the same token, a member of species 1 needs half the niche space of an individual of species 2, and β = 1/2. In other words, $K_2 = K_1/a$ and $K_1 = K_2/\beta$, which means that the lines $dN_1/dt = 0 = dN_2/dt$ will exactly coincide. It can also be seen from the equations in the previous sentence that $a = K_1/K_2$ and $\beta = K_2/K_1$ — that is, that a and β are reciprocals of one another, and thus, the ecological requirements of the two species must be identical.

However, as pointed out previously, two different species (almost by definition) will not occupy the same ecological niche. Therefore, the equalities dealt with above are so improbable that one must look to the inequalities to learn more about interspecific competition. These inequalities may be of four different types:

1. In the first case (Figure 16-3), the zero-growth curve, $dN_1/dt = 0$, lies above the zero-growth curve, $dN_2/dt = 0$. In the area K_2, 0, K_2/β, both N_1 and N_2 will increase, but in the area K_1/a, K_2, K_2/β, K_1, even though N_1 will continue to increase, N_2 will decrease. The horizontal arrows indicate the direction of changes in population size of species 1; the vertical arrows, the direction of changes in species 2. Considered jointly, the arrows provide a vector that indicates the direction of change in the composition of the combined populations. Therefore, in time, the joint changes in population number favor species 1 at the expense of species 2, and eventually the species 2 population will go to extinction (N_2 = 0), and N_1 will equilibrate at K_1, as shown in the figure.

In this instance, it can be seen in the figure that $K_1/a > K_2$ and $K_2/\beta < K_1$, $a/K_1 < 1/K_2$ and $\beta/K_2 > 1/K_1$ and from these, the following holds, $a < K_1/K_2$ and $\beta > K_2/K_1$. Therefore, when these relationships hold between the competition coefficients and K_1 and K_2, species 1 can be expected to win the competition.

2. In the second case (Figure 16-4), the zero-growth curve $dN_2/dt = 0$ lies above the zero-growth curve $dN_1/dt = 0$, and species 2 will eventually win out over species 1, with the equilibrium values N_1 = 0 and $N_2 = K_2$.

In this case, it can be seen that $K_1/a < K_2$ and $K_2/\beta > K_1$. Then $a/K_1 > 1/K_2$ and $\beta/K_2 < 1/K_1$, and $a > K_1/K_2$ and $\beta < K_2/K_1$.

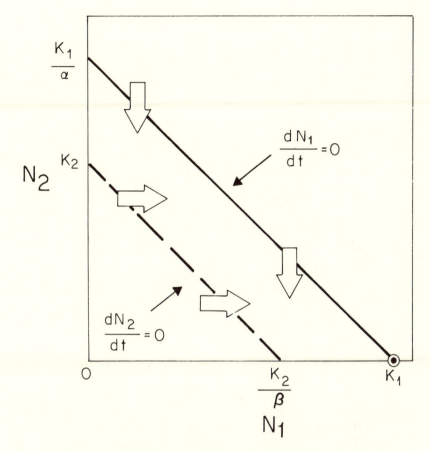

Figure 16-3. Conditions of competition between species 1 and 2 under which species 1 wins and species 2 goes to extinction. Horizontal arrows, N_1; vertical arrows, N_2; $\odot$ = stable equlibrium point.

3. In the third case (Figure 16-5), the lines for the zero-growth curves intersect at point U. In the area K_1/a, 0, K_2/β, U, both N_1 and N_2 will increase. Point U represents an equilibrium point, but it is an unstable equilibrium. If the population numbers drift into the area K_2, K_1/a, U, species 2 will continue to increase in numbers, as shown by the arrows, but species 1 will decrease, and N_1 will go to zero while N_2 goes to K_2. On the other hand, if the population numbers drift into the area K_1, K_2/β, U, then N_1 will go to K_1 and N_2 to zero. It is unlikely that the unstable equilibrium would persist very long if it were ever achieved, and the outcome of the competition in this case

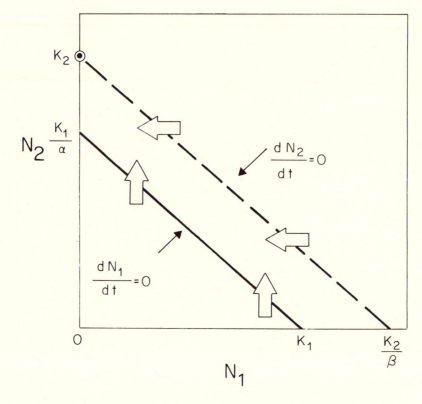

Figure 16-4. Conditions of competition between species 1 and 2 under which species 2 wins and species 1 goes to extinction. Horizontal arrows, N_1; vertical arrows, N_2; ⊙ = stable equilibrium point.

is dependent on the relative numbers of the two species. If N_1/N_2 becomes greater than the equilibrium ratio (shown in Figure 16-5 as the line 0, U) species 1 should win out; if N_1/N_2 becomes less than that value, Species 2 should win out.

In this case, it can be seen that $K_1/a < K_2$ and $K_2/\beta < K_1$. These convert to $a/K_1 > 1/K_2$ and $\beta/K_2 > 1/K_1$ and then to $a > K_1/K_2$ and $\beta > K_2/K_1$.

4. In the fourth and final case (Figure 16-6), the lines for the zero-growth curves intersect at point S. In the area K_2, 0, K_1, S, both N_1 and N_2 increase; in the area K_1, K_2/β, S, species 1 decreases but species 2 increases; in the area K_1/a, K_2, S, species 1 increases but species 2 decreases; beyond K_1/a, S, K_2/β both species will decrease in number. The net result is that in all four sectors, the vector arrows converge

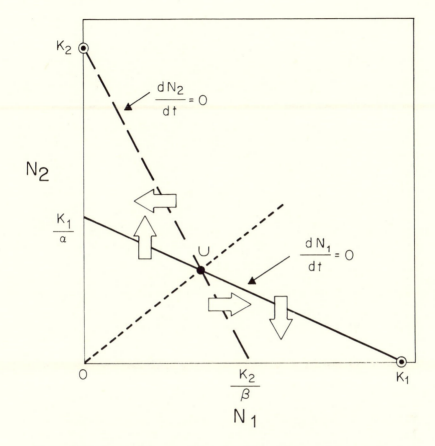

Figure 16-5. Competitive conditions in which either species may win, depending on their relative numbers, while the other goes to extinction. An equilibrium is possible, but it is unstable and unlikely to persist. Horizontal arrows, N_1; vertical arrows, N_2; $\odot$ = stable equilibrium; $\cdot$ = unstable equilibrium.

on S, which is, therefore, a stable equilibrium. Any perturbation away from S will be counteracted in time so that the relative population numbers return to S.

In Figure 16-6 it can be seen that $K_1/a > K_2$ and $K_2/\beta > K_1$. These relations convert to $a/K_1 < 1/K_2$ and $\beta/K_2 < 1/K_1$ and finally to $a < K_1/K_2$ and $\beta < K_2/K_1$.

Early in the discussions of competitive exclusion, it was pointed out that the inhibitory effect of one individual of species 1 on the population growth of species 1 was $1/K_1$ while the effect of one

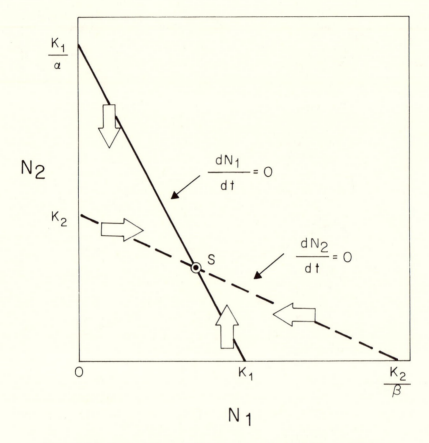

Figure 16-6. Competitive conditions generating a stable equilibrium between species 1 and 2. In essence, their niches differ sufficiently to permit their coexistence. Horizontal arrows, N_1; vertical arrows, N_2; $\odot$ = stable equilibrium point.

individual of species 2 on the population growth of species 1 was a/K_1. Conversely, the inhibitory effect of an individual of species 2 on the population growth of species 2 is $1/K_2$ while the effect of one individual of species 1 on the growth of species 2 is β/K_2.

The reason for the stable equilibrium as the expected outcome in the fourth case is very simply that, as shown by the relations $a/K_1 < 1/K_2$ and $\beta/K_2 < 1/K_1$, any individual inhibits the population growth of its own species more than it inhibits the population growth of the other species. In other words, the two species must occupy somewhat different niches, and are not in such direct competition that they can not coexist. (This method of reasoning can also be applied to cases 1, 2, and 3 and may help one to understand the outcome better in those cases as well.) Furthermore, the less similar the species, the smaller the competition coefficients, α and β, will become, and the more likely coexistence will be. If they do not affect one another at all, α and β will equal zero.

The "principle" of competitive exclusion has received so much attention in the literature that it is difficult not to take it seriously. Nevertheless, unless the principle can be formulated in such a way that it can be used to make predictions, it hardly merits the attention it has received. What must be done now is to consider whether it is possible to make predictions based on the principle of competitive exclusion.

Even those not particularly enamored of the concept have often gone on to develop complex mathematical treatments of the ideas involved (Armstrong and McGehee, 1980). The difficulty with the models is that, of necessity, they require such simplistic assumptions that they fail to reflect the realities in natural populations.

Consider the nature of some of these assumptions. In the first place, the competitive exclusion equations are based on the logistic curve of population growth and thus are subject to the same assumptions as the logistic curve. Among these are the following:

1. All the individuals in the population are ecologically identical—that is, they all have the same genotype and phenotype, and all are equally likely to die, give birth, and so on. Such an assumption eliminates age structure entirely from consideration although, alternatively, it can be assumed that the population has a stable age distribution initially and maintains it, which is equally unlikely. The assumption of genotypic and phenotypic uniformity also is highly improbable.

2. The population size and density have been measured in appropriate units. For example, if individuals in crowded populations are smaller than in uncrowded populations, number alone may not be an appropriate measure.

3. There is a real attribute of the population corresponding to r, the intrinsic rate of natural increase, and r remains constant as the population grows.

4. There is an inverse linear relationship between population density and the rate of population growth. In other words, each individual added to the population decreases the rate of population growth by a constant amount independent of the population density.

5. There is no time lag associated with the influence of density on the rate of increase. In other words, the population responds instantan-

eously to the addition of each new individual, an unlikely assumption in view of the built-in time lags in almost any life history where a period of development is required before the new individual becomes an adult.

6. There is a constant upper limit to population size, K, the carrying capacity of the environment, for a given set of conditions, and the rate of increase at any given time is linearly proportional to the difference between N at that time and K, the upper limit.

Obviously, all these assumptions about the logistic curve are open to question. In addition, in the competition equations, a and β are assumed to be constant—that is, the effect of an individual of one species on the population growth of the other species is constant and independent of population density. Moreover, given the differences between members of different species, the very idea of a competition coefficient itself is open to question. It is doubtful that merely multiplying the number of one species by a coefficient can convert them into an equivalent number of individuals of another species.

Recall that the concept of competitive exclusion has been stated in various ways. In one version, exclusion occurred if the competitors occupied "identical ecological niches"; in another, the competitors occupied "similar ecological niches"; in a third form, growing out of Birch's definition of competition, they competed for a common resource in limited supply. It was pointed out above that no two individuals, let alone two species, occupy identical niches. Therefore, all the statements of competitive replacement that refer to "identical niches," "complete competitors" and the like are clearly invalid because no two species occupy the same ecological niche.

With those versions that refer to "similar ecological niches," the question becomes, How similar must the niches be for competitive exclusion to occur? The answer, of course, is that if one species population becomes extinct, then their niches are presumed to be so similar that they can not coexist. If both species persist, their niches are presumed to differ. In other words, the frequently stated corollary of competitive exclusion that, if two species persist in a particular region, it can be taken as axiomatic that their ecological niches do not coincide, is not predictive, but merely descriptive. What is axiomatic is that, because the gene pools of two different species are and must be different, their ecological niches must also be different whether the two species are able to coexist or not. The most sensible statement possible is merely that two species with similar ecological requirements may or may not be able to coexist.

Finally, if the concept is formulated in terms of a common limiting resource, the evidence given above makes clear that even though both species may subsist on a common resource, they will not utilize it in exactly the same way, and they may or may not coexist.

Ayala (1969, 1970a, 1970b, 1972a, Ayala et al., 1973) claimed to have invalidated the principle of competitive exclusion experimentally because he demonstrated that two species of *Drosophila* were able to coexist even when $a > K_1/K_2$ and $\beta > K_2/K_1$, contrary to the Lotka-Volterra conditions for coexistence, which are that $a < K_1/K_2$ and $\beta < K_2/K_1$. Gause (1970) criticized Ayala's results because "clearly this case represents competition between two species belonging to different ecological niches," and also because Ayala's data were not amenable to mathematical treatment since he only counted adults and not larvae. Gause then added, "Ayala's data in no way contradict or reject the principle of competitive exclusion between two species, which I demonstrated experimentally in a system of two species belonging to the same ecological niche." Ayala (1970b), however, stated as his conclusion that "Two species competing for limiting resources may coexist." Thus they were using different definitions of competitive exclusion. In his experiments with *D. pseudoobscura* and *D. serrata, pseudoobscura* eliminated *serrata* at 19°C and *serrata* eliminated *pseudoobscura* at 25°C. At 23.5°C the two species could coexist. Thus a slight change in conditions reversed the outcome of the competition in a manner parallel to Merrell (1951). It would be interesting to know the estimates of a, β, K_1, and K_2 at these other temperatures. However, in view of the many assumptions listed above in relation to the Lotka-Volterra equations, it is doubtful that any experiments can be devised to test competitive exclusion adequately.

The main difficulty with the "principle" of competitive exclusion is that it cannot be used to make predictions. The primary value of a principle in biology is that it makes predictions possible, as in the case of Mendel's laws. With competitive exclusion, the assumptions involved are so simplified that the model bears little relation to actual populations and thus cannot really be subjected to experimental test. Rather than generating predictions, just the reverse is true. The "principle" and its corollary are used to explain, but not predict, the experimental results. If two species coexist, they must occupy different ecological niches or utilize the same limiting resources differently. If one species population becomes extinct, it is cited as an example of competitive exclusion. At this point, the following quotation from Ewens (1969) seems apropos:

It must be understood that, because the biological world is infinitely more complex than our mathematical models, it is impossible to expect that mathematics can play in the biological sciences the fundamental and ubiquitous role which it plays in the physical sciences. We must therefore expect that such principles as do emerge must be treated with some measure of caution and with full appreciation of the difficulty of forming workable and meaningful mathematical models in the biological world.

Competitive exclusion has stimulated a great deal of thought about the relationships between closely related species, and in that may lie its greatest value.

Although the validity or utility of competitive exclusion may be doubtful, the following conclusions are possible from the foregoing discussion:

1. Every species — indeed, every individual — will have its own unique ecological niche in terms of the multi-dimensional niche-space concept.

2. Because this niche has an infinite number of dimensions, our understanding of it will be determined and limited by the dimensions we choose to study.

3. The niche of members of the same species will more nearly coincide than the niches of different species. Therefore, intraspecific competition is apt to be more severe than interspecific competition.

4. The evidence for interspecific competition is clear in many cases, but natural selection will tend to operate in such a way as to minimize interspecific competition. Where two ecologically similar species are in competition, they may or may not be able to coexist depending, apparently, on the degree of niche overlap under the prevailing conditions.

L'ENVOI

At the outset I pointed out that the two major problems in evolution are adaptation and speciation. The methodology of ecological genetics permits the study of the adaptation of populations to their environments and the mechanisms by which they respond to environmental change. In essence, ecological genetics is the direct study of evolution as it occurs. To study the interplay between a species population and its environment, one must learn something about both the species and its environment. Thus, one seeks to learn something about the distribution of the species and, hence, the nature of its ecological requirements. Estimates of population size and density, birth and death rates, and immigration and emigration rates provide insight into the population dynamics of the species. Size and density estimates may also reveal something about the most favorable habitat for the species and, hence, about its ecological niche. Estimates of population size and of migration rates also have evolutionary and genetic implications, for the effective population size is related to the potential for random genetic drift, and gene flow between populations is dependent on some form of migration.

To study genetic and evolutionary changes in populations, gene-frequency data are necessary. Ideally, one might wish to monitor changes throughout the entire genome; in practice, this is not possible. Instead, there has been what Lewontin has called "a struggle to measure variation." At first, a few loci affecting visible traits were studied. Subsequently, viability mutants, especially lethals and semi-lethals, were observed. More recently, protein electromorphs detected by gel electrophoresis have been monitored. Although this approach represents a considerable advance over previous methods—for example, in the number of loci studied—it, too, has its limitations. Only a relatively small sample of the structural gene portion of the genome can be assayed and not all variants can be detected with the usual techniques. Thus the struggle to measure variation continues.

The factors responsible for gene-frequency change and, thus, ultimately for evolution are mutation, selection, gene flow, and random genetic drift. Therefore, if one is to study adaptation and evolution in populations, one must not only estimate changes in gene frequency, but also estimate mutation rates, selection coefficients, migration coefficients, and the effective size of populations. Only thus will it be possible to determine what role each of these factors plays in causing the observed gene-frequency changes. Perhaps it should be added that the struggle to measure variation may seem like child's play compared to the struggle needed to measure these factors in natural populations. For more than a decade, population geneticists have been busily exploiting electrophoresis as a new tool for measuring variation in species of every description. The success of this new technique in measuring variation led to a new dilemma: The observed levels of heterozygosity and polymorphism were so high that they were difficult to explain under the prevailing theories. This dilemma, in turn, led to the development of the neutralist theory of evolution, the concept that the observed variation at the molecular level is neutral with respect to fitness.

The controversy between the neutralists and the selectionists has gone unresolved for more than a decade. Because it is impossible to prove neutrality directly, the neutralists have been forced to seek indirect evidence, for example by the comparison of observed gene-frequency distributions with those predicted under the neutral theory. The selectionists, on the other hand, have had difficulty demonstrating the adaptive significance of electrophoretic variants because the functional significance of these proteins is generally not well understood. Only recently have efforts begun to study directly the relation between electromorphs and fitness. As the pathway of events between variation at the molecular level and reproductive fitness is apt to be long and complex, such studies are not easy even when the functional role of the protein is known. Nevertheless, the point has been reached where one must go beyond cataloging the distribution and frequency of allelic variants in natural populations. Further understanding of the evolutionary process requires efforts to determine the adaptive significance of the observed variation and to study the interplay among mutation, selection, migration, and drift, not just in a theoretical way, but in actual populations.

Even though all species appear to have a common genetic material and to be governed by the same genetic principles, genetic systems differ in different species and there is no reason to suppose that the evolutionary process is identical in all species. A great need at present is for comparative studies both of closely related species and of distinctively different species. Such studies provide a means for assessing the relative importance of mutation, selection, migration, and drift in the adaptation and evolution of diverse groups. Until recently, the concept of gradual allopatric speciation had practically reached the status of dogma, but now the possibility of other modes of speciation is accepted and is being explored.

Ecological genetics can contribute to the resolution of a number of other fundamental questions in addition to neutralism versus selectionism and the modes of speciation. As heterosis became a less plausible explanation for the high levels of genic polymorphism observed in wild populations, other explanations were sought such as frequency-dependent selection, density-related selection, and selection heterogeneous in either time or space. Even though heterosis may seem an insufficient explanation for genic polymorphisms, chromosomal polymorphisms, as exemplified by translocation heterozygotes and inversion polymorphisms, seem to persist because of the heterozygous advantage they confer on the heterozygotes. Similarly, allopolyploids in plants and asexually reproducing groups in both plants and animals generally have a hybrid origin and, thus, serve as a means to capture and perpetuate heterosis. Studies of these and other genetic systems should reveal whether their adaptive significance is tied primarily to heterosis or whether other factors are also involved. A related question is the role of hybridization and gene flow as a source of genetic variation in evolving populations. Even though both mutation and gene flow are recognized as sources of new variation in populations, less is known about the role of gene flow than mutation in this respect.

The crucial step in the process of speciation is the origin of isolating mechanisms, which limit hybridization and gene flow between diverging populations. The methodology of ecological genetics is a useful approach to the study of the origin and evolution of reproductive isolation.

Related to this question is the nature of the genetic changes involved in the genetic divergence. Quantitative polygenic variability has generally been thought to be the basis for the differences between diverging populations as in Wright's shifting balance theory of evolution. However, recent studies have revealed that distinctive chromosomal rearrangements have originated in closely related groups far more often than had previously been suspected. Furthermore, some of the better-studied examples of evolutionary change (industrial melanism, pesticide resistance, disease resistance, host race shifts) appear to be mediated by allelic substitutions at just a few major gene loci. Further studies should reveal the relative importance of chromosomal rearrangements, major gene substitutions, and gradual changes in polygenically controlled quantitative traits in the speciation process. It seems inevitable that the rapidly increasing understanding of the nature and organization of the genetic material and of the regulation of gene expression will influence the development of evolutionary theory more in the future than has been the case in the past.

A nagging question is the relative importance of unique events in determining the course of evolution. Much of the theory is based on recurrent events such as recurrent mutation, recurrent migration, constant selection pressure, etc. However, the possible role of unique events, a unique mutation or recombination or migration or selective event, needs to be explored.

Although studies in ecological genetics have focused almost exclusively on one species at a time, it seems inevitable that sooner or later studies in ecological genetics with two or more species will be attempted. Such studies require the monitoring of the environment as well as the genetic changes in both species. A few such studies of competition and of parasitism have been carried out, as noted earlier. Studies of the coevolution and, in some cases, the coadaptation of host and parasite are probably the most common type of multispecies study of this sort, but the amount of work involved in monitoring the genotypes of more than one species has probably inhibited the number of such studies. The evolution of Batesian mimicry presents an intriguing problem, but attention has been directed primarily to the genetic changes in the mimic while little attention has been paid to the genetics of the model or, for that matter, of the predators that prey on them. The same can be said for most studies of predation where attention is directed far more often to the effects of predation on the prey than on the predator.

Implicit in the discussion of species interactions is the extent to which the behavior of individuals of the same or of different species toward one another affects the course of evolution. There are many indications that it does, tracing all the way back to Darwin's *Descent of Man and Selection in Relation to Sex*. But much remains to be learned in this area.

At a more mundane, practical level, the conceptual framework of ecological genetics has numerous applications. Animal and plant breeders are practical ecological geneticists in the sense that their breeding programs are directed toward developing breeds or varieties of animals and plants well adapted to the environments in which they will live. Similarly, species introductions, whether of wild turkeys into new areas or trees in reforestation projects, will have a greater chance of success if some thought is given to the genotypes of the individuals being introduced, and also, especially in the case of the turkeys, to the amount of genetic variation in the introduced population on which selection can act. The efforts to save threatened and endangered species should be directed not just to building up the numbers of individuals, but also toward preserving as much of the genetic variability of the species as possible. Finally, populations today are facing a variety of new and often unprecedented environmental changes, ranging from habitat destruction to exposure to pesticides, antibiotics, herbicides, and other chemical and physical agents (many of them carcinogenic or mutagenic or both), as well as to alien species or diseases to which they have never been exposed. These changes not only threaten the existence of individuals, but have genetic implications for the populations that can no longer be safely ignored. These problems can best be studied and understood as problems in ecological genetics.

References

References

Alexander, M. L. 1960. Radiosensitivity at specific autosomal loci in mature sperm and spermatogonial cells of *Drosophila melanogaster*. Genetics 45:1019-1022.

Alexander, R. D. 1960. Sound communication in Orthoptera and Cicadidae. In *Animal Sounds and Communication*. W. E. Lanyon and W. N. Tavolga, Eds. AIBS Publ. No. 7:38-92.

Alexander, R. D. 1968. Life cycle origins, speciation, and related phenomena in crickets. Quart. Rev. Biol. 43:1-41.

Alexander, R. D., and R. S. Bigelow. 1960. Allochronic speciation in field crickets, and a new species, *Acheta veletis*. Evol. 14:334-346.

Alexander, R. D., and T. E. Moore. 1962. The evolutionary relationships of 17-year and 13-year cicadas, and three new species (Homoptera, Cicadidae, *Magicicada*). Misc. Publ. Mus. Zool., Univ. Michigan 121:1-59.

Allard, R. W. 1960. *Principles of Plant Breeding*. Wiley. New York.

Allard, R. W. and A. L. Kahler. 1971. Allozyme polymorphisms in plant populations. Stadler Genetics Symp. 3:9-24.

Allard, R. W., and A. L. Kahler. 1972. Patterns of molecular variation in plant populations. Proc. 6th Berk. Symp. Math. Stat. Prob. Vol. 5:237-254.

Allee, W. C., A. E. Emerson, O. Park, T. Park, and K. P. Schmidt. 1949. *Principles of Animal Ecology*. Saunders. Philadelphia.

Allen, A. C. 1966. The effects of recombination on quasinormal second and third chromosomes of *Drosophila melanogaster* from a natural population. Genetics 54:1409-1422.

Allison, A. C. 1964. Polymorphism and natural selection in human populations. Cold Spring Harbor Symp. Quant. Biol. 29:137-149.

Ancona, V. d'. 1954. *The Struggle for Existence*. Brill. Leiden.

Anderson, E. 1948. Hybridization of the habitat. Evol. 2:1-9.

Anderson, E. 1949. *Introgressive Hybridization*. Wiley. New York.

Anderson, E. 1952. *Plants, Man, and Life*. University of California Press. Berkeley.

Anderson, E. 1953. Introgressive hybridization. Biol. Rev. Camb. Phil. Soc. 28:280-307.

Anderson, E., and L. Hubricht. 1938. Hybridization in *Tradescantia*. III. The evidence for introgressive hybridization. Amer. J. Botany 25:396-402.

Anderson, E., and G. L. Stebbins. 1954. Hybridization as an evolutionary stimulus. Evol. 8:378-388.

Anderson, W. W. 1971. Genetic equilibrium and population growth under density-regulated selection. Amer. Nat. 105:489-498.

Anderson, W. W., T. Dobzhansky, and O. Pavlovsky. 1967. Selection and inversion polymorphism in experimental populations of *Drosophila pseudoobscura* initiated with chromosomal constitutions of natural populations. Evol. 21:664-671.

Anderson, W. W., T. Dobzhansky, O. Pavlovsky, J. Powell, and D. Yardley. 1975. Genetics of natural populations. XLII. Three decades of genetic change in *Drosophila pseudoobscura*. Evol. 29:24-36.

Anderson, W. W., C. Oshima, T. Watanabe, T. Dobzhansky, and O. Pavlovsky. 1968. Genetics of natural populations. XXXIX. A test of the possible influence of two insecticides on the chromosomal polymorphism in *Drosophila pseudoobscura*. Genetics 58:423-434.

Andrewartha, H. G. 1961. *Introduction to the Study of Animal Populations*. University of Chicago Press. Chicago.

Andrewartha, H. G., and L. C. Birch. 1954. *The Distribution and Abundance of Animals*. University of Chicage Press. Chicago.

Andrewartha, H. G., and T. O. Browning. 1958. Williamson's theory of interspecific competition. Nature 181:1415.

Antonovics, J. 1968. Evolution in closely adjacent plant populations. V. The evolution of self-fertility. Heredity 23:219-328.

Antonovics. J. 1971. The effects of a heterogeneous environment on the genetics of natural populations. Amer. Sci. 59:593-599.

Antonovics, J., and A. D. Bradshaw. 1970. Evolution in closely adjacent plant populations. VIII. Clinal patterns at a mine boundary. Heredity 25:349-362.

Antonovics, J., A. D. Bradshaw, and R. G. Turner. 1971. Heavy metal tolerance in plants. Adv. Ecol. Res. 7:1-58.

Armstrong, R. A. and R. McGehee. 1980. Competitive exclusion. Amer. Nat. 115:151-170.

Aston, J. L., and A. D. Bradshaw. 1966. Evolution in closely adjacent plant populations. II. *Agrostis stolonifera* in maritime habitats. Heredity 21:649-664.

Auerbach, C. and B. J. Kilbey. 1971. Mutation in eukaryotes. Ann. Rev. Genetics 5:163-218.

Avery, A. G., S. Satina, and J. Rietsema. 1959. *Blakeslee: The Genus Datura*. Ronald Press. New York.

Avise, J. C., and R. K. Selander. 1972. Evolutionary genetics of cave-dwelling fishes of the genus *Astyanax*. Evol. 26:1-19.

Avise, J. C., M. H. Smith, R. K. Selander, T. E. Taylor, and P. R. Ramsey. 1974. Biochemical polymorphism and systematics in the genus *Peromyscus*. V. Insular and mainland species of the subgenus *Haplomylomys*. Syst. Zool. 23:226-238.

Ayala, F. J. 1979. Experimental invalidation of the principle of competitive exclusion. Nature 224:1076-1079.

Ayala, F. J. 1970a. Invalidation of competitive exclusion defended. Nature 227:89-90.

Ayala, F. J. 1970b. Competition, coexistence, and evolution. In *Essays in Evolution and*

Genetics in Honor of Theodosius Dobzhansky. M. K. Hecht and W. C. Steere, Eds. Appleton-Century-Crofts. New York.

Ayala, F. J. 1972a. Competition between species. Amer. Sci. 60:348-357.

Ayala, F. J. 1972b. Darwinian versus non-Darwinian evolution in natural populations of *Drosophila*. Proc. 6th Berk. Symp. Math. Stat. and Prob. Vol. 5:211-236.

Ayala, F. J. 1974, Biological evolution: natural selection or random walk. Amer. Sci. 62:692-701.

Ayala, F. J. 1975a. Genetic differentiation during the speciation process. Evol. Biol. 8: 1-78.

Ayala, F. J. 1975b. Scientific hypotheses, natural selection, and the neutrality theory of protein evolution. In *The Role of Natural Selection in Human Evolution*. F. M. Salzano, Ed. North-Holland. Amsterdam.

Ayala, F. J., Ed. 1976. *Molecular Evolution*. Sinauer. Sunderland, Mass.

Ayala, F. J., and T. Dobzhansky. 1974. A new subspecies of *Drosophila pseudoobscura*. Pan Pacific Entomologist 50:211-219.

Ayala, F. J., and M. E. Gilpin. 1974. Gene frequency comparisons between taxa: Support for the natural selection of protein polymorphisms. Proc. Nat. Acad. Sci. 71:4847-4849.

Ayala, F. J., M. E.Gilpin, and J. G. Ehrenfeld. 1973. Competition between species: Theoretical models and experimental tests. Theoret. Pop. Biol. 4:331-356.

Ayala, F. J., J. R. Powell, and T. Dobzhansky. 1971. Enzyme variability in the *Drosophila willistoni* group. II. Polymorphisms in continental and island populations of *Drosophila willistoni*. Proc. Nat. Acad. Sci. 68:2480-2483.

Ayala, F. J., J. R. Powell, M. L. Tracey, C. A. Mourão, and S. Pérez-Salas. 1972. Enzyme variability in the *Drosophila willistoni* group. IV. Genic variation in natural populations of *Drosophila willistoni*. Genetics 70:113-139.

Ayala, F. J., M. L. Tracey, L. G. Barr, J. F. McDonald, and S. Pérez-Salas. 1974. Genetic variation in natural populations of five *Drosophila* species and the hypothesis of the selective neutrality of protein polymorphisms. Genetics 77:343-384.

Ayala, F. J., M. L. Tracey, D. Hedgecock, and R. C. Richmond. 1974. Genetic differentiation during the speciation process in *Drosophila*. Evol. 28:576-592.

Baerends, G., C. Beer, and A. Manning, Eds. 1976. *Function and Evolution in Behaviour*. Clarendon Press. New York.

Baker, H. G. 1951. Hybridization and natural gene-flow between higher plants. Biol. Rev. Camb. Phil. Soc. 26:302-337.

Baker, H. G., and G. L. Stebbins, Eds. 1965. *The Genetics of Colonizing Species*. Academic Press. New York.

Band, H. T., and P. T. Ives. 1963. Comparison of lethal and semilethal frequencies in second and third chromosomes from a natural population of *Drosophila melanogaster*. Canad. J. Genet. Cytol. 5:351-357.

Barker, J. S. F., and L. J. Cummins. 1969. Disruptive selection for sternopleural bristle number in *Drosophila melanogaster*. Genetics 61:697-712.

Bateman, A. J. 1950. Is gene dispersion normal? Heredity 4:353-363.

Bateman, K. G. 1959a. The genetic assimilation of the dumpy phenocopy. J. Genetics 56:341-351.

Bateman, K. G. 1959b. Genetic assimilation of four venation phenocopies. J. Genet. 56: 443-474.

Beckenbach, A. T., and S. Prakash. 1977. Examination of allelic variation at the hexokinase loci of *Drosophila pseudoobscura* and *D. persimilis* by different methods. Genetics 87:743-761.

Berger, E. M. 1971. A temporal survey of allelic variation in natural and laboratory populations of *Drosophila melanogaster*. Genetics 67:121-136.

Bernstein, S. C., L. H. Throckmorton, and J. L. Hubby. 1973. Still more genetic variability in natural populations. Proc. Nat. Acad. Sci. 70:3928-3931.

Berry, R. J., and J. Peters. 1976. Genes, survival, and adjustment in an island population of the house mouse. In *Population Genetics and Ecology*. S. Karlin and E. Nevo, Eds. Academic Press. New York.

Bijlsma, R., and W. Van Delden. 1977. Polymorphism at the G6PD and 6PGD loci in *Drosophila melanogaster*. I. Evidence for selection in experimental populations. Genet. Res. 30:221-236.

Bijlsma-Meeles, E., and W. Van Delden. 1974. Intra- and interpopulation selection concerning the alcohol dehydrogenase locus in *Drosophila melanogaster*. Nature 247: 369-371.

Birch, L. C. 1948. The intrinsic rate of natural increase of an insect population. J. Animal Ecol. 17:15-26.

Birch, L. C. 1955. Selection in *Drosophila pseudoobscura* in relation to crowding. Evol. 9:389-399.

Birch, L. C. 1957. The meanings of competition. Amer. Nat. 91:5-18.

Birch, L. C. 1960. The genetic factor in population ecology. Amer. Nat. 94:5-24.

Bishop, J. A., and L. M. Cook. 1975. Moths, melanism and clean air. Sci. Amer. 232: (1):90-99.

Blair, W. F. 1964. Isolating mechanisms and interspecies interactions in anuran amphibians. Quart. Rev. Biol. 39:334-344.

Bock, W. J., and G. von Wahlert. 1965. Adaptation and the form-function complex. Evol. 19:269-299.

Bodmer, W. F. and L. L. Cavalli-Sforza. 1976. *Genetics, Evolution, and Man*. Freeman. San Francisco.

Bodmer, W. F., and A. W. F. Edwards. 1960. Natural selection and the sex ratio. Ann. Human Genet. 24:239-244.

Bonnell, M. L., and R. K. Selander. 1974. Elephant seals: Genetic variation and near extinction. Science 184:908-909.

Bonnier, G., and K. G. Lüning. 1949. Studies on X-ray mutations in the white and forked loci of *Drosophila melanogaster*. I. A statistical analysis of mutation frequencies. Hereditas. 35:163-189.

Bradshaw, A. D. 1971. Plant evolution in extreme environments. In *Ecological Genetics and Evolution*. R. Creed, Ed. Blackwell. Oxford.

Bradshaw, A. D. 1972. Some of the evolutionary consequences of being a plant. Evol. Biol. 5:25-47.

Breckenridge, W. J. 1944. *Reptiles and Amphibians of Minnesota*. University of Minnesota Press. Minneapolis.

Briscoe, D. A., A. Robertson, and J. Malpica. 1975. Dominance at ADH locus in response of adult *Drosophila melanogaster* to environmental alcohol. Nature 255:148-149.

British Ecological Society. 1944. Easter Meeting 1944. Symposium on "The ecology of closely allied species." J. Animal Ecol. 13:176-177.

Britten, R. J., and E. H. Davidson. 1969. Gene regulation for higher cells: a theory. Science 165:349-357.

Brncic, D. 1954. Heterosis and the integration of the genotype in geographic populations of *Drosophila pseudoobscura*. Genetics 39:77-88.

Brncic, D. 1961. Integration of the genotype in geographic populations of *Drosophila pavani*. Evol. 15:92-97.

Brower, J. V. Z. 1958. Experimental studies of mimicry in some North American butterflies. I. The Monarch, *Danaus plexippus*, and Viceroy, *Limenitis archippus archippus*. II. *Battus philenor* and *Papilio troilus, P. polyxenes,* and *P. glaucus.* III. *Danaus gilippus berenice* and *Limenitis archippus floridensis.* Evol. 12:32-47; 123-136; 273-285.

Brower, J. V. Z. 1960. Experimental studies of mimicry. IV. The reactions of starlings to different proportions of models and mimics. Amer. Nat. 94:271-282.

Brower, J. V. Z., and L. P. Brower. 1962. Experimental studies of mimicry. VI. The reaction of toads (*Bufo terrestris*) to honeybees (*Apis mellifera*) and their dronefly mimics (*Eristalis vinetorum*). Amer. Nat. 96:297-308.

Brower, J. V. Z., and L. P. Brower. 1966. Experimental evidence of the effects of mimicry. Amer. Nat. 100:173-187.

Brower, L. P., J. V. Z. Brower, F. G. Stiles, H. J. Croze, and A. S. Horner. 1964. Mimcry: differential advantage of color patterns in the natural environment. Science 144: 183-185.

Brower, L. P., J. V. Z. Brower, and P. W. Westcott. 1960. Experimental studies of mimicry. V. The reaction of toads (*Bufo terrestris*) to bumblebees (*Bombus americanorum*) and their robberfly mimics (*Mallophora bomboides*), with a discussion of aggressive mimicry. Amer. Nat. 94:343-356.

Brower, L. P., F. H. Pough, and H. R. Meck. 1970. Theoretical investigations of automimicry. I. Single trial learning. Proc. Nat. Acad. Sci. 66:1059-1066.

Brown, A. W. A., and R. Pal. 1971. *Insecticide Resistance in Arthropods.* 2nd Ed. W. H. O. Geneva. Monogr. Ser. No. 38.

Brown, W. L. 1957. Centrifugal speciation. Quart. Rev. Biol. 32:247-277.

Brown, W. L., and E. O. Wilson. 1956. Character displacement. Syst. Zool. 5:49-64.

Brues, A. M. 1964. The cost of evolution vs. the cost of not evolving. Evol. 18:379-383.

Brues, A. M. 1969. Genetic load and its varieties. Science 164:1130-1136.

Buri, P. 1956. Gene frequency in small populations of mutant *Drosophila.* Evol. 10:367-402.

Burla, H., A. B. da Cunha, A. G. F. Cavalcanti, T. Dobzhansky, and C. Pavan. 1950. Population density and dispersal rates in Brazilian *Drosophila willistoni.* Ecology 31: 393-404.

Bush, G. L. 1969. Sympatric host race formation and speciation in frugivorous flies of the genus *Rhagoletis* (Diptera, Tephritidae). Evol. 23:237-251.

Bush, G. L. 1974. The mechanism of sympatric host race formation in the true fruit flies (Tephritidae). In *Genetic Mechanisms of Speciation in Insects.* M. J. D. White, Ed. Australia and New Zealand Book Co. Sydney.

Bush, G. L. 1975. Modes of animal speciation. Ann. Rev. Ecol. Syst. 6:339-364.

Cain, A. J., and P. M. Sheppard. 1954. Natural selection in *Cepaea.* Genetics 39:89-116.

Cain, A. J., P. M. Sheppard, J. M. B. King, M. A. Carter, J. D. Currey, B. Clarke, C. Diver, J. Murray, and R. W. Arnold. 1968. Studies on *Cepaea.* Philos. Trans. Roy. Soc. London (B) 253:383-595.

Calhoun, J. B. 1952. The social aspects of population dynamics. J. Mammology 33: 139-159.

Calhoun, J. B. 1962. *The Ecology and Sociology of the Norway Rat.* U. S. Dept. of Health, Education, and Welfare. Public Health Service Document No. 1008.

Camin, J. H., and P. R. Ehrlich. 1958. Natural selection in water snakes (*Natrix sipedon,* L.) on islands in Lake Erie. Evol. 12:504-511.

Cannon, W. B. 1929. Organization for physiological homeostasis. Physiol. Rev. 9:399-431.

Cannon, W. B. 1932. *The Wisdom of the Body.* Norton. New York.

Carlquist, S. 1966. The biota of long-distance dispersal. I. Principles of dispersal and evolution. Quart. Rev. Biol. 41:247-270.

Carlson, A. D., and J. Copeland. 1978. Behavioral plasticity in the flash communication system of fireflies. Amer. Sci. 66:340-346.

Carpenter, C. R. 1958. Territoriality: A review of concepts and problems. In *Behavior and Evolution*. A. Roe and G. G. Simpson, Eds. Yale University Press. New Haven.

Carson, H. L. 1958a. Response to selection under different conditions of recombination in *Drosophila*. Cold Spring Harbor Symp. Quant. Biol. 23:291-306.

Carson, H. L. 1958b. The population genetics of *Drosophila robusta*. Adv. Genetics 9: 1-40.

Carson, H. L. 1959. Genetic conditions which promote or retard the formation of species. Cold Spring Harbor Symp. Quant. Biol. 24:87-105.

Carson, H. L. 1965. Chromosomal morphism in geographically widespread species of *Drosophila*. In *The Genetics of Colonizing Species*. H. G. Baker and G. L. Stebbins, Eds. Academic Press. New York.

Carson, H. L. 1968. The population flush and its genetic consequences. In *Population Biology and Evolution*. R. C. Lewontin, Ed. Syracuse University Press. Syracuse.

Carson, H. L. 1971. Speciation and the founder principle. Stadler Genet. Symp. 3:51-70.

Carson, H. L. 1973. Reorganization of the gene pool during speciation. In *Genetic Structure of Populations*. N. E. Morton, Ed. *Population Genetics Monographs*. Vol. 3. University of Hawaii Press.

Carson, H. L. 1975. The genetics of speciation at the diploid level. Amer. Nat. 109:83-92.

Carson, H. L., and W. B. Heed. 1964. Structural homozygosity in marginal populations of nearctic and neotropical species of *Drosophila* in Florida. Proc. Nat. Acad. Sci. 52: 427-430.

Castle, W. E. 1903. The laws of heredity of Galton and Mendel, and some laws governing race improvement by selection. Proc. Amer. Acad. Arts Sci. 39:223-242.

Castle, W. E. 1919. Piebald rats and the theory of genes. Proc. Nat. Acad. Sci. 5:126-130.

Cavalli-Sforza, L. L. 1963. Genetic drift for blood groups. In *Genetics of Migrant and Isolate Populations*. E. Goldschmidt, Ed. Williams and Wilkins. Baltimore.

Cavalli-Sforza, L. L. 1969. Genetic drift in an Italian population. Sci. Amer. 221(2): 30-37.

Cavalli-Sforza, L. L., I. Barrai, and A. W. F. Edwards. 1964. Analysis of human evolution under random genetic drift. Cold Spring Harbor Symp. Quant. Biol. 29:9-20.

Cavalli-Sforza, L. L., and W. F. Bodmer. 1971. *The Genetics of Human Populations*. Freeman. San Francisco.

Chabora, H. J. 1968. Disruptive selection for sternopleural chaeta number in various strains of *Drosophila melanogaster*. Amer. Nat. 102:525-532.

Charlesworth, B. 1971. Selection in density-regulated populations. Ecology 52:469-474.

Cherry, L. M., S. M. Case, and A. C. Wilson. 1978. Frog perspective on the morphological difference between humans and chimpanzees. Science 200:209-211.

Chetverikov, S. S. 1926. On certain aspects of the evolutionary process from the standpoint of modern genetics. Proc. Amer. Phil. Soc. 105:167-195. (1961) (Translation). M. Barker, Tr. I. M. Lerner, Ed. Original in Zh. Eksp. noi Biol. A2:3-54.

Chitty, D. 1957. Self-regulation of numbers through changes in viability. Cold Spring Harbor Symp. Quant. Biol. 22:277-280.

Chitty, D. 1960. Population processes in the vole and their relevance to general theory. Canad. J. Zool. 38:99-113.

Chitty, D. 1967. The natural selection of self-regulatory behavior in animal populations. Proc. Ecol. Soc. Australia 2:51-78.

Chitty, D. 1970. Variation and population density. Symp. Zool. Soc. London 26:327-333.

Christian, J. J. 1961. Phenomena associated with population density. Proc. Nat. Acad. Sci. 47:428-449.

Christian, J. J., and D. E. Davis. 1964. Endocrines, behavior, and population. Science 146:1550-1560.

Christiansen, F. B. 1974. Sufficient conditions for protected polymorphism in a subdivided population. Amer. Nat. 108:157-166.

Christiansen, F. B. 1975. Hard and soft selection in a subdivided population. Amer. Nat. 109:11-16.

Clarke, B. 1972. Density-dependent selection. Amer. Nat. 106:1-13.

Clarke, B. 1975. The contribution of ecological genetics to evolutionary theory: Detecting the direct effects of natural selection on particular polymorphic loci. Genetics 79(Suppl):101-113.

Clarke, C. A., and P. M. Sheppard. 1955. A preliminary report on the genetics of the Machaon group of swallowtail butterflies. Evol. 9:182-201.

Clarke, C. A., and P. M. Sheppard. 1959a. The genetics of some mimetic forms of *Papilio dardanus*, Brown, and *Papilio glaucus*, Linn. J. Genet. 56:236-260.

Clarke, C. A., and P. M. Sheppard. 1959b. The genetics of *Papilio dardanus*, Brown. I. Race Cenea from South Africa. Genetics 44:1347-1358.

Clarke, C. A., and P. M. Sheppard. 1960a. The genetics of *Papilio dardanus*, Brown. II. Races Dardanus, Polytrophus, Meseres, and Tibullus. Genetics 45:439-457.

Clarke, C. A., and P. M. Sheppard. 1960b. The evolution of dominance under disruptive selection. Heredity 14:73-88.

Clarke, C. A., and P. M. Sheppard. 1960c. Super-genes and mimicry. Heredity 14:175-185.

Clarke, C. A., and P. M. Sheppard. 1960d. The evolution of mimicry in the butterfly, *Papilio dardanus*. Heredity 14:163-173.

Clarke, C. A., and P. M. Sheppard. 1962. The genetics of the mimetic butterfly *Papilio glaucus*. Ecology 43:159-161.

Clarke, C. A., and P. M. Sheppard. 1964. Genetic control of the melanic form *insularia* of the moth *Biston betularia*. Nature 202:215-216.

Clarke, C. A., and P. M. Sheppard. 1966. A local survey of the distribution of industrial melanic forms in the moth *Biston betularia* and estimates of the selective values of these in an industrial environment. Proc. Roy. Soc. London (B) 165:424-439.

Clarke, C. A., P. M. Sheppard, and I. W. B. Thornton. 1968. The genetics of the mimetic butterfly *Papilio memnon*. Phil. Trans. Roy. Soc. London (B)254:37-89.

Clausen, J. 1951. *Stages in the Evolution of Plant Species*. Cornell University Press. Ithaca.

Clausen, J., and W. M. Hiesey. 1958. Experimental studies on the nature of species. IV. Genetic structure of ecological races. Carnegie Inst. Wash. Publ. 615:1-312.

Clausen, J., D. D. Keck, and W. M. Hiesey. 1940. Experimental studies on the nature of species. I. The effect of varied environments on western North American plants. Carnegie Inst. Wash. Publ. 520:1-452.

Clausen, J., D. D. Keck , and W. M. Hiesey. 1945. Experimental studies on the nature of species. II. Plant evolution through amphiploidy and autoploidy, with examples from the Madiinae. Carnegie Inst. Wash. Publ. 564:1-174.

Clausen, J., D. D. Keck, and W. M. Hiesey. 1947. Heredity of geographically and ecologically isolated races. Amer. Nat. 81:114-133.

Clausen, J., D. D. Keck, and W. M. Hiesey. 1948. Experimental studies on the nature of species. III. Environmental responses of climatic races of *Achillea*. Carnegie Inst. Wash. Publ. 581:1-129.

Clegg, M. T., and R. W. Allard. 1972. Patterns of genetic differentiation in the slender wild oat species, *Avena barbata*. Proc. Nat. Acad. Sci. 69:1820-1824.

Cleland, R. E. 1962. The cytogenetics of *Oenothera*. Adv. Genetics 11:147-237.

Clements, F. E., and V. E. Shelford. 1939. *Bio-ecology*. Wiley. New York.

Cobbs, G. 1976. Polymorphism for dimerizing ability at the esterase-5 locus in *Drosophila pseudoobscura*. Genetics 82:53-62.

Cobbs, G., and S. Prakash. 1977a. A comparative study of the esterase-5 locus in *Drosophila pseudoobscura*, *D. persimilis*, and *D. miranda*. Genetics 85:697-711.

Cobbs, G., and S. Prakash. 1977b. An experimental investigation of the unit charge model of protein polymorphism and its relation to the esterase-5 locus of *Drosophila pseudoobscura*, *D. persimilis*, and *D. miranda*. Genetics 87:717-742.

Cochrane, B. J. 1976. Heat stability variants of esterase-6 in *Drosophila melanogaster*. Nature 263:131-132.

Cockrum, E. L. 1952. A check-list and bibliography of hybrid birds in North America north of Mexico. Wilson Bull. 64:140-159.

Cody, M. S. 1966. A general theory of clutch size. Evol. 20:174-184.

Colbert, E. H. 1972. *Lystrosaurus* and Gondwanaland. Evol. Biol. 6:157-177.

Cole, L. C. 1954. Some features of random population cycles. J. Wildlife Mgt. 18:2-24.

Cole, L. C. 1960. Competitive exclusion. Science 132:348-349.

Cook. L. M., R. R. Askew, and J. A. Bishop. 1970. Increasing frequency of the typical form of the Peppered Moth in Manchester. Nature 227:1155.

Cordeiro, A. R. 1952. Experiments on the effects in heterozygous condition of second chromosomes for natural populations of *Drosophila willistoni*. Proc. Nat. Acad. Sci. 38:471-478.

Cory, L., P. Fjeld, and W. Serat. 1971. Environmental DDT and the genetics of natural populations. Nature 229:128-130.

Cory, L., and J. J. Manion. 1955. Ecology and hybridization in the genus *Bufo* in the Michigan-Indiana region. Evol. 9:42-51.

Cott, H. B. 1940. *Adaptive Coloration in Animals*. Methuen. London.

Coyne, J. A. 1976. Lack of genic similarity between sibling species of *Drosophila* as revealed by varied techniques. Genetics 84:593-607.

Coyne, J. A., and A. A. Felton. 1977. Genic heterogeneity at two alcohol dehydrogenase loci in *Drosophila pseudoobscura* and *Drosophila persimilis*. Genetics 87:285-304.

Crew, F. A. E. 1937. The sex ratio. Amer. Nat. 71:529-559.

Crombie, A. C. 1945. On competition between different species of graminivorous insects. Proc. Roy. Soc. London (B)132:362-395.

Crombie, A. C. 1947. Interspecific competition. J. Animal Ecol. 16:44-73.

Crosby, J. L. 1963. The evolution and nature of dominance. J. Theoret. Biol. 5:35-51.

Crow, J. F. 1954. Breeding structure of populations. II. Effective population number. In *Statistics and Mathematics in Biology*. O. Kempthorne et al., Eds. Iowa State College Press. Ames. Iowa.

Crow, J. F. 1955. General theory of population genetics: Synthesis. Cold Spring Harbor Symp. Quant. Biol. 20:54-59.

Crow, J. F. 1957. Genetics of insect resistance to chemicals. Ann. Rev. Entomol. 2:227-246.

Crow, J. F. 1961. Mutation in man. In *Progress in Medical Genetics*. Vol. 1. A. G. Steinberg, Ed. Grune and Stratton. New York.

Crow, J. F. 1963. The concept of genetic load: A reply. Amer. J. Human Genetics 15: 310-315.

Crow, J. F. 1966. Evolution of resistance in hosts and pests. In *Scientific Aspects of Pest Control*. NAS/NRC. Washington, D.C.

Crow, J. F. 1970. Genetic loads and the costs of natural selection. In *Mathematical Topics in Population Genetics*. K. Kojima, Ed. Springer-Verlag. New York.

Crow, J. F. 1972. Darwinian and non-Darwinian evolution. Proc. 6th Berkeley Symp. Math. Stat. Prob. Vol. 5:1-22.

Crow, J. F., and C. Denniston, Eds. 1974. *Genetic Distance*. Plenum Press. New York.

Crow, J. F., and M. Kimura. 1970. *An Introduction to Population Genetics Theory*. Harper and Row. New York.

Crow, J. F., and N. Morton. 1955. Measurement of gene frequency drift in small populations. Evol. 9:202-214.

Crow. J. F., and R. G. Temin. 1964. Evidence for the partial dominance of recessive lethal genes in natural populations of *Drosophila*. Amer. Nat. 98:21-33.

Crozier, R. H. 1973. Apparent differential selection at an isozyme locus between queens and workers of the ant *Aphaenogaster rudis*. Genetics 73:313-318.

Crumpacker, D. W. 1967. Genetic loads in Maize (*Zea mays* L.) and other crossfertilized plants and animals. Evol. Biol. 1:306-424.

Crumpacker, D. W., and J. S. Williams. 1973. Density, dispersion, and population structure in *Drosophila pseudoobscura*. Ecol. Monog. 43:499-538.

Cunha, A. B. da, T. Dobzhansky, O. Pavlovsky, and B. Spassky. 1959. Genetics of natural populations. XXVIII. Supplementary data on the chromosomal polymorphism in *Drosophila willistoni* in relation to its environment. Evol. 13:389-404.

Dapkus, D. C. 1976. Differential survival involving the Burnsi phenotype in the northern leopard frog, *Rana pipiens*. Herpetologica 32:325-327.

Darlington, C. D. 1956. Natural populations and the breakdown of classical genetics. Proc. Roy. Soc. London (B) 145:350-364.

Darlington, C. D., and K. Mather. 1949. *The Elements of Genetics*. Allen and Unwin. London.

Darwin, C. R. 1859. *On the Origin of Species by Means of Natural Selection*. John Murray. London.

Darwin, C. R. 1872. *The Origin of Species*. 6th Ed. (With additions and corrections, 1890). Appleton. New York.

Darwin, C. R. 1871. *The Descent of Man and Selection in Relation to Sex*. John Murray. London.

Dayhoff, M. O., R. V. Eck, and C. M. Park. 1972. A model of evolutionary change in proteins. Atlas of Protein Sequence and Structure 5:89-100.

DeBach, P. 1966. The competitive displacement and coexistence principles. Ann. Rev. Ent. 11:183-212.

DeJong, G., and W. Scharloo. 1976. Environmental determination of selective significance or neutrality of amylase variants in *Drosophila melanogaster*. Genetics 84:77-94.

Dempster, E. R. 1955. Maintenance of genetic heterogeneity. Cold Spring Harbor Symp. Quant. Biol. 20:25-32.

Dickinson, H., and J. Antonovics. 1973. Theoretical considerations of sympatric divergence. Amer. Nat. 107:256-274.

Dingle, H. 1972. Migration strategies of insects. Science 175:1327-1335.

Dobzhansky, T. 1940. Speciation as a stage of evolutionary divergence. Amer. Nat. 74: 312-321.

Dobzhansky, T. 1941. *Genetics and the Origin of Species*. 2nd Ed. Columbia University Press. New York.

Dobzhansky, T. 1943. Genetics of natural populations. IX. Temporal changes in the composition of populations of *Drosophila pseudoobscura*. Genetics 28:162-186.

Dobzhansky, T. 1944. Experiments on sexual isolation in *Drosophila*. III. Geographic strains of *Drosophila sturtevanti*. Proc. Nat. Acad. Sci. 30:335-339.

Dobzhansky, T. 1946. Genetics of natural populations. XIII. Recombination and variability in populations of *Drosophila pseudoobscura*. Genetics 31:269-290.

Dobzhansky, T. 1948. Genetics of natural populations. XVI. Altitudinal and seasonal changes produced by natural selection in certain populations of *Drosophila pseudoobscura* and *Drosophila persimilis*. Genetics 33:158-176.

Dobzhansky, T. 1950. Genetics of natural populations. XIX. Origin of heterosis through natural selection in populations of *Drosophila pseudoobscura*. Genetics 35:288-302.

Dobzhansky, T. 1951. *Genetics and the Origin of Species*. 3rd Ed. Columbia University Press. New York.

Dobzhansky, T. 1952a. Nature and origin of heterosis. In *Heterosis*. J. W. Gowen, Ed. Iowa State College Press. Ames.

Dobzhansky, T. 1952b. Genetics of natural populations. XX. Changes induced by drought in *Drosophila pseudoobscura* and *Drosophila persimilis*. Evol. 6:234-243.

Dobzhansky, T. 1955. A review of some fundamental concepts and problems of population genetics. Cold Spring Harbor Symp. Quant. Biol. 20:1-15.

Dobzhansky, T. 1956. Genetics of natural populations. XXV. Genetic changes in populations of *Drosophila pseudoobscura* and *Drosophila persimilis* in some localities in California. Evol. 10:82-92.

Dobzhansky, T. 1957. Genetics of natural populations. XXVI. Chromosomal variability in island and continental populations of *Drosophila willistoni* from Central America and the West Indies. Evol. 11:280-293.

Dobzhansky, T. 1962a. *Mankind Evolving*. Yale University Press. New Haven.

Dobzhansky, T. 1962b. Rigid vs. flexible chromosomal polymorphisms in *Drosophila*. Amer. Nat. 96:321-328.

Dobzhansky, T. 1968a. On some fundamental concepts of Darwinian biology. Evol. Biol. 2:1-34.

Dobzhansky, T. 1968b. Adaptedness and fitness. In *Population Biology and Evolution*. R. C. Lewontin, Ed. Syracuse University Press. Syracuse, New York.

Dobzhansky, T. 1970. *Genetics of the Evolutionary Process*. Columbia University Press. New York.

Dobzhansky, T. 1971. Evolutionary oscillations in *Drosophila pseudoobscura*. In *Ecological Genetics and Evolution*. R. Creed, Ed. Blackwell. Oxford.

Dobzhansky, T. 1973. Active dispersal and passive transport in *Drosophila*. Evol. 27:565-575.

Dobzhansky, T. 1974. Genetic analysis of hybrid sterility within the species *Drosophila pseudoobscura*. Hereditas 77:81-88.

Dobzhansky, T., W. W. Anderson, and O. Pavlovsky. 1966. Genetics of natural populations. XXXVIII. Continuity and change in populations of *Drosophila pseudoobscura* in western United States. Evol. 20:418-427.

Dobzhansky, T., W. W. Anderson, O. Pavlovsky, B. Spassky, and C. J. Wills. 1964. Genetics of natural populations. XXXV. A progress report on genetic changes in populations of *Drosophila pseudoobscura* in the American Southwest. Evol. 18:164-176.

Dobzhansky, T., F. J. Ayala, G. L. Stebbins, and J. W. Valentine. 1977. *Evolution*. Freeman. San Francisco.

Dobzhansky, T. and C. Epling. 1944. Taxonomy, geographic distribution, and ecology of *Drosophila pseudoobscura* and its relatives. Carnegie Inst. Wash. Publ. 554:1-46.

Dobzhansky, T., A. S. Hunter, O. Pavlovsky, B. Spassky, and B. Wallace. 1963. Genetics of natural populations. XXXI. Genetics of an isolated marginal population of *D. pseudoobscura*. Genetics 48:91-103.

Dobzhansky, T. and H. Levene. 1948. Genetics of natural populations. XVII. Proof of operation of natural selection in wild populations of *Drosophila pseudoobscura*. Genetics 33:537-547.

Dobzhansky, T., and E. Mayr. 1944. Experiments on sexual isolation in *Drosophila*. I. Geographic strains of *Drosophila willistoni*. Proc. Nat. Acad. Sci. 30:238-244.

Dobzhansky, T., and O. Pavlovsky. 1953. Indeterminate outcome of certain experiments on *Drosophila* populations. Evol. 7:198-210.

Dobzhansky, T., and O. Pavlovsky. 1955. An extreme case of heterosis in a Central American population of *Drosophila tropicalis*. Proc. Nat. Acad. Sci. 41:289-295.

Dobzhansky, T., and O. Pavlovsky. 1957. An experimental study of interaction between genetic drift and natural selection. Evol. 11:311-319.

Dobzhansky, T., and O. Pavlovsky. 1958. Interracial hybridization and breakdown of co-adapted gene complexes in *Drosophila paulistorum* and *Drosophila willistoni*. Proc. Nat. Acad. Sci. 44:622-629.

Dobzhansky, T., and J. R. Powell. 1974. Rates of dispersal of *Drosophila pseudoobscura* and its relatives. Proc. Royal Soc. Lond. (B) 187:281-298.

Dobzhansky, T. and B. Spassky. 1944. Genetics of natural populations. XI. Manifestation of genetic variants in *Drosophila pseudoobscura* in different environments. Genetics 29:270-290.

Dobzhansky, T. and B. Spassky. 1953. Genetics of natural populations. XXI. Concealed variability in two sympatric species of *Drosophila*. Genetics 38:471-484.

Dobzhansky, T. and B. Spassky. 1954. Genetics of natural populations. XXII. A comparison of the concealed variability in *Drosophila prosaltans* with that in other species. Genetics. 39:472-487.

Dobzhansky, T., B. Spassky, and N. Spassky. 1952. A comparative study of mutation rates in two ecologically diverse species of *Drosophila*. Genetics 37:650-664.

Dobzhansky, T., B. Spassky, and N. Spassky. 1954. Rates of spontaneous mutation in the second chromosomes of the sibling species, *Drosophila pseudoobscura* and *Drosophila persimilis*. Genetics 39:899-907.

Dobzhansky, T. and N. P. Spassky. 1962. Genetic drift and natural selection in experimental populations of *Drosophila pseudoobscura*. Proc. Nat. Acad. Sci. 48:148-156.

Dobzhansky, T. and G. Streisinger. 1944. Experiments on sexual isolation in *Drosophila*. II. Geographic strains of *Drosophila prosaltens*. Proc. Nat. Acad. Sci. 30:340-345.

Dobzhansky, T. and S. Wright. 1941. Genetics of natural populations. V. Relations between mutation rate and accumulation of lethals in populations of *Drosophila pseudoobscura*. Genetics 26:23-51.

Dobzhansky, T. and S. Wright. 1943. Genetics of natural populations. X. Dispersion rates in *Drosophila pseudoobscura*. Genetics 28:304-340.

Dobzhansky, T. and S. Wright. 1947. Genetics of natural populations. XV. Rate of diffusion of a mutant gene through a population of *Drosophila pseudoobscura*. Genetics 32:303-339.

Dowdeswell, W. H. 1961. Experimental studies on natural selection in the butterfly *Maniola jurtina*. Heredity 16:39-52.

Drake, J. W. Mutagenic mechanisms. Ann. Rev. Genetics 3:247-268.

Dubinin, N. P. 1946. On lethal mutations in natural populations. Genetics 31:21-38.

Dubinin, N. P., D. D. Romashov, M. A. Heptner, and Z. A. Demidova. 1937. Aberrant polymorphism in *Drosophila fasciata* Meig (Syn. - *melanogaster* Meig) (In Russian) Biol. Zhurnal 6:311-354.

Dubinin, N. P., and D. D. Tiniakov. 1945. Seasonal cycles and the concentration of inversions in populations of *Drosophila funebris*. Amer. Nat. 79:570-572.

Dubinin, N. P., and G. G. Tiniakov. 1946a. Inversion gradients and natural selection in ecological races of *Drosophila funebris*. Genetics 31:537-545.

Dubinin, N. P., and G. G. Tiniakov. 1946b. Structural chromosome variability in urban and rural populations of *Drosophila funebris*. Amer. Nat. 80:393-396.

Dubinin, N. P., and fourteen collaborators. 1934. Experimental study of the ecogenotypes of *Drosophila melanogaster*. (In Russian). Biol. Zhurnal 3:166-216.

Dunn, L. C. 1955. Widespread distribution of mutant alleles (t-alleles) in populations of wild house mice. Nature 176:1275-1276.

Dunn, L. C. 1956. Analysis of a complex gene in the house mouse. Cold Spring Harbor Symp. Quant. Biol. 21:187-195.

Dunn, L. C. 1957a. Evidence of evolutionary forces leading to the spread of lethal genes in wild populations of the house mouse. Proc. Nat. Acad. Sci. 43:158-163.

Dunn, L. C. 1957b. Studies of the genetic variability in populations of wild house mice. II. Analysis of eight additional alleles at locus T. Genetics 42:299-311.

Dunn, L. C., and S. Gluecksohn-Schoenheimer. 1945. Dominance modification and physiological effects of genes. Proc. Nat. Acad. Sci. 31:82-84.

DuToit, A. L. 1937. *Our Wandering Continents*. Oliver and Boyd. Edinburgh.

East, E. M. 1910. A Mendelian interpretation of variation that is apparently continuous. Amer. Nat. 44:65-82.

East. E. M. 1915. Studies on size inheritance in *Nicotiana*. Genetics 1:164-176.

East, E. M. 1936. Heterosis. Genetics 21:375-397.

Ehrlich, P. R. 1961. Intrinsic barriers to dispersal in the checkerspot butterfly, *Euphydryas editha*. Science 134:108-109.

Ehrich, P. R. 1965. The population biology of the butterfly, *Euphydryas editha*. II. The structure of the Jasper Ridge Colony. Evol. 19:327-336.

Ehrlich, P. R., and R. W. Holm. 1962. Patterns and populations. Science 137:652-657.

Ehrlich, P. R., and R. W. Holm. 1963. *The Process of Evolution*. McGraw-Hill. New York.

Ehrlich, P. R., and P. H. Raven. 1969. Differentiation of populations. Science 165:1228-1232.

Ehrlich, P. R., and R. R. White. 1980. Colorado checkerspot butterflies: isolation, neutrality, and the biospecies. Amer. Nat. 115:328-341.

Ehrlich, P. R., R. R. White, M. C. Singer, S. W. McKechnie, and L. E. Gilbert. 1975. Checkerspot butterflies: an historical perspective. Science 188:221-228.

Ehrman, L., and C. Petit. 1968. Genotype frequency and mating success in the *willistoni* species group of *Drosophila*. Evol. 22:649-658.

Ehrman, L., and J. Probber. 1978. Rare *Drosophila* males: The mysterious matter of choice. Amer. Sci. 66:216-222.

Elton, C. S. 1927. *Animal Ecology*. Sidgwick and Jackson. London.

Elton, C. S. 1942. *Voles, Mice, and Lemmings: Problems in Population Dynamics*. Clarendon Press. Oxford.

Elton, C. S. 1946. Competition and the structure of animal communities. J. Animal Ecol. 15:54-68,

Elton, C. S. 1958. *The Ecology of Invasions by Animals and Plants*. Methuen. London.

Elton, C. S. 1966. *The Pattern of Animal Communities*. Methuen. London.

Elton, C. S., and R. S. Miller. 1954. The ecological survey of animal communities with a practical system of classifying habitats by structural characters. J. Ecol. 42:460-496.

Emlen, J. M. 1968a. A note on natural selection and the sex-ratio. Amer. Nat. 102:94-95.

Emlen, J. M. 1968b. Selection for the sex ratio. Amer. Nat. 102:589-591.

Emlen, J. M. 1973. *Ecology: An Evolutionary Approach*. Addison-Wesley. Reading, Massachusetts.

Emlen, J. M. 1978. Density anomalies and regulatory mechanisms in land bird populations on the Florida peninsula. Amer. Nat. 112:265-286.

Endler, J. A. 1973. Gene flow and population differentiation. Science 179:243-250.

Endler, J. A. 1977. *Geographic Variation, Speciation, and Clines*. Princeton University Press. Princeton, N.J.

Epling, C., D. F. Mitchell, and R. H. T. Mattoni. 1953. On the role of inversions in wild populations of *Drosophila pseudoobscura*. Evol. 7:342-365.

Epling, C., D. F. Mitchell, and R. H. T. Mattoni. 1955. Frequencies of inversion combinations in the third chromosome of wild males of *Drosophila pseudoobscura*. Proc. Nat. Acad. Sci. 41:915-921.

Epling, C., D. F. Mitchell, and R. H. T. Mattoni. 1957. The relation of an inversion system to recombination in wild populations. Evol. 11:225-247.

Evans, F. C., and F. E. Smith. 1952. The intrinsic rate of natural increase for the human louse, *Pediculus humanus* L. Amer. Nat. 86:299-310.

Ewens, W. J. 1965a. A note on Fisher's theory of the evolution of dominance. Ann. Human Genet. 29:85-88.

Ewens, W. J. 1965b. Further notes on the evolution of dominance. Heredity 20:443-450.

Ewens, W. J. 1966. Linkage and the evolution of dominance. Heredity 21:363-370.

Ewens, W. J. 1967. A note on the mathematical theory of the evolution of dominance. Amer. Nat. 101:35-40.

Ewens, W. J. 1969. *Population Genetics*. Methuen. London.

Ewens, W. J. 1973. Comments on Dr. Kimura's paper. Genetics (Suppl.) 73:36-38.

Ewens, W. J., and M. W. Feldman. 1976. The theoretical assessment of selective neutrality. In *Population Genetics and Ecology*. S. Karlin and E. Nevo, Eds. Academic Press. New York.

Falconer, D. S. 1960. *Introduction to Quantitative Genetics*. Ronald. New York.

Feller, W. 1966. On the influence of natural selection on population size. Proc. Nat. Acad. Sci. 55:733-738.

Feller, W. 1967. On fitness and the cost of natural selection. Genet. Res. 9:1-15.

Fisher, R. A. 1918. The correlation between relatives on the supposition of Mendelian inheritance. Trans. Roy. Soc. Edinburgh 52:399-433.

Fisher, R. A. 1927. On some objections to mimicry theory: statistical and genetic. Trans. Roy. Ent. Soc. London 75:269-278.

Fisher, R. A. 1928a. The possible modification of the response of the wild type to recurrent mutations. Amer. Nat. 62:115-126.

Fisher, R. A. 1928b. Two further notes on the origin of dominance. Amer. Nat. 62:571-574.

Fisher, R. A. 1930. *The Genetical Theory of Natural Selection*. Clarendon. Oxford. (2nd Ed. Dover. New York. 1958).

Fisher, R. A. 1931. The evolution of dominance. Biol. Rev. Cambridge Phil. Soc. 6:345-368.

Fisher, R. A. 1935. Dominance in poultry. Phil. Trans. Roy. Soc. London (B) 225:195-226.

Fisher, R. A. 1938. Dominance in poultry. Feathered feet, rose comb, internal pigment, and pile. Proc. Roy. Soc. London (B) 125:25-48.

Fisher, R. A. 1954. Retrospect of the criticisms of the theory of natural selection. In *Evolution as a Process*. J. Huxley, A. C. Hardy, and E. B. Ford, Eds. Allen and Unwin. London.

Fisher, R. A., and E. B. Ford. 1947. The spread of a gene in natural conditions in a colony of the moth, *Panaxia dominula*. Heredity 1:143-174.

Fisher, R. A., and E. B. Ford. 1950. The "Sewall Wright effect." Heredity 4:117-119.

Fisher, R. A., and S. B. Holt. 1944. The experimental modification of dominance in Danforth's short-tailed mutant mice. Ann. Eugenics 12:102-120.

Fisher, R. A., F. R. Immer, and O. Tedin. 1932. The genetical interpretation of statistics of the third degree in the study of quantitative inheritance. Genetics 17:107-124.

Fitch, W. M. 1975. Evolutionary rates in proteins and the cost of natural selection: implications for neutral mutations. In *The Role of Natural Selection in Human Evolution*. F. M. Salzano, Ed. North-Holland. Amsterdam.

Fitch, W. M., and E. Margoliash. 1967. Construction of phylogenetic trees. Science 155: 279-284.

Fitch, W. M., and E. Markowitz. 1970. An improved method for determining codon variability in a gene and its application to the rate of fixation of mutations in evolution. Biochem. Genet. 4:579-593.

Flor, H. H. 1971. Current status of the gene-for-gene concept. Ann. Rev. Phytopathol. 9:275-296.

Ford, E. B. 1937. Problems of heredity in the Lepidoptera. Biol. Rev. Cambridge Philos. Soc. 12:461-503.

Ford, E. B. 1940a. Genetic research in the Lepidoptera. Ann. Eugenics 10:227-252.

Ford, E. B. 1940b. Polymorphism and taxonomy. In *The New Systematics*. J. Huxley, Ed. Clarendon Press. Oxford.

Ford, E. B. 1953. The genetics of polymorphism in the Lepidoptera. Adv. Genetics 5: 43-87.

Ford, E. B. 1960. Evolution in progress. In *Evolution after Darwin. Vol. I. The Evolution of Life*. S. Tax, Ed. University of Chicago. Chicago.

Ford, E. B. 1964. *Ecological Genetics*. 1st Ed. Methuen. London.

Ford, E. B. 1971. *Ecological Genetics*. 3rd Ed. Chapman and Hall. London.

Ford, E. B. 1975. *Ecological Genetics*. 4th Ed. Chapman and Hall. London.

Frydenberg, O. 1963. Population studies of a lethal mutant in *Drosophila melanogaster*. I. Behavior in populations with discrete generations. Hereditas 50:89-116.

Frydenberg, O. 1964. Long-term instability of an ebony polymorphism in artificial populations of *Drosophila melanogaster*. Hereditas 51:198-206.

Frye, S. H. 1961. Drosophila Information Service 35:82-83.

Fuerst, P. A., R. Chakraborty, and M. Nei. 1977. Statistical studies on protein polymorphism in natural populations. I. Distribution of single locus heterozygosity. Genetics 86:455-483.

Futuyma, D. J. 1979. *Evolutionary Biology*. Sinauer. Sunderland, Mass.

Gaines, M. S., and C. J. Krebs. 1971. Genetic changes in fluctuating vole populations. Evol. 25:702-723.

Gardner, C. O. 1963. Estimates of genetic parameters in cross-fertilizing plants and their implications in plant breeding. In *Statistical Genetics and Plant Breeding*. W. D. Hanson and H. F. Robinson, Eds. NAS/NRC Publ. 982:225-252.

Gardner, C. O. 1977. Quantitative genetic research in plants: Past accomplishments and research needs. In *Proceedings of the International Conference on Quantitative Genetics*. E. Pollack, O. Kempthorne, and T. B. Bailey, Eds. Iowa State University Press. Ames.

Gause, G. F. 1934. *The Struggle for Existence*. Williams and Wilkins. Baltimore.

Gause, G. F. 1970. Criticism of invalidation of principle of competitive exclusion. Nature 227:89.

Gause, G. F., and A. A. Witt. 1935. Behavior of mixed populations and the problem of natural selection. Amer. Nat. 69:596-609.

Georghiou, G. P. 1969. Genetics of resistance to insecticides in houseflies and mosquitoes. Exp. Parasitol 26:224-255.

Gershenson, S. 1928. A new sex ratio abnormality in *Drosophila obscura*. Genetics 13: 488-507.

Gershenson, S. 1945. Evolutionary studies on the distribution and dynamics of melanism in the hamster (*Cricetus cricetus* L.). I. Distribution of black hamsters in the Ukrainian and Bashkirian Soviet Socialist Republics (U.S.S.R.). Genetics 30:207-251.

Ghiselin, M. T. 1966. On semantic pitfalls of biological adaptation. Philos. Science 33: 147-153.

Gibson, J. 1970. Enzyme flexibility in *Drosophila melanogaster*. Nature 227:959-960.

Gibson, J. B., and R. Miklovich. 1971. Modes of variation in alcohol dehydrogenase in *Drosophila melanogaster*. Experientia 27:99-100.

Gilbert, O., T. B. Reynoldson, and J. Hobart. 1952. Gause's hypothesis: an examination. J. Animal Ecol. 21:310-312.

Gillespie, J. H. 1974a. Polymorphism in patchy environments. Amer. Nat. 108:145-151.

Gillespie, J. H. 1974b. The role of environmental grain in the maintenance of genetic variation. Amer. Nat. 108:831-836.

Gillespie, J. H. 1975. The role of migration in the genetic structure of populations in temporally and spatially varying environments. I. Conditions for polymorphism. Amer. Nat. 109:127-135.

Gillespie, J. H., and K. Kojima. 1968. The degree of polymorphism in enzymes involved in energy production compared to that in nonspecific enzymes in two *Drosophila ananassae* populations. Proc. Nat. Acad. Sci. 61:582-585.

Glass, B., M. S. Sacks, E. F. Jahn, and C. Hess. 1952. Genetic drift in a religious isolate: an analysis of the causes of variation in blood groups and other gene frequencies in a small population. Amer. Nat. 86:145-159.

Glen, W. 1975. *Continental Drift and Plate Tectonics*. Merrill. Columbus, Ohio.

Goldschmidt, R. 1935. *Lymantria*. Bibliogr. Genet. 11:1-186.

Goldschmidt, R. 1935. Gen und Ausseneigenschaft. I. Untersuchungen an *Drosophila*. Z. indukt. Abstamm. -u. Vererb.-Lehre 69:38-69.

Goldschmidt, R. 1938. *Physiological Genetics*. McGraw-Hill. New York.

Goldschmidt, R. 1940. *The Material Basis of Evolution*. Yale University Press. New Haven.

Goldschmidt, R. 1945. Mimetic polymorphism, a controversial chapter of Darwinism. Quart. Rev. Biol. 20:147-164; 205-230.

Goodhart, C. B. 1962. Variation in a colony of the snail *Cepaea nemoralis* (L.) J. Animal Ecol. 31:207-237.

Goodhart, C. B. 1963. "Area effects" and non-adaptive variation between populations of *Cepaea*. Heredity 18:459-465.

Grant, V. 1949. Pollination systems as isolating mechanisms in flowering plants. Evol. 3:82-97.

Grant, V. 1950. The flower constancy of bees. Bot. Rev. 16:379-398.

Grant, V. 1963. *The Origin of Adaptations*. Columbia University Press. New York.

Grant, V. 1971. *Plant Speciation*. Columbia University Press. New York.

Grant, V. 1975. *Genetics of Flowering Plants*. Columbia University Press. New York.

Gray, A. P. 1954. *Mammalian Hybrids*. Commonwealth Agricultural Bureaux. Farnham Royal. England.

Gray, A. P. 1958. *Bird Hybrids*. Commonwealth Agricultural Bureaux. Farnham Royal. England.

Greenberg, R., and J. F. Crow. 1960. A comparison of the effect of lethal and detrimental chromosomes from *Drosophila* populations. Genetics 45:1153-1168.

Grinnell, J. 1904. The origin and distribution of the chesnut-backed chickadee. Auk 21: 364-382.

Grinnell, J. 1917. The niche-relationships of the California thrasher. Auk 34:427-433.

Grinnell, J. 1928. The presence and absence of animals. Univ. Calif. Chronicle 30:429-450.

Hairston, N. G., D. W. Tinkle, and H. M. Wilbur. 1970. Natural selection and the parameters of population growth. J. Wildl. Mgmt. 34:681-690.

Haldane, J. B. S. 1924a. A mathematical theory of natural and artificial selection. I. Trans. Cambridge Phil. Soc. 23:19-41.

Haldane, J. B. S. 1924b. A mathematical theory of natural and artificial selection. II. Proc. Cambridge. Phil. Soc. Biol. Sci. 1:158-163.

Haldane, J. B. S. 1924c. A mathematical theory of natural and artificial selection. III. Proc. Cambridge Phil. Soc. 23:363-372.

Haldane, J. B. S. 1924d. A mathematical theory of natural and artificial selection. IV. Proc. Cambridge Phil. Soc. 23:607-615.

Haldane, J. B. S. 1927. A mathematical theory of natural and artificial selection. V. Proc. Cambridge Phil. Soc. 23:838-844.

Haldane, J. B. S. 1930a. A mathematical theory of natural and artificial selection. VI. Proc. Cambridge Phil. Soc. 26:220-230.

Haldane, J. B. S. 1930b. A note on Fisher's theory of the origin of dominance and on a correlation between dominance and linkage. Amer. Nat. 64:87-90.

Haldane, J. B. S. 1931a. A mathematical theory of natural and artificial selection. VII. Proc. Cambridge Phil. Soc. 27:131-136.

Haldane, J. B. S. 1931b. A mathematial theory of natural and artificial selection. VIII. Proc. Cambridge Phil. Soc. 27:137-142.

Haldane, J. B. S. 1932a. A mathematical theory of natural and artificial selection. IX. Proc. Cambridge Phil. Soc. 28:244-248.

Haldane, J. B. S. 1932b. *The Causes of Evolution*. Longmans, Green. London. (Reprint. 1966. Cornell University Press. Ithaca).

Haldane, J. B. S. 1934. A mathematical theory of natural and artificial selection. X. Genetics 19:412-429.

Haldane, J. B. S. 1937. The effect of variation on fitness. Amer. Nat. 71:337-349.

Haldane, J. B. S. 1939. The theory of the evolution of dominance. J. Genet. 37:365-374.

Haldane, J. B. S. 1956. The relation between density regulation and natural selection. Proc. Roy. Soc. London (B) 145:306-308.

Haldane, J. B. S. 1957. The cost of natural selection. J. Genet. 55:511-524.

Haldane, J. B. S. 1960. More precise expressions for the cost of natural selection. J. Genet. 57:351-360.

Haldane, J. B. S. 1964. A defense of beanbag genetics. Perspect. Biol. Med. 7:343-359.

Haldane, J. B. S., and S. D. Jayakar. 1963. Polymorphism due to selection in varying directions. J. Genet. 58:237-242.

Haldane, J. B. S., and S. D. Jayakar. 1965. The nature of human genetic loads. J. Genet. 59:53-59.

Halkka, O. 1964. Geographical, spatial and temporal variability in the balanced polymorphisms of *Philaenus spumarius*. Heredity 19:383-401.

Halkka, O. 1967. Ecology and ecological genetics of *Philaenus spumarius* L. (Homoptera). Ann. Zool. Fenn. 4:1-18.

Hamilton, W. D. 1967. Extraordinary sex ratios. Science 156:477-488.

Hamrick, J. L., and R. W. Allard. 1972. Microgeographical variation in allozyme frequencies in *Avena barbata*. Proc. Nat. Acad. Sci. 69:2100-2104.

Hardin, G. 1960. The competitive exclusion principle. Science 131:1292-1297.

Hardy, G. H. 1908. Mendelian proportions in a mixed population. Science 28:49-50.

Harris, H. 1966. Enzyme polymorphisms in man. Proc. Roy. Soc. London (B) 164: 298-310.

Hartl, D. L. 1975. Genetic dissection of segregation distortion. II. Mechanism of suppression of distortion by certain inversions. Genetics 80:539-547.

Hartl, D. L., and Y. Hiraizumi. 1975. Segregation distortion after fifteen years. In *The Genetics of Drosophila melanogaster*. Vol 1. E. Novitski and M. Ashburner, Eds. Academic Press. New York.

Hartl, D. L., Y. Hiraizumi, and J. F. Crow. 1967. Evidence for sperm dysfunction as the mechanism of segregation distortion in *Drosophila melanogaster*. Proc. Nat. Acad. Sci. 58:2240-2245.

Hedrick, P. W. 1974. Genetic variation in a heterogeneous environment. I. Temporal heterogeneity and the absolute dominance model. Genetics 78:757-770.

Hedrick, P. W., M. E. Ginevan, and E. P. Ewing. 1976. Genetic polymorphism in heterogeneous environments. Ann. Rev. Ecol. Syst. 7:1-32.

Heiser, C. B. 1949. Natural hybridization with particular reference to introgression. Bot. Rev. 15:645-687.

Heiser, C. B. 1973. Introgression reexamined. Bot. Rev. 39:347-366.

Hernstein, R. J. 1971. I. Q. Atlantic 228:44-64.

Hexter, W. M. 1955. A population analysis of heterozygote frequencies in *Drosophila*. Genetics 40:444-459.

Hickey, D. A. 1977. Selection for amylase allozymes in *Drosophila melanogaster*. Evol. 31:800-804.

Hinegardner, R. 1976. Evolution of genome size. In *Molecular Evolution*. F. J. Ayala, Ed. Sinauer. Sunderland, Mass.

Hiraizumi, Y. 1962. Distorted segregation and genetic load. Japan. J. Genet. 37:147-154.

Hiraizumi, Y., and J. F. Crow. 1960. Heterozygous effects on viability, fertility, rate of development, and longevity of *Drosophila* chromosomes that are lethal when homozygous. Genetics 45:1071-1083.

Hiraizumi, Y., L. Sandler, and J. F. Crow. 1960. Meiotic drive in natural populations of *Drosophila melanogaster*. III. Populational implications with special application to the segregation distorter locus. Evol. 14:433-444.

Hjort, J. 1926. Fluctuations in the year classes of important food fishes. J. Conseil Perm. Internat. L'Exploration de la Mer 1:1-38.

Hoekstra, R. F. 1975. A deterministic model of cyclical selection. Genet. Res. 25:1-15.

Huang, S. L., M. Singh, and K. Kojima. 1971. A study of frequency-dependent selection observed in the esterase-6 locus of *Drosophila melanogaster* using a conditional medium method. Genetics 68:97-104.

Hubbs, C. L. 1955. Hybridization between fish species in nature. Syst. Zool. 4:1-20.

Hubby, J. L., and R. C. Lewontin. 1966. A molecular approach to the study of genic heterozygosity in natural populations. I. The number of alleles at different loci in *Drosophila pseudoobscura*. Genetics 54:577-594.

Hubby, J. L., and L. H. Throckmorton. 1965. Protein differences in *Drosophila*. II. Comparative species genetics and evolutionary problems. Genetics 52:203-215.

Huettel, M. D., and G. L. Bush. 1972. The genetics of host selection and its bearing on sympatric speciation in *Procecidochares* (Diptera: Tephritidae). Entomol. Exp. Appl. 15:465-480.

Hull, F. H. 1945. Recurrent selection for specific combining ability in corn. J. Amer. Soc. Agron. 37:134-145.

Hull, F. H. 1946. Regression analyses of corn yield data (Abstr.) Genetics 31:219.

Hull, F. H. 1952. Recurrent selection and overdominance. In *Heterosis*. J. W. Gowen, Ed. Iowa State College Press. Ames.

Hutchinson, G. E. Concluding remarks. Cold Spring Harbor Symp. Quant. Biol. 22:415-427.

Hutchinson, G. E. 1965. *The Ecological Theater and the Evolutionary Play*. Yale University Press. New Haven.

Hutchinson, G. E., and E. S. Deevey, Jr. 1949. Ecological studies on populations. Survey of Biol. Prog. 1:325-359.

Huxley, J. S. 1939. Clines: an auxiliary method in taxonomy. Bijdr. Dierk. 27:491-520.

Huxley, J. S. 1943. *Evolution: The Modern Synthesis*. Harper. New York.

Ingram, V. M. 1963. *The Hemoglobins in Genetics and Evolution*. Columbia University Press. New York.

Ives, P. T. 1945. The genetic structure of American populations of *Drosophila melanogaster*. Genetics 30:167-196.

Ives, P. T. 1950. The importance of mutation rate genes in evolution. Evol. 4:236-252.

Ives, P. T. 1954. Genetic changes in American populations of *Drosophila melanogaster*. Proc. Nat. Acad. Sci. 40:87-92.

Jaenike, J. 1978. Ecological genetics in *Drosophila athabasca*: Its effect on local abundance. Amer. Nat. 112:287-299.

Jain, S. K., and A. D. Bradshaw. 1966. Evolutionary divergence among adjacent plant populations. I. The evidence and its theoretical analysis. Heredity 21:407-441.

James, J. W. 1965. Simultaneous selection for dominant and recessive mutants. Heredity 20:142-144.

Jensen, A. R. 1969. How much can we boost I.Q. and scholastic achievement. Harvard Educ. Rev. 39:1-123.

Johnson, C. G. 1969. *Migration and Dispersal of Insects by Flight*. Methuen. London.

Johnson, F. M., C. Kanapi, R. H. Richardson, M. R. Wheeler, and W. S. Stone. 1966. An analysis of polymorphisms among isozyme loci in dark and light *Drosophila ananassae* strains from American and Western Samoa. Proc. Nat. Acad. Sci. 56:119-125.

Johnson, F. M., and A. Powell. 1974. The alcohol-dehydrogenases of *Drosophila melanogaster*: Frequency changes associated with heat and cold shock. Proc. Nat. Acad. Sci. 71:1783-1784.

Johnson, F. M., and H. E. Schaffer. 1973. Isozyme variability in species of the genus *Drosophila*. VII. Genotype-environment relationships in populations of *D. melanogaster* from the eastern United States. Biochem. Genet. 10:149-163.

Johnson, F. M., H. E. Schaffer, J. E. Gillaspy, and E. S. Rockwood. 1969. Isozyme genotype-environmental relationships in natural populations of the harvester ant, *Pogonomyrmex barbatus*, from Texas. Biochem. Genet. 3:429-450.

Johnson, G. B. 1973a. Importance of substrate variability to enzyme polymorphisms. Nature 243:151-153.

Johnson, G. B. 1973b. Enzyme polymorphism and biosystematics: The hypothesis of selective neutrality. Ann. Rev. Ecol. Syst. 4:93-116.

Johnson, G. B. 1974. Enzyme polymorphism and metabolism. Science 184:28-37.

Johnson, G. B. 1975. Enzyme polymorphism and adaptation. Stadler Genet. Symp. 7:91-116.

Johnson, G. B. 1976a. Genetic polymorphism and enzyme function. In *Molecular Evolution*. F. J. Ayala, Ed. Sinauer. Sunderland, Mass.

Johnson, G. B. 1976b. Hidden alleles at the α-glycero-phosphate dehydrogenase locus in the *Colias* butterflies. Genetics 83:149-167.

Johnson, G. B. 1977a. Assessing electrophoretic similarity: The problem of hidden heterogeneity. Ann. Rev. Ecol. Syst. 8:309-328.

Johnson, G. B. 1977b. Evaluation of the stepwise mutation model of electrophoretic mobility: Comparison of the gel sieving behavior of alleles at the esterase-5 locus of *Drosophila pseudoobscura*. Genetics 87:139-157.

Johnson, M. S. 1971. Adaptive lactate dehydrogenase variation in the crested blenny, *Anoplarchus*. Heredity 27:205-226.

Jones, D. F. 1917. Dominance of linked factors as a means of accounting for heterosis. Genetics 2:466-479.

Kalmus, H., and C. A. B. Smith. 1960. Evolutionary origin of sexual differentiation and the sex ratio. Nature 186:1004-1006.

Kamping, H., and W. Van Delden. 1978. The alcohol dehydrogenase polymorphism in populations of *Drosophila melanogaster*. II. The relation between ADH activity and adult mortality. Biochem. Genet. 16:541-551.

Keith, L. B. 1963. *Wildlife's Ten-year Cycle*. University of Wisconsin Press. Madison.

Kempthorne, O. 1977a. Status of quantitative genetic theory. In *Proceedings of the International Conference on Quantitative Genetics*. E. Pollack, O. Kempthorne, and T. B. Bailey, Eds. Iowa State University Press. Ames.

Kempthorne, O. 1977b. The International Conference on Quantitative Genetics: Introduction. In *Proceedings of the International Conference on Quantitative Genetics*. E. Pollack, O. Kempthorne, and T. B. Bailey, Eds. Iowa State University Press. Ames.

Kerr, W. E., and S. Wright. 1954a. Experimental studies of the distribution of gene frequencies in very small populations of *Drosophila melanogaster*. I. Forked. Evol. 8:172-177.

Kerr, W. E., and S. Wright. 1954b. Experimental studies of the distribution of gene frequencies in very small populations of *Drosophila melanogaster*. III. Aristapedia and spineless. Evol. 8:293-302.

Kerster, H. W. 1964. Neighborhood size in the Rusty Lizard, *Sceloporus olivaceus*. Evol. 18:445-457.

Kerster, H. W., and D. A. Levin. 1968. Neighborhood size in *Lithospermum carolinense*. Genetics 60:577-583.

Kessler, S. 1966. Selection for and against ethological isolation between *Drosophila pseudoobscura* and *Drosophila persimilis*. Evol. 20:634-645.

Kettlewell, H. B. D. 1956. Industrial melanism in the Lepidoptera and its contribution to our knowledge of evolution. Proc. 10th Internat. Cong. Ent. 2:831-841.

Kettlewell, H. B. D. 1958. A survey of the frequencies of *Biston betularia* (L.) (Lep.) and its melanic forms in Great Britain. Heredity 12:51-72.

Kettlewell, H. B. D. 1961. The phenomenon of industrial melanism in Lepidoptera. Ann. Rev. Entomol. 6:245-262.

Kettlewell, H. B. D. 1965. Insect survival and selection for pattern. Science 148:1290-1296.

Kettlewell, H. B. D. 1973. *The Evolution of Melanism*. Clarendon Press. Oxford.

Kidwell, J. F., M. T. Clegg, F. M. Stewart, and T. Prout. 1977. Regions of stable equilibria for models of differential selection in the two sexes under random mating. Genetics 85:171-183.

Kimura, M. 1955. Stochastic processes and distribution of gene frequencies under natural selection. Cold Spring Harbor Symp. Quant. Biol. 20:33-53.

Kimura, M. 1968. Evolutionary rate at the molecular level. Nature 217:624-626.

Kimura, M. 1969. The rate of molecular evolution considered from the standpoint of population genetics. Proc. Nat. Acad. Sci. 63:1181-1188.

Kimura, M. 1979a. Model of effectively neutral mutations in which selective constraint is incorporated. Proc. Nat. Acad. Sci. 76:3440-3444.

Kimura, M. 1979b. The neutral theory of molecular evolution. Sci. Amer. 241(5):98-126.

Kimura, M., and J. F. Crow. 1963. The measurement of effective population number. Evol. 17:279-288.

Kimura, M., and J. F. Crow. 1964. The number of alleles that can be maintained in a finite population. Genetics 49:725-738.

Kimura, M., and T. Ohta. 1969. The average number of generations until fixation of a mutant gene in a finite population. Genetics 61:763-771.

Kimura, M., and T. Ohta. 1971a. Protein polymorphism as a phase of molecular evolution. Nature. 229:467-469.

Kimura, M., and T. Ohta. 1971b. *Theoretical Aspects of Population Genetics*. Princeton University Press. Princeton, N.J.

Kimura, M., and T. Ohta. 1971c. On the rate of molecular evolution. J. Mol. Evol. 1:1-17.

Kimura, M., and T. Ohta. 1972. Population genetics, molecular biometry, and evolution. Proc. 6th Berkeley Symp. Math. Stat. Prob. Vol. 5:43-68.

Kimura, M., and T. Ohta. 1973. Mutation and evolution at the molecular level. Genetics (Suppl.) 73:19-35.

Kimura, M., and T. Ohta. 1974. On some principles governing molecular evolution. Proc. Nat. Acad. Sci. 71:2848-2852.

King, C. E., and W. W. Anderson. 1971. Age-specific selection. II. The interaction between r and K during population growth. Amer. Nat. 105:137-156.

King, J. L. 1967. Continuously distributed factors affecting fitness. Genetics 55:483-492.

King, J. L. 1974. Isoallele frequencies in very large populations. Genetics 76:607-613.

King, J. L., and T. H. Jukes. 1969. Non-Darwinian evolution: Random fixation of selectively neutral mutations. Science 164:788-798.

King, J. L., and T. Ohta. 1975. Polyallelic mutational equilibria. Genetics 79:681-691.

King, M. -C., and A. C. Wilson. 1975. Evolution at two levels in humans and chimpanzees. Science 188:107-116.

Kitagawa, O. 1967. Genetic divergence in M. Vetukhiv's experimental populations of *Drosophila pseudoobscura*. Genet. Res. 10:303-312.

Knerer, G., and C. E. Atwood. 1973. Diprionid sawflies: Polymorphism and speciation. Science 179:1090-1099.

Koehn, R. K. 1969. Esterase heterogeneity: Dynamics of a polymorphism. Science 163:943-944.

Koehn, R. K. 1970. Functional and evolutionary dynamics of polymorphic esterases in Catastomid fishes. Trans. Amer. Fish. Soc. 99:219-228.

Koehn, R. K., R. Milkman, and J. B. Mitton. 1976. Population genetics of marine pelecypods. IV. Selection, migration, and genetic differentiation in the blue mussel *Mytilus edulis*. Evol. 30:2-32.

Koehn, R. K., and J. B. Mitton. 1972. Population genetics of marine pelecypods. I. Ecological heterogeneity and evolutionary strategy at an enzyme locus. Amer. Nat. 106:47-56.

Koehn, R. K., J. E. Perez, and R. B. Merritt. 1971. Esterase enzyme function and genetical structure of populations of the freshwater fish *Notropis stramineus*. Amer. Nat. 105:51-69.

Kojima, K. 1971. Is there a constant fitness for a given genotype? No! Evol. 25:281-285.

Kojima, K., J. Gillespie, and Y. N. Tobari. 1970. A profile of *Drosophila* species' enzymes assayed by electrophoresis. I. Number of alleles, heterozygosities, and linkage disequilibrium in glucose-metabolizing systems and some other enzymes. Biochem. Genet. 4:627-637.

Kojima, K., P. Smouse, S. Yang, P. S. Nair, and D. Brncic. 1972. Isozyme frequency patterns in *Drosophila pavani* associated with geographical and seasonal variables. Genetics 72:721-731.

Kojima, K., and Y. N. Tobari. 1969. The pattern of viability changes associated with genotype frequency at the alcohol dehydrogenase locus in a population of *Drosophila melanogaster*. Genetics 61:201-209.

Kojima, K., and K. M. Yarbrough. 1967. Frequency dependent selection at the Esterase-6 locus in *Drosophila melanogaster*. Proc. Nat. Acad. Sci. 57:645-649.

Kolman, W. 1960. The mechanism of natural selection for the sex ratio. Amer. Nat. 94: 373-377.

Koopman, K. F. 1950. Natural selection for reproductive isolation between *Drosophila pseudoobscura* and *Drosophila persimilis*. Evol. 4:135-145.

Krebs, C. J. 1970. *Microtus* population biology. IV. Behavioral changes associated with the population cycle in *M. ochrogaster* and *M. pennsylvanicus*. Ecology 51:34-52.

Krebs, C. J., M. S. Gaines, B. L. Keller, J. H. Myers, and R. H. Tamarin. 1973. Population cycles in small rodents. Science 179:35-44.

Kruse, K. C., and D. G. Dunlap. 1976. Serum albumins and hybridization in two species of the *Rana pipiens* complex in the North Central United States. Copeia 1976(2): 394-396.

Lack, D. L. 1943. *The Life of the Robin*. Harmondsworth. Penguin. London.

Lack, D. L. 1945. The ecology of closely related species with special reference to cormorant (*Phalacrocorax carbo*) and shag (*P. aristotelis*). J. Animal Ecol. 14:12-16.

Lack, D. L 1946. Competition for food by birds of prey. J. Animal Ecol. 15:123-129.

Lack, D. L. 1947. *Darwin's Finches*. Cambridge University Press. Cambridge.

Lack, D. L. 1954. *The Natural Regulation of Animal Numbers*. Clarendon Press. Oxford.

Lack, D. L. 1966. *Population Studies of Birds*. Clarendon Press. Oxford.

Laibach, F. 1925. Das Taubwerden von Bastardsamen und die Künstliche Aufzucht früh absterbender Bastardembryonen. Zeit. Botanik 17:417-459.

Lamotte, M. 1951. Recherches sur la structure génétique des populations naturelles de *Cepaea nemoralis* (L.). Bull. Biol. (France) Suppl. 35:1-239.

Lamotte, M. 1959. Polymorphism of natural populations of *Cepaea nemoralis*. Cold Spring Harbor Symp. Quant. Biol. 24:65-84.

Latimer, H. 1958. A study of the breeding barrier between *Gilia australis* and *Gilia splendens*. Ph.D. Thesis. Claremont Graduate School. Claremont, California.

Latter, B. D. H. 1976. The intensity of selection for electrophoretic variants in natural populations of *Drosophila*. In *Population Genetics and Ecology*. S. Karlin and E. Nevo, Eds. Academic Press. New York.

Lawler, G. H. 1965. Fluctuations in the success of year-classes of whitefish populations with special reference to Lake Erie. J. Fisheries Res. Bd. Canada 22:1197-1227.

Leigh, E. G. 1970. Sex ratio and differential mortality between the sexes. Amer. Nat. 104:205-210.

Lerner, I. M. 1954. *Genetic Homeostasis*. Oliver and Boyd. Edinburgh.

Lerner, I. M. 1958. *The Genetic Basis of Selection*. Wiley. New York.

Lerner, I. M. 1965. Ecological genetics: Synthesis. Proc. XI Internat. Cong. Genetics 2: 489-494.

Lerner, I. M. 1968. *Heredity, Evolution, and Society*. Freeman. San Francisco.

Lerner, I. M. 1972. Polygenic inheritance and human intelligence. Evol. Biol. 6:399-414.

Lerner, I. M., and E. R. Dempster. 1962. Indeterminism in interspecific competition. Proc. Nat. Acad. Sci. 48:821-826.

Lerner, I. M., and F. K. Ho. 1961. Genotype and competitive ability of *Tribolium* species. Amer. Nat. 95:329-343.

Leslie, P. H. 1945. On the use of matrices in certain population mathematics. Biometrika 33:183-212.

Leslie, P. H., and T. Park. 1949. The intrinsic rate of natural increase of *Tribolium castaneum* Herbst. Ecology 30:469-477.

Leslie, P. H., and R. M. Ranson. 1940. The mortality, fertility, and rate of natural increase of the Vole (*Microtus agrestis*) as observed in the laboratory. J. Animal Ecol. 9:27-52.

Levene, H. 1949. A new measure of sexual isolation. Evol. 3:315-321.

Levene, H. 1953. Genetic equilibrium when more than one ecological niche is available. Amer. Nat. 87:331-333.

Levin, D. A. 1979. The nature of plant species. Science 204:381-384.

Levin, D. A., and H. W. Kerster. 1967. Natural selection for reproductive isolation in *Phlox*. Evol. 21:679-687.

Levin, D. A., and H. W. Kerster. 1968. Local gene dispersal in *Phlox*. Evol. 22:130-139.

Levin, D. A., and H. W. Kerster. 1974. Gene flow in seed plants. Evol. Biol. 7:139-220.

Levin, S. A. 1978. On the evolution of ecological parameters. In *Ecological Genetics: The Interface*. P. F. Brussard, Ed. Springer-Verlag. New York.

Levins, R. 1968. *Evolution in Changing Environments*. Princeton University Press. Princeton, N.J.

Levins, R., and R. H. MacArthur. 1966. Maintenance of genetic polymorphism in a heterogeneous environment: Variations on a Theme by Howard Levene. Amer. Nat. 100: 585-590.

Levitan, M. 1951a. Selective differences between males and females in *Drosophila robusta*. Amer. Nat. 85:385-388.

Levitan, M. 1951b. Experiments on chromosomal variability in *Drosophila robusta*. Genetics 36:285-305.

Levitan, M. 1957. Natural selection for linked gene arrangements. Anat. Rec. 127:430.

Lewis, H. 1966. Speciation in flowering plants. Science 152:167-172.

Lewis, H. 1973. The origin of diploid neospecies in *Clarkia*. Amer. Nat. 107:161-170.

Lewontin, R. C. 1955. The effects of population density and composition on viability in *Drosophila melanogaster*. Evol. 9:27-41.

Lewontin, R. C. 1957. The adaptations of populations to varying environments. Cold Spring Harbor Symp. Quant. Biol. 22:395-408.

Lewontin, R. C. 1961. Evolution and the theory of games. J. Theor. Biol. 1:382-403.

Lewontin, R. C. 1965. The role of linkage in natural selection. Proc. XI Internat. Cong. Genetics 3:517-525.

Lewontin, R. C. 1967. Population genetics. Ann Rev. Genetics 1:37-70.

Lewontin, R. C. 1968. Introduction. *Population Biology and Evolution*. Syracuse University Press. Syracuse, N.Y.

Lewontin, R. C. 1972. The apportionment of human diversity. Evol. Biol. 6:381-398.

Lewontin, R. C. 1974. *The Genetic Basis of Evolutionary Change*. Columbia University Press. New York.

Lewontin, R. C. 1975. Genetic aspects of intelligence. Ann. Rev. Genetics 9:387-405.

Lewontin, R. C. 1979. Fitness, survival, and optimality. In *Analysis of Ecological Systems*. D. J. Horn, G. R. Stairs, and R. D. Mitchell, Eds. Ohio State. Columbus.

Lewontin, R. C., and L. C. Birch. 1966. Hybridization as a source of variation for adaptation to new environments. Evol. 20:315-336.

Lewontin, R. C., and L. C. Dunn. 1960. The evolutionary dynamics of a polymorphism in the house mouse. Genetics 45:705-722.

Lewontin, R. C., L. R. Ginzburg, and S. D. Tuljapurkar. 1978. Heterosis as an explanation for large amounts of genic polymorphism. Genetics 88:149-169.

Lewontin, R. C., and J. L. Hubby. 1966. A molecular approach to the study of genic hetero-zygosity in natural populations. II. Amount of variation and degree of heterozygosity in natural populations of *Drosophila pseudoobscura*. Genetics 54:595-609.

Lewontin, R. C., and Y. Matsuo. 1963. Interaction of genotypes determining viability in *Drosophila busckii*. Proc. Nat. Acad. Sci. 49:270-278.

Li, C. C. 1955. The stability of an equilibrium and the average fitness of a population. Amer. Nat. 89:281-296.

Li, C. C. 1963a. Decrease of population fitness upon inbreeding. Proc. Nat. Acad. Sci. 49:439-445.

Li, C. C. 1963b. The way the load ratio works. Amer. J. Human Genet. 15:316-321.

Li, C. C. 1976. *First Course in Population Genetics*. Boxwood Press. Pacific Grove, Cal-ifornia.

Li, W. -H. 1978. Maintenance of genetic variability under the joint effect of mutation, selection and random drift. Genetics 90:349-382.

Lillie, F. R. 1921. Studies of fertilization. VIII. On the measure of specificity in fertili-zation between two associated species of the sea-urchin genus, *Strongylocentrotus*. Biol. Bull. 40:1-22.

Lim, H. C., V. R. Vickery, and D. K. M. Kevan. 1973. Cytogenetic studies in relation to taxonomy within the family Gryllidae (Orthoptera). I. Subfamily Gryllinae. Canad. J. Zool. 51:179-186.

Littlejohn, M. J. 1969. The systematic significance of isolating mechanisms. In *Systematic Biology*. Nat. Acad. Sci. Publ. No. 1692.

Livingstone, F. B. 1964. On the nonexistence of human races. In *The Concept of Race*. A. Montague, Ed. Macmillan. New York.

Lloyd, M., and H. S. Dybas. 1966a. The periodical cicada problem. I. Population ecology. Evol. 20:133-149.

Lloyd, M., and H. S. Dybas. 1966b. The periodical cicada problem. II. Evolution. Evol. 20:466-505.

Loehlin, J. C., G. Lindzey, and J. N. Spuhler. 1975. *Race Differences in Intelligence*. Free-man. San Francisco.

Lotka, A. J. 1925. *Elements of Physical Biology*. Williams and Wilkins. Baltimore.

Lotka, A. J. 1932. The growth of mixed populations: Two species competing for a common food supply. J. Wash. Acad. Sci. 22:461-469.

Lotsy, J. P. 1916. *Evolution by Means of Hybridization*. Nijhoff. The Hague.

Lotsy, J. P. 1932. On the species of the taxonomist in relation to evolution. Genetica 13:1-16.

Ludwig, W. 1950. Zur Theorie der Konkurrenz. Die Annidation (Einnischung) als fünfter Evolutionsfaktor. Neue Ergeb. Probleme Zoologie. Klatt-Festschrift. 1950:516-537.

Lyon, M., and R. Meredith. 1964. Investigation of the nature of t-alleles in the mouse. I. Genetic analysis of a series of mutants derived from a lethal allele. II. Genetic analysis of an unusual mutant allele and its derivative. III. Short tests of some further mutant alleles. Heredity 19:301-312; 313-325; 327-330.

Lyon, M. F., and T. Morris. 1966. Mutation rates at a new set of specific loci in mice. Genet. Res. 7:12-17.

MacArthur, R. H. 1961. Population effects of natural selection. Amer. Nat. 95:195-199.

MacArthur, R. H. 1962. Some generalized theorems of natural selection. Proc. Nat. Acad. Sci. 48:1893-1897.

MacArthur, R. H. 1965. Ecological consequences of natural selection. In *Theoretical and Mathematical Biology*. T. H. Waterman and H. J. Morowitz, Eds. Blaisdell. New York.

MacArthur, R. H. 1968. The theory of the niche. In *Population Biology and Evolution*. R. C. Lewontin, Ed. Syracuse University Press. Syracuse, N.Y.

MacArthur, R. M., and E. O. Wilson. 1967. *The Theory of Island Biogeography.* Princeton University Press. Princeton, N.J.

MacFayden, A. 1963. *Animal Ecology.* 2nd Ed. Pitman. London.

MacLulich, D. A. 1937. Fluctuations in the numbers of the varying hare (*Lepus americanus*). Univ. Toronto Studies, Biol. Ser. No. 43.

Malécot, G. 1948. Les mathematiques de l'heredite. Masson et Cie. Paris. (Translated by D. M. Yermanos. 1969. Freeman. San Francisco).

Malogolowkin-Cohen, C., A. Simmons, and H. Levene. 1965. A study of sexual isolation between certain strains of *Drosophila paulistorum*. Evol. 19:95-103.

Mangelsdorf, A. J. 1952. Gene interaction in heterosis. In *Heterosis.* J. W. Gowen, Ed. Iowa State College Press. Ames.

Manwell, C., and C. M. A. Baker. 1970. *Molecular Biology and the Origin of Species.* University of Washington Press. Seattle.

Manwell, C., and C. M. A. Baker. 1976. Protein polymorphisms in domesticated species: Evidence for hybrid origin? In *Population Genetics and Ecology.* S. Karlin and E. Nevo, Eds. Academic Press. New York.

Marinković, D., and F. J. Ayala. 1975a. Fitness of allozyme variants in *Drosophila pseudoobscura*. I. Selection at the *Pgm-1* and *Me-2* loci. Genetics 79:85-95.

Marinković, D., and F. J. Ayala. 1975b. Fitness of allozyme variants in *Drosophila pseudoobscura*. II. Selection at the *Est-5*, *Odh*, and *Mdh-2* loci. Genet. Res. 24:137-149.

Markert, C. L., and F. Møller. 1959. Multiple forms of enzymes: Tissue, ontogenetic, and species-specific patterns. Proc. Nat. Acad. Sci. 45:753-763.

Marshall, G. A. K. 1908. On diaposematism, with reference to some limitations of the Müllerian hypothesis of mimicry. Trans. Entomol. Soc. London (1908):93-142.

Mather, K. 1943. Polygenic inheritance and natural selection. Biol. Rev. Cambridge Phil. Soc. 18:32-64.

Mather, K. 1944. The genetical activity of heterochromatin. Proc. Roy. Soc. London (B)132:308-332.

Mather, K. 1955. Polymorphism as an outcome of disruptive selection. Evol. 9:52-61.

Mather, K., and J. L. Jinks. 1971. *Biometrical Genetics.* Cornell University Press. Ithaca.

Matthew, W. D. 1939. *Climate and Evolution.* 2nd Ed. New York Acad. Sci. New York.

Maxson, L., E. Pepper, and R. D. Maxson. 1977. Immunological resolution of a diploid-tetraploid species complex of tree frogs. Science 197:1012-1013.

May, R. M., G. R. Conway, M. P. Hassell, and T. R. E. Southwood. 1974. Time delays, density-dependence, and single-species oscillations. J. Animal Ecol. 43:747-770.

Maynard Smith, J. 1962. Disruptive selection, polymorphism, and sympatric speciation. Nature 195:60-62.

Maynard Smith, J. 1966. Sympatric speciation. Amer. Nat. 100:637-650.

Maynard Smith, J. 1968. "Haldane's dilemma" and the rate of evolution. Nature 219:1114-1116.

Maynard Smith, J. 1970. Genetic polymorphism in a varied environment. Amer. Nat. 104:487-490.

Maynard Smith, J., and J. Haigh. 1974. The hitch-hiking effect of a favourable gene. Genet. Res. 23:23-35.

Mayr, E. 1942. *Systematics and the Origin of Species.* Columbia University Press. New York.

Mayr, E. 1950. Taxonomic categories in fossil hominids. Cold Spring Harbor Symp. Quant. Biol. 15:109-118.

Mayr, E. 1954. Change of genetic environment and evolution. In *Evolution as a Process.* J. Huxley, A. C. Hardy, and E. B. Ford, Eds. Allen and Unwin. London.

Mayr, E. 1959. Where are we? Genetics and twentieth century Darwinism. Cold Spring Harbor Symp. Quant. Biol. 24:1-14.

Mayr, E. 1963. *Animal Species and Evolution*. Harvard University Press. Cambridge, Mass.

Mayr, E. 1970. *Populations, Species, and Evolution*. Harvard University Press. Cambridge, Mass.

Mayr, E., E. G. Linsley, and R. L. Usinger. 1953. *Methods and Principles of Systematic Zoology*. McGraw-Hill. New York.

McDonald, J., and F. J. Ayala. 1974. Genetic response to environmental heterogeneity. Nature. 250:572-574.

McDowell, R. E., and S. Prakash. 1976. Allelic heterogeneity within allozymes separated by electrophoresis in *Drosophila pseudoobscura*. Proc. Nat. Acad. Sci. 73:4150-4153.

McKechnie, S. W., P. R. Ehrlich, and R. P. White. 1975. Population genetics of *Euphydryas* butterflies. I. Genetic variation and the neutrality hypothesis. Genetics 81:571-594.

McMillan, I., and A. Robertson. 1974. The power of methods for the detection of major genes affecting quantitative characters. Heredity 32:349-356.

McNeilly, T. 1968. Evolution in closely adjacent plant populations. III. *Agrostis tenuis* on a small copper mine. Heredity 23:99-108.

McNeilly, T. S., and J. Antonovics. 1968. Evolution in closely adjacent plant populations. IV. Barriers to gene flow. Heredity 23:205-218.

McNeilly, T., and A. D. Bradshaw. 1968. Evolutionary processes in populations of copper tolerant *Agrostis tenuis* Sibth. Evol. 22:108-118.

Mecham, J. S. 1961. Isolating mechanisms in anuran amphibians. In *Vertebrate Speciation*. W. F. Blair, Ed. University of Texas Press. Austin.

Medawar, P. B. 1951. Problems of adaptation. New Biology 11:10-26.

Mendel, G. 1866. Versuche über Pflanzen-Hybriden. Verh. Naturforsch. Ver. in Brünn. 4:3-47. Also available in German in J. Heredity 42:1-47. Translations in W. Bateson, 1909, *Mendel's Principles of Heredity*, Cambridge University Press, London, and in C. Stern and E. R. Sherwood, 1966, *The Origin of Genetics*, Freeman, San Francisco.

Merrell, D. J. 1950. Measurement of sexual isolation and selective mating. Evol. 4:326-331.

Merrell, D. J. 1951. Interspecific competition between *Drosophila funebris* and *Drosophila melanogaster*. Amer. Nat. 85:159-169.

Merrell, D. J. 1953. Gene frequency changes in small laboratory populations of *Drosophila melanogaster*. Evol. 7:95-101.

Merrell, D. J. 1954. Sexual isolation between *Drosophila persimilis* and *Drosophila pseudoobscura*. Amer. Nat. 88:93-99.

Merrell, D. J. 1960. Mating preferences in *Drosophila*. Evol. 14:525-526.

Merrell, D. J. 1963. "Heterosis" in *Drosophila*. Evol. 17:481-485.

Merrell, D. J. 1965a. Lethal frequency and allelism in DDT-resistant populations and their controls. Amer. Nat. 99:411-417.

Merrell, D. J. 1965b. The distribution of the dominant Burnsi gene in the leopard frog, *Rana pipiens*. Evol. 19:69-85.

Merrell, D. J. 1965c. Competition involving dominant mutants in experimental populations of *Drosophila melanogaster*. Genetics 52:165-189.

Merrell, D. J. 1968. A comparison of the estimated size and the "effective size" of breeding populations of the leopard frog, *Rana pipiens*. Evol. 22:274-283.

Merrell, D. J. 1969a. The evolutionary role of dominant genes. In *Genetics Lectures*. Vol. 1:169-194. R. Bogart, Ed. Oregon State University Press. Corvallis.

Merrell, D. J. 1969b. Limits on heterozygous advantage as an explanation of polymorphism. J. Heredity 60:180-182.

Merrell, D. J. 1970. Migration and gene dispersal in *Rana pipiens*. Amer. Zool. 10:47-52.

Merrell, D. J. 1972. Laboratory studies bearing on pigment pattern polymorphisms in wild populations of *Rana pipiens*. Genetics 70:141-161.

Merrell, D. J. 1975a. *An Introduction to Genetics*. Norton. New York.

Merrell, D. J. 1975b. In defense of frogs. Science 189:838.

Merrell, D. J. 1977. The life history of the leopard frog, *Rana pipiens*, in Minnesota. Occ. Papers Bell Mus. Nat. Hist. Univ. Minnesota 15:1-23.

Merrell, D. J., and C. F. Rodell. 1968. Seasonal selection in the leopard frog, *Rana pipiens*. Evol. 22:284-288.

Merrell, D. J., and J. C. Underhill. 1956. Selection for DDT resistance in inbred, laboratory, and wild stocks of *Drosophila melanogaster*. J. Econ. Ent. 49:300-306.

Mettler, L. E., and T. G. Gregg. 1969. *Population Genetics and Evolution*. Prentice-Hall. Englewood Cliffs, New Jersey.

Metz, C. W. 1947. Duplications of chromosome parts as a factor in evolution. Amer. Nat. 81:81-103.

Michaelis, P. 1951. Plasmavererbung und Heterosis. Z. Pflanzen 30:250-275.

Mickey, G. H. 1954. Visible and lethal mutations in *Drosophila*. Amer. Nat. 88:241-255.

Milkman, R. D. 1967. Heterosis as a major cause of heterozygosity in nature. Genetics 55:493-495.

Milkman, R., and R. Koehler. 1976. Isoelectric focusing of MDH and 6-PGDH from *Escherichia coli* of diverse natural origins. Biochem. Genet. 14:517-522.

Miller, R. S. 1967. Pattern and process in competition. Adv. Ecol. Res. 4:1-74.

Milne, A. 1961. Definition of competition among animals. Symp. Soc. Exptl. Biol. 15:40-61.

Mitton, J. B., and R. K. Koehn. 1975. Genetic organization and adaptive response of allozymes to ecological variables in *Fundulus heteroclitus*. Genetics 79:97-111.

Moore, W. S. 1977. An evaluation of narrow hybrid zones in vertebrates. Quart. Rev. Biol. 52:263-277.

Morgan, P. 1975. Selection acting directly on an enzyme polymorphism. Heredity 34:124-127.

Morgan, T. H. 1929. The variability of eyeless. Publ. Carnegie Inst. Wash. 399:139-168.

Morris, R. F., Ed. 1963. The dynamics of epidemic spruce budworm populations. Mem. Entomol. Soc. Canada 31:1-332.

Morton, N. E. 1960. The mutational load due to detrimental genes in man. Amer. J. Human Genet. 12:348-364.

Morton, N. E., J. F. Crow, and H. J. Muller. 1956. An estimate of the mutational damage in man from data on consanguineous marriages. Proc. Nat. Acad. Sci. 42:855-863.

Moynihan, M. 1968. Social mimicry: Character convergence versus character displacement. Evol. 22:315-331.

Muller, H. J. 1927. Artificial transmutation of the gene. Science 66:84-87.

Muller, H. J. 1932. Further studies on the nature and causes of gene mutations. Proc. 6th Internat. Cong. Genetics 1:213-255.

Muller, H. J. 1940. Bearings of the "*Drosophila*" work on systematics. In *The New Systematics*. J. S. Huxley, Ed. Oxford University Press. Oxford.

Muller, H. J. 1942. Isolating mechanisms, evolution, and temperature. Biol. Symp. 6:71-125.

Muller, H. J. 1948. Evidence of the precision of genetic adaptation. Harvey Lectures 43:165-229.

Muller, H. J. 1950. Our load of mutations. Amer. J. Human Genet. 2:111-176.

Muller, H. J. 1958. Evolution by mutations. Bull. Amer. Math. Soc. 64:137-160.

Muller, H. J., and I. I. Oster. 1963. Some mutational techniques in *Drosophila*. In *Methodology in Basic Genetics*. W. J. Burdette, Ed. Holden-Day. San Francisco.

Müntzing, A. 1930. Über Chromosomenvermehrung in *Galeopsis*-Kreuzungen und ihre phylogenetische Bedeutung. Hereditas 14:153-172.

Müntzing, A. 1932. Cyto-genetic investigations on synthetic *Galeopsis tetrahit*. Hereditas 16:105-154.

Müntzing, A. 1974. Accessory chromosomes. Ann. Rev. Genetics 8:243-266.

Nei, M. 1972. Genetic distance between populations. Amer. Nat. 106:283-292.

Nei, M. 1975. *Molecular Population Genetics and Evolution*. North-Holland. Amsterdam.

Nei, M. 1976. Mathematical models of speciation and genetic distance. In *Population Genetics and Ecology*. S. Karlin and E. Nevo, Eds. Academic Press. New York.

Nei, M., and A. K. Roychoudhury. 1974. Genic variation within and between the three major races of man, Caucasoids, Negroids, and Mongoloids. Amer. J. Human Genet. 26:421-443.

Nichols, W. W. 1975. Somatic mutation in biologic research. Hereditas 81:225-236.

Nicholson, A. J. 1933. The balance of animal populations. J. Animal Ecol. 2:132-178.

Nicholson, A. J. 1954. An outline of the dynamics of animal populations. Austral. J. Zool. 2:9-65.

Nicholson, A. J. 1957. Self-adjustment of populations to change. Cold Spring Harbor Symp. Quant. Biol. 22:153-173.

Nilsson-Ehle, H. 1909. Kreuzungsuntersuchungen an Hafer und Weizen. Lunds. Univ. Aarsk. N. F. 5:1-122.

Novitski, E. 1951. Non-random disjunction in *Drosophila*. Genetics 36:267-280.

Novitski, E. 1967. Nonrandom disjunction in *Drosophila*. Ann. Rev. Genetics 1:71-86.

Novitski, E. and E. R. Dempster. 1958. The analysis of data from laboratory populations of *Drosophila melanogaster*. Genetics 43:470-479.

Novitski, E., W. J. Peacock, and J. Engel. 1965. Cytological basis of sex ratio in *Drosophila pseudoobscura*. Science 148:516-517.

Nozawa, K. 1972. Population genetics of Japanese monkeys. I. Estimation of effective troop size. Primates 13:381-393.

Oakeshott, J. G. 1976. Selection at the alcohol dehydrogenase locus in *Drosophila melanogaster* imposed by environmental ethanol. Genet. Res. 26:265-274.

Odum, E. P. 1971. *Fundamentals of Ecology*. 3rd Ed. Saunders. Philadelphia.

Ohno, S. 1970. *Evolution by Gene Duplication*. Springer-Verlag. New York.

Ohta, T. 1973. Slightly deleterious mutant substitutions in evolution. Nature 246:96-98.

Ohta, T. 1974. Mutational pressure as the main cause of molecular evolution and polymorphism. Nature 252:351-354.

Ohta, T. 1976. Role of very slightly deleterious mutations in molecular evolution and polymorphism. Theor. Pop. Biol. 10:254-275.

Ohta, T., and M. Kimura. 1970. Development of associative overdominance through linkage disequilibrium in finite populations. Genet. Res. 16:165-177.

Ohta, T., and M. Kimura. 1973. A model of mutation appropriate to estimate the number of electrophoretically detectable alleles in a finite population. Genet. Res. 22:201-204.

Ohta, T., and M. Kimura. 1975. Theoretical analysis of electrophoretically detectable polymorphisms: models of very slightly deleterious mutations. Amer. Nat. 109:137-145.

Oppenoorth, F. J. 1965. Biochemical genetics of insecticide resistance. Ann. Rev. Entomol. 10:185-206.

Owen, A. R. G. 1953. A genetical system admitting of two distinct stable equilibria under natural selection. Heredity 7:97-102.

Owen, D. F. 1964. Density effects in polymorphic land snails. Heredity 20:312-315.

Park, T. 1954. Experimental studies of interspecies competition. II. Temperature, humidity, and competition in two species of *Tribolium*. Physiol. Zool. 27:177-238.

Patten, B. C. 1961. Competitive exclusion. Science 134:1599-1601.

Patterson, J. T., and W. S. Stone. 1952. *Evolution in the Genus Drosophila*. Macmillan. New York.

Pavan, C., A. R. Cordeiro, N. Dobzhansky, T. Dobzhansky, C. Malogolowkin, B. Spassky, and M. Wedel. 1951. Concealed genic variability in Brazilian populations of *Drosophila willistoni*. Genetics 36:13-30.

Peacock, W. J., and G. L. G. Miklos. 1973. Meiotic drive in *Drosophila*: New interpretations of the segregation distorter and sex chromosome systems. Adv. Genetics 17: 361-409.

Petit, C., and L. Ehrman. 1969. Sexual selection in *Drosophila*. Evol. Biol. 3:177-223.

Petras, M. L. 1967. Studies of natural populations of *Mus*. I. Biochemical polymorphisms and their bearing on breeding structure. Evol. 21:259-274.

Petrides, G. A. 1950. The determination of sex and age ratios in fur animals. Amer. Midl. Nat. 43:355-382.

Pianka, E. R. 1970. On r- and K-selection. Amer. Nat. 104:592-597.

Pianka, E. R. 1972. r and K selection or b and d selection? Amer. Nat. 106:581-588.

Pimentel, D. 1961. Animal population regulation by the genetic feedback mechanism. Amer. Nat. 95:65-79.

Pimentel, D. 1965. Population ecology and the genetic feedback mechanism. Proc. 11th Internat. Cong. Genetics 2:483-488.

Pimentel, D. 1968. Population regulation and genetic feedback. Science 159:1432-1437.

Pitelka, F. A. 1957. Some aspects of population structure in the short-term cycle of the brown lemming in northern Alaska. Cold Spring Harbor Symp. Quant. Biol. 22:237-251.

Pittendrigh, C. S. 1958. Adaptation, natural selection, and behavior. In *Behavior and Evolution*. A. Roe and G. G. Simpson, Eds. Yale University Press. New Haven.

Plapp, F. W. 1976. Biochemical genetics of insecticide resistance. Ann. Rev. Entomol. 21:179-197.

Platz, J. E. 1972. Sympatric interaction between two forms of leopard frog (*Rana pipiens* complex) in Texas. Copeia 1972(2):232-240.

Plunkett, C. R. 1932. Temperature as a tool of research in phenogenetics: Methods and results. Proc. 6th Internat. Cong. Genetics 2:158-160.

Plunkett, C. R. 1933. A contribution to the theory of dominance. Amer. Nat. 67:84-85. (Abstr.).

Pollak, E., O. Kempthorne, and T. B. Bailey, Eds. 1977. *Proceedings of the International Conference on Quantitative Genetics*. Iowa State University Press. Ames.

Poulson, D. F. 1963. Cytoplasmic inheritance and hereditary infections in *Drosophila*. In *Methodology in Basic Genetics*. W. J. Burdette, Ed. Holden-Day. San Francisco.

Poulson, D. F., and B. Sakaguchi. 1961. Nature of sex-ratio agent in *Drosophila*. Science 133:1489-1490.

Powell, J. R. 1971. Genetic polymorphisms in varied environments. Science 174:1035-1036.

Powell, J. R. 1973. Apparent selection of enzyme alleles in laboratory populations of *Drosophila*. Genetics 75:557-570.

Powell, J. R. 1975a. Protein variation in natural populations of animals. Evol. Biol. 8: 79-119.

Powell, J. R. 1975b. Isozymes and non-Darwinian evolution. In *Isozymes. Vol. 4. Genetics and Evolution*. C. L. Markert, Ed. Academic Press. New York.

Powell, J., and T. Dobzhansky. 1976. How far do flies fly? Amer. Sci. 64:179-185.

Powers, L. 1944. An expansion of Jones's Theory for the explanation of heterosis. Amer. Nat. 78:275-280.

Prakash, S. 1969. Genic variation in a natural population of *Drosophila persimilis*. Proc. Nat. Acad. Sci. 62:778-784.

Prakash, S. 1972. Origin of reproductive isolation in the absence of apparent genic differentiation in a geographic isolate of *Drosophila pseudoobscura*. Genetics 72:143-155.

Prakash, S. 1973. Patterns of gene variation in central and marginal populations of *Drosophila robusta*. Genetics 75:347-369.

Prakash, S. 1977. Allelic variants at the xanthine dehydrogenase locus affecting enzyme activity in *Drosophila pseudoobscura*. Genetics 87:159-168.

Prakash, S., R. C. Lewontin, and J. L. Hubby. 1969. A molecular approach to the study of genic heterozygosity in natural populations. IV. Patterns of genic variation in central, marginal and isolated populations of *Drosophila pseudoobscura*. Genetics 61: 841-858.

Prout, T. 1952. Selection against heterozygotes for autosomal lethals in natural populations of *Drosophila willistoni*. Proc. Nat. Acad. Sci. 38:478-481.

Prout, T. 1968. Sufficient conditions for multiple niche polymorphism. Amer. Nat. 102: 493-496.

Punnett, R. C. 1911. *Mendelism*. 3rd Ed. Macmillan. New York.

Punnett, R. C. 1915. *Mimicry in Butterflies*. Cambridge University Press. Cambridge.

Rasmusson, J. M. 1935. Studies on the inheritance of quantitative characters in *Pisum*. I. Preliminary note on the genetics of flowering. Hereditas 20:161-180.

Raven, P. H. 1962. Interspecific hybridization as an evolutionary stimulus in *Oenothera*. Proc. Linnaean Soc. London 173:92-98.

Reed, S. C., and E. W. Reed. 1948. Natural selection in laboratory populations of *Drosophila*. Evol. 2:176-186.

Remington, C. L. 1968. Suture-zones of hybrid interaction between recently joined biotas. Evol. Biol. 2:321-428.

Rendel, J. M. 1959. Canalization of the scute phenotype of *Drosophila*. Evol. 13:425-439.

Rendel, J. M. 1962. Evolution of dominance. In *The Evolution of Living Organisms*. G. W. Leeper, Ed. Melbourne University Press. Melbourne.

Rensch, B. 1960. *Evolution above the Species Level*. Columbia University Press. New York.

Rhoades, M. 1941. The genetic control of mutability in maize. Cold Spring Harbor Symp. Quant. Biol. 9:138-144.

Rhoades, M. M. 1942. Preferential segregation in maize. Genetics 27:395-407.

Rhoades, M. M. 1952. Preferential segregation in maize. In *Heterosis*. J. W. Gowen, Ed. Iowa State College Press. Ames.

Richardson, R. H. 1969. Migration and enzyme polymorphisms in natural populations of *Drosophila*. Jap. J. Genet. 44(Suppl. 1):172-179.

Richardson, R. H. 1970. Models and analyses of dispersal patterns. In *Mathematical Topics in Population Genetics*. K. Kojima, Ed. Springer-Verlag. New York.

Richardson, R. H., R. J. Wallace, S. J. Gage, G. D. Bouchey, and M. Denell. 1969. Neutron activation techniques for labeling *Drosophila* in natural populations. Studies in Genetics. V. Univ. Texas Publ. 6918:171-186.

Richmond, R. C. 1970. Non-Darwinian evolution: a critique. Nature 225:1025-1028.

Ricklefs, R. E. 1973. *Ecology*. Chiron Press. Newton, Mass.

Rizki, M. T. 1951. Morphological differences between two sibling species, *Drosophila pseudoobscura* and *Drosophila persimilis*. Proc. Nat. Acad. Sci. 37:156-159.

Roberts, H. F. 1929. *Plant Hybridization before Mendel*. Princeton University Press. Princeton. (Reprinted by Hafner, New York, 1965.)

Robertson, A. 1967. The nature of quantitative genetic variation. In *Heritage from Mendel*. R. A. Brink, Ed. University of Wisconsin Press. Madison.

Robertson, A. 1970. A note on disruptive selection experiments in *Drosophila*. Amer. Nat. 104:561-569.

Rogers, J. S. 1972. Measures of genetic similarity and distance. Studies in Genetics. VII. Univ. Texas Publ. 7213:145-153.

Rosenzweig, M. L., and R. H. MacArthur. 1963. Graphical representation and stability conditions of predator-prey interactions. Amer. Nat. 97:209-223.

Ross, H. H. 1957. Principles of natural coexistence indicated by leaf-hopper populations. Evol. 11:113-129.

Roughgarden, J. 1971. Density-dependent natural selection. Ecology 52:453-468.

Russell, W. L. 1962. An augmenting effect of dose fractionation on radiation-induced mutation rate in mice. Proc. Nat. Acad. Sci. 48:1724-1727.

Russell, W. L., L. B. Russell, and M. B. Cupp. 1959. Dependence of mutation frequency on radiation dose rate in female mice. Proc. Nat. Acad. Sci. 45:18-23.

Ryan, F. J. 1963. Mutation and population genetics. In *Methodology in Basic Genetics*. W. J. Burdette, Ed. Holden-Day. San Francisco.

Salthe, S. N. 1969. Geographic variation of the lactate dehydrogenases of *Rana pipiens* and *Rana palustris*. Biochem. Genet. 2:271-303.

Sandler, L., Y. Hiraizumi, and I. Sandler. 1959. Meiotic drive in natural populations of *Drosophila melanogaster*. I. The cytogenetic basis of Segregation-Distortion. Genetics 44:233-250.

Sandler, L., and E. Novitski. 1957. Meiotic drive as an evolutionary force. Amer. Nat. 91:105-110.

Sanghvi, L. D. 1963. The concept of genetic load: A critique. Amer. J. Human Genet. 15:298-309.

Savage, J. M. 1958. The concept of ecological niche with reference to the theory of natural coexistence. Evol. 12:111-112.

Sax, K. 1923. The association of size differences with seed-coat pattern and pigmentation in *Phaseolus vulgaris*. Genetics 8:552-560.

Scarr-Salapatek, S. 1971. Race, social class, and I.Q. Science 174:1285-1295.

Scarr-Salapatek, S. 1974. Genetics and the development of intelligence. In *Review of Child Development*. IV. F. Horowitz, E. M. Hetherington, S. Scarr-Salapatek, and J. Siegal, Eds. University of Chicago Press. Chicago.

Scattergood, L. W. 1954. Estimating fish and wildlife populations: A survey of methods. In *Statistics and Mathematics in Biology*. O. Kempthorne, T. A. Bancroft, J. W. Gowen, and J. L. Bush, Eds. Iowa State College Press. Ames.

Schaffer, H. E., and F. M. Johnson. 1974. Isozyme allelic frequencies related to selection and gene-flow hypotheses. Genetics 77:163-168.

Scharloo, W. 1964. The effect of disruptive and stabilizing selection on a *cubitus interruptus* mutant in *Drosophila*. Genetics 50:553-562.

Scharloo, W. 1971. Reproductive isolation by disruptive selection: Did it occur? Amer. Nat. 105:83-86.

Scharloo, W., M. den Boer, and M. S. Hoogmoed. 1967. Disruptive selection on sternopleural chaetae number. Genet. Res. 9:115-118.

Scharloo, W., M. S. Hoogmoed, and A. ter Kuile. 1967. Stabilizing and disruptive selection on a mutant character in *Drosophila*. I. The phenotypic variance and its components. Genetics 56:709-726.

Schlager, G., and M. M. Dickie. 1966. Spontaneous mutation rates at five coat-color loci in mice. Science 151:205-206.

Schlager, G., and M. M. Dickie. 1971. Natural mutation rates in the house mouse. Estimates for five specific loci and dominant mutations. Mutation Research 11:89-96.

Schmalhausen, I. I. 1949. *Factors of Evolution*. Blakiston. Philadelphia.

Schultz, J., and H. Redfield. 1951. Interchromosomal effects on crossing over in *Drosophila*. Cold Spring Harbor Symp. Quant. Biol. 16:175-197.

Scudder, G. G. E. 1974. Species concepts and speciation. Canad. J. Zool. 52:1121-1134.

Selander, R. K. 1970. Behavior and genetic variation in natural populations. Amer. Zool. 10:53-66.

Selander, R. K. 1972. Sexual selection and dimorphism in birds. In *Sexual Selection and the Descent of Man. 1871-1971*. B. Campbell, Ed. Aldine. Chicago.

Selander, R. K. 1976. Genic variation in natural populations. In *Molecular Evolution*. F. J. Ayala, Ed. Sinauer. Sunderland, Mass.

Selander, R. K., and W. E. Johnson. 1973. Genetic variation among vertebrate species. Ann. Rev. Ecol. Syst. 4:75-91.

Selander, R. K., and D. W. Kaufman. 1973a. Self-fertilization and genetic population structure in a colonizing land snail. Proc. Nat. Acad. Sci. 70:1186-1190.

Selander, R. K., and D. W. Kaufman. 1973b. Genic variability and strategies of adaptation in animals. Proc. Nat. Acad. Sci. 70:1875-1877.

Selander, R. K., M. H. Smith, S. Y. Yang, W. E. Johnson, and J. B. Gentry. 1971. Biochemical polymorphism and systematics in the genus *Peromyscus*. I. Variation in the old-field mouse (*Peromyscus polionotus*). Studies in Genetics VI. Univ. Texas Publ. 7103:49-90.

Semeonoff, R., and F. W. Robertson. 1968. A biochemical and ecological study of plasma esterase polymorphism in natural populations in the field vole, *Microtus agrestis* L. Biochem. Genet. 1:205-227.

Shaw, R. F., and J. D. Mohler. 1953. The selective significance of the sex ratio. Amer. Nat. 87:337-342.

Sheppard, P. M. 1958. *Natural Selection and Heredity*. Hutchinson. London.

Sheppard, P. M. 1959. The evolution of mimicry: A problem in ecology and genetics. Cold Spring Harbor Symp. Quant. Biol. 24:131-140.

Sheppard, P. M. 1961a. Recent genetical work on polymorphic mimetic *Papilios*. Symp. Roy. Ent. Soc. 1:23-30.

Sheppard, P. M. 1961b. Some contributions to population genetics resulting from the study of the *Lepidoptera*. Adv. Genetics 10:165-216.

Sheppard, P. M. 1969. Evolutionary genetics of animal populations: The study of natural populations. Proc. 12th Inter. Congr. Genetics 3:261-279.

Sheppard, P. M. 1975. *Natural Selection and Heredity*. 4th Ed. Hutchinson. London.

Sheppard, P. M., and E. B. Ford. 1966. Natural selection and the evolution of dominance. Heredity 21:139-147.

Shorey, H. H. 1973. Behavioral responses to insect pheromones. Ann. Rev. Ent. 18:349-380.

Shull, G. H. 1948. What is "heterosis"? Genetics 33:439-446.

Sierts-Roth, U. 1953. Geburts- und Aufzuchtgewichte von Rassehunden. Z. Hundeforsch. 20:1-122.

Simmons, M. J. 1976. Heterozygous effects of irradiated chromosomes on viability in *Drosophila melanogaster*. Genetics 84:353-374.

Simpson, G. G. 1953. *Evolution and Geography*. Condon Lectures. Oregon State University Press. Corvallis.

Simpson, G. G. 1961. *Principles of Animal Taxonomy*. Columbia University Press. New York.

Singh, R. S., J. L. Hubby, and L. H. Throckmorton. 1975. The study of genic variation by electrophoretic and heat denaturation techniques at the octanol dehydrogenase locus in members of the *Drosophila virilis* group. Genetics 80:637-650.

Singh, R. S., R. C. Lewontin, and A. A. Felton. 1976. Genetic heterogeneity within electrophoretic "alleles" of xanthine dehydrogenase in *Drosophila pseudoobscura*. Genetics 84:609-629.

Slobodkin, L. B. 1961. *Growth and Regulation of Animal Populations*. Holt, Rinehart and Winston. New York. (2nd Ed. Dover. New York. 1980).

Smith, F. H., and Q. D. Clarkson. 1956. Cytological studies of interspecific hybridization in *Iris*, subsection Californicae. Amer. J. Botany 43:582-588.

Sneath, P. H. A., and R. R. Sokal. 1973. *Numerical Taxonomy*. Freeman. San Francisco.

Snyder, T. P. 1974. Lack of allozymic variability in three bee species. Evol. 28:687-689.

Sokal, R. R., and T. J. Crovello. 1970. The biological species concept: A critical evaluation. Amer. Nat. 104:127-153.

Sokal, R. R., and P. H. A. Sneath. 1963. *Principles of Numerical Taxonomy*. Freeman. San Francisco.

Soulé, M. 1973. The epistasis cycle: A theory of marginal populations. Ann. Rev. Ecol. Syst. 4:165-187.

Southwood, T. R. E., R. M. May, M. P. Hassell, and G. R. Conway. 1974. Ecological strategies and natural selection. Amer. Nat. 108:791-804.

Spassky, B., R. C. Richmond, S. Pérez-Salas, O. Pavlovsky, C. A. Mourão, A. S. Hunter, H. Hoenigsberg, T. Dobzhansky, and F. J. Ayala. 1971. Geography of the sibling species related to *Drosophila willistoni*, and of the semi-species of the *Drosophila paulistorum* complex. Evol. 25:129-143.

Spiess, E. B. 1970. Mating propensity and its genetic basis in *Drosophila*. Evol. Biol. (Suppl.): 315-379.

Spiess, E. B. 1977. *Genes in Populations*. Wiley. New York.

Spieth, H. J. 1952. Mating behavior within the genus *Drosophila* (Diptera). Bull. Amer. Mus. Nat. Hist. 99:395-474.

Spieth, H. T. 1958. Behavior and isolating mechanisms. In *Behavior and Evolution*. A. Roe and G. G. Simpson, Eds. Yale University Press. New Haven.

Spieth, P. 1974. Gene flow and genetic differentiation. Genetics 78:961-965.

Sprague, E. F. 1962. Pollination and evolution in *Pedicularis* (Scrophulariaceae). Aliso 5: 181-209.

Stadler, L. J. 1942. Some observations on gene variability and spontaneous mutation. Spragg Memorial Lectures on Plant Breeding (Third Series) Michigan State. Pp. 3-15.

Stadler, L. J. 1948. Spontaneous mutation at the R locus in maize. I. Race differences in mutation rate. Amer. Nat. 82:289-314.

Stalker, H. D. 1942. Sexual isolation in the species complex *Drosophila virilis*. Genetics 27:238-257.

Stalker, H. D. 1961. The genetic systems modifying meiotic drive in *Drosophila paramelanica*. Genetics 46:177-202.

Stebbins, G. L. 1950. *Variation and Evolution in Plants*. Columbia University Press. New York.

Stebbins, G. L. 1958. The inviability, weakness, and sterility of interspecific hybrids. Adv. Genetics 9:147-215.

Stebbins, G. L. 1959. The role of hybridization in evolution. Proc. Amer. Phil. Soc. 103: 231-251.

Stebbins, G. L. 1963. Perspectives. I. Amer. Sci. 51:362-370.

Stebbins, G. L. 1969. The significance of hybridization for plant taxonomy and evolution. Taxon 18:26-35.

Stebbins, G. L. 1971a. *Chromosomal Evolution in Higher Plants.* Addison-Wesley. Reading, Mass.

Stebbins, G. L. 1971b. *Processes of Organic Evolution.* 2nd Ed. Prentice-Hall. Englewood Cliffs, New Jersey.

Stebbins, G. L. 1974. *Flowering Plants. Evolution above the Species Level.* Belknap Press. Cambridge, Mass.

Stebbins, G. L., and R. C. Lewontin. 1972. Comparative evolution at the levels of molecules, organisms, and populations. Proc. 6th Berkeley Symp. Math. Stat. Prob. 5:23-42.

Steinberg, A. G., H. K. Bleibtreu, T. W. Kurczynski, A. O. Martin, and E. M. Kurczynski. 1966. Genetic studies on an inbred human isolate. Proc. Third Internat. Cong. Human Genetics:267-289.

Stephens, S. G. 1950. The internal mechanisms of speciation in *Gossypium.* Bot. Rev. 16:115-149.

Stephens, S. G. 1951. Possible significance of duplication in evolution. Adv. Genet. 4: 247-265.

Stern, C. 1958. Selection for subthreshold differences and the origin of pseudoexogenous adaptations. Amer. Nat. 92:313-316.

Stern, C., G. Carson, M. Kinst, E. Novitski, and D. Uphoff. 1952. The viability of heterozygotes for lethals. Genetics 37:413-450.

Stern, C., and E. W. Schaeffer. 1943. On wild-type iso-alleles in *Drosophila melanogaster.* Proc. Nat. Acad. Sci. 29:361-367.

Stern, J. T. 1970. The meaning of "Adaptation" and its relation to the phenomenon of natural selection. Evol. Biol. 4:39-66.

Stern, V. M., and A. Mueller. 1968. Techniques of marking insects with micronized fluorescent dust with especial emphasis on marking millions of *Lygus hesperus* for dispersal studies. J. Econ. Entomol. 61:1232-1237.

Stevenson, A. C., and C. B. Kerr. 1967. On the distribution of frequencies of mutation to genes determining harmful traits in man. Mutation Res. 4:339-352.

Streisinger, G. 1948. Experiments on sexual isolation in *Drosophila.* IX. Behavior of males with etherized females. Evol. 2:187-188.

Strickberger, M. W., and C. J. Wills. 1966. Monthly frequency changes in *Drosophila pseudoobscura* third chromosome gene arrangements in a California locality. Evol. 20: 592-602.

Sturtevant, A. H. 1942. The classification of the genus *Drosophila,* with descriptions of nine new species. Univ. Texas Publ. 4213:5-51.

Sturtevant, A. H., and T. Dobzhansky. 1936. Geographical distribution and cytology of "sex-ratio" in *Drosophila pseudoobscura* and related species. Genetics 21:473-490.

Sullivan, W. 1974. *Continents in Motion.* McGraw-Hill. New York.

Sved, J. A., and O. Mayo. 1970. The evolution of dominance. In *Mathematical Topics in Population Genetics.* K. Kojima, Ed. Springer-Verlag. New York.

Sved, J. A., T. E. Reed, and W. F. Bodmer. 1967. The number of balanced polymorphisms that can be maintained in a natural population. Genetics 55:469-481.

Tamarin, R. H., and C. J. Krebs. 1969. *Microtus* population biology. II. Genetic changes at the transferrin locus in fluctuating populations of two vole species. Evol. 23:183-211.

Tauber, C. A., and M. J. Tauber. 1977a. Sympatric speciation based on allelic changes at three loci: Evidence from natural populations in two habitats. Science 197:1298-1299.

Tauber, C. A., and M. J. Tauber. 1977b. A genetic model for sympatric speciation through habitat diversification and seasonal isolation. Nature 268:702-705.

Tauber, C. A., M. J. Tauber, and J. R. Nechols. 1977. Two genes control seasonal isolation in sibling species. Science 197:592-593.

Thoday, J. M. 1953. Components of fitness. Symp. Soc. Exptl. Biol. 7:96-113.

Thoday, J. M. 1958a. Natural selection and biological process. In *A Century of Darwin*. S. A. Barnett, Ed. Heinemann. London.

Thoday, J. M. 1958b. Homeostasis in a selection experiment. Heredity 12:401-415.

Thoday, J. M. 1961. Location of polygenes. Nature 191:368-370.

Thoday, J. M. 1965. Effects of selection for genetic diversity. Proc. 11th Intern. Congr. Genet. 3:533-540.

Thoday, J. M. 1972. Disruptive selection. Proc. Royal Soc. Lond. (B) 182:109-143.

Thoday, J. M. 1977. Effects of specific genes. In *Proceedings of the International Conference on Quantitative Genetics*. E. Pollack, O. Kempthorne, and T. B. Bailey, Eds. Iowa State University Press. Ames.

Thoday, J. M., and T. B. Boam. 1959. Effects of disruptive selection. III. Polymorphism and divergence without isolation. Heredity 13:205-218.

Thoday, J. M., and J. B. Gibson. 1962. Isolation by disruptive selection. Nature 193: 1164-1166.

Thoday, J. M., and J. B. Gibson. 1970. The probability of isolation by disruptive selection. Amer. Nat. 104:219-230.

Thoday, J. M., and J. B. Gibson. 1971. Reply to Scharloo. Amer. Nat. 105:86-88.

Thompson, J. N., and T. N. Kaiser. 1977. Selection acting upon slow-migrating ADH alleles differing in enzyme activity. Heredity 38:191-195.

Thompson, J. H., and J. M. Thoday. 1975. Genetic assimilation of part of a mutant phenotype. Genet. Res. 26:149-162.

Thomson, G. 1977. The effect of a selected locus on linked neutral loci. Genetics 85: 753-788.

Throckmorton, L. H. 1977. *Drosophila* systematics and biochemical evolution. Ann. Rev. Ecol. Syst. 8:235-254.

Timoféeff-Ressovsky, N. W. 1940a. Mutations and geographical variation. In *The New Systematics*. J. Huxley, Ed. Clarendon Press. Oxford.

Timoféeff-Ressovsky, N. W. 1940b. Zur Analyse des Polymorphismus bei *Adalia bipunctata*. Biol. Zentr. 60:130-137.

Timoféeff-Ressovsky, N. W., and E. A. Timoféeff-Ressovsky. 1940. Populations-genetische Versuche an *Drosophila*. Z. indukt. Abstamm. Vererbungsl. 79:28-49.

Tinbergen, N. 1951. *The Study of Instinct*. Oxford University Press. Oxford.

Tinkle, D. W. 1965. Population structure and effective size of a lizard population. Evol. 19:569-573.

Toda, M. J. 1974. A preliminary study on microdistribution and dispersal in Drosophilid natural populations. J. Fac. Sci. Hokkaido Univ. Ser. VI. Zool. 19:641-656.

Trippa, G., A. De Marco, A. Micheli, and B. Nicoletti. 1974. Recovery of SD chromosomes from *Drosophila melanogaster* males when heterozygous with structurally different second chromosomes. Canad. J. Genet. Cytol. 16:257-266.

Trivers, R. L. 1972. Parental investment and sexual selection. In *Sexual Selection and the Descent of Man. 1871-1971*. B. Campbell, Ed. Aldine. Chicago.

Trivers, R. L., and D. E. Willard. 1973. Natural selection of parental ability to vary the sex ratio of offspring. Science 179:90-92.

Turesson, G. 1922a. The species and the variety as ecological units. Hereditas 3:100-113.

Turesson, G. 1922b. The genotypical response of the plant species to the habitat. Hereditas 3:211-350.

Turesson, G. 1923. The scope and import of genecology. Hereditas 4:171-176.

Turesson, G. 1925. The plant species in relation to habitat and climate. Hereditas 6:147-236.

Turesson, G. 1930. The selective effect of climate upon the plant species. Hereditas 14: 99-152.

Turner, J. R. G., and M. H. Williamson. 1968. Population size, natural selection and the genetic load. Nature 218:700.

Turrill, W. B. 1936. Natural selection and the distribution of plants. Proc. Royal Soc. London (B) 121:49-52.

Turrill, W. B. 1938. Taxonomy and genetics. J. Botany 76:33-39.

Turrill, W. B. 1940. Experimental and synthetic plant taxonomy. In *The New Systematics*. J. Huxley, Ed. Clarendon Press. Oxford.

Turrill, W. B. 1946. The ecotype concept: A consideration with appreciation and criticism, especially of recent trends. New Phytologist 45:34-43.

Twitty, V. C. 1959. Migration and speciation in newts. Science 139:1735-1743.

Twitty, V. C. 1966. *Of Scientists and Salamanders*. Freeman. San Francisco.

Udvardy, M. D. F. 1959. Notes on the ecological concepts of habitat, biotope, and niche. Ecology 40:725-728.

Underhill, J. C., and D. J. Merrell. 1966. Fecundity, fertility, and longevity of DDT-resistant and susceptible populations of *Drosophila melanogaster*. Ecology 47:140-142.

Utida, S. 1957. Population fluctuation, an experimental and theoretical approach. Cold Srping Harbor Symp. Quant. Biol. 22:139-151.

Uzzell, T. 1970. Meiotic mechanisms of naturally occurring unisexual vertebrates. Amer. Nat. 104:433-445.

Valentine, J. W. 1976. Genetic strategies of adaptation. In *Molecular Evolution*. F. J. Ayala, Ed. Sinauer. Sunderland, Mass.

Van Delden, W., A. C. Boerema, and A. Kamping. 1978. The alcohol dehydrogenase polymorphisms in populations of *Drosophila melanogaster*. I. Selection in varying environments. Genetics 90:161-191.

Van Delden, W., A. Kamping, and H. Van Dijk. 1975. Selection at the alcohol dehydrogenase locus in *Drosophila melanogaster*. Experientia 31:418-419.

Van Valen, L. 1960. Further competitive exclusion. Science 132:1674-1675.

Van Valen, L. 1963. Haldane's dilemma, evolutionary rates and heterosis. Amer. Nat. 97:185-190.

Varley, G. C. 1949. Population changes in German forest pests. J. Animal Ecol. 18:117-122.

Vavilov, N. I. 1926. *Studies on the Origin of Cultivated Plants*. Institute de Botanique Appliquée et D'Amelioration des Plantes. Leningrad.

Verner, J. 1965. Selection for sex ratio. Amer. Nat. 99:419-421.

Vetukhiv, M. 1953. Viability of hybrids between local populations of *Drosophila pseudoobscura*. Proc. Nat. Acad. Sci. 39:30-34.

Vetukhiv, M. 1954. Integration of the genotype in local populations of three species of *Drosophila*. Evol. 8:241-251.

Vetukhiv, M. 1956. Fecundity of hybrids between geographic populations of *Drosophila pseudoobscura*. Evol. 10:139-146.

Vetukhiv, M. 1957. Longevity of hybrids between geographic populations of *Drosophila pseudoobscura*. Evol. 11:348-360.

Vigue, C. L., and F. M. Johnson. 1973. Isozyme variability in species of the genus *Drosophila*. VI. Frequency-property-environment relationships of allelic alcohol dehydrogenases in *Drosophila melanogaster*. Biochem. Genet. 9:213-227.

Volpe, E. P. 1952. Physiological evidence for natural hybridization of *Bufo americanus* and *Bufo fowleri*. Evol. 6:393-406.

Volterra, V. 1926a. Variazioni e fluttuazioni del numero d'individui in specie animali conviventi. Mem. Acad. Lincei. Ser. 6, Vol. 2:31-113. (Translation: Variations and fluctuations of the number of individuals in animal species living together. In *Animal Ecology*. 1931. R. N. Chapman. McGraw-Hill. New York.).

Volterra, V. 1926b. Fluctuations in the abundance of a species considered mathematically. Nature 118:558-560.

Volterra, V. 1931. *Leçons sur la Théorie Mathématique de la Lutte pour la Vie*. Gauthier-Villars. Paris.

Vries, H. de. 1901. *The Mutation Theory*. 2 vol. Tr. J. B. Farmer and A. D. Darbishire. Open Court. Chicago. (1909).

Waddington, C. H. 1942. Canalization of development and the inheritance of acquired characters. Nature 150:563-565.

Waddington, C. H. 1953a. Epigenetics and evolution. Symp. Soc. Exp. Biol. 7:186-199.

Waddington, C. H. 1953b. Genetic assimilation of an acquired character. Evol. 7:118-126.

Waddington, C. H. 1956. Genetic assimilation of the bithorax phenotype. Evol. 10:1-13.

Waddington, C. H. 1957a. The genetic basis of the 'assimilated bithorax' stock. J. Genetics 55:241-245.

Waddington, C. H. 1957b. *The Strategy of the Genes*. Allen and Unwin. London.

Waddington, C. H. 1959a. Canalization of development and genetic assimilation of acquired characters. Nature 183:1654-1655.

Waddington, C. H. 1959b. Evolutionary adaptation. In *Evolution after Darwin. I. The Evolution of Life*. S. Tax, Ed. University of Chicago Press. Chicago.

Waddington, C. H. 1961. Genetic assimilation. Adv. Genetics 10:257-294.

Waddington, C. H. 1968. The paradigm for the evolutionary process. In *Population Biology and Evolution*. R. C. Lewontin, Ed. Syracuse University Press. Syracuse.

Wagner, R. P. 1944. Nutritional differences in the *mulleri* group. Univ. Texas Publ. 4920:39-41.

Wahlund, S. 1928. Zusammensetzung von Populationen und Korrelations-erscheinungen vom Standpunkt der Vererbungslehre aus betrachtet. Hereditas 11:65-106.

Walker, T. J. 1974. Character displacement and acoustic insects. Amer. Zool. 14:1137-1150.

Wallace, A. R. 1876. *The Geographical Distribution of Animals*. 2 vols. Macmillan. London.

Wallace, A. R. 1889. *Darwinism: An Exposition of the Theory of Natural Selection*. Macmillan. London.

Wallace, B. 1955. Interpopulation hybrids in *Drosophila melanogaster*. Evol. 9:302-316.

Wallace, B. 1958a. The average effect of radiation-induced mutations on viability in *Drosophila melanogaster*. Evol. 12:532-552.

Wallace, B. 1958b. The role of heterozygosity in *Drosophila* populations. Proc. 10th Internat. Cong. Genetics 1:408-419.

Wallace, B. 1958c. The comparison of observed and calculated zygotic distributions. Evol. 12:113-115.

Wallace, B. 1966a. On the dispersal of *Drosophila*. Amer. Nat. 100:551-563.

Wallace, B. 1966b. Distance and the allelism of lethals in a tropical population of *Drosophila melanogaster*. Amer. Nat. 100:565-578.

Wallace, B. 1968a. *Topics in Population Genetics*. Norton. New York.

Wallace, B. 1968b. On the dispersal of *Drosophila*. Amer. Nat. 102:85-87.

Wallace, B. 1970a. *Genetic Load*. Prentice-Hall. Englewood Cliffs, New Jersey.

Wallace, B. 1970b. Observations on the microdispersion of *Drosophila melanogaster*. Evol. Biol. (Suppl.):381-399.

Wallace, B. 1975. Hard and soft selection revisited. Evol. 29:465-473.

Wallace, B., and J. C. King. 1952. A genetic analysis of the adaptive values of populations. Proc. Nat. Acad. Sci. 38:706-715.

Wallace, B., and C. Madden. 1953. The frequencies of sub- and supervitals in experimental populations of *Drosophila melanogaster*. Genetics 38:456-470.

Wallace, B., and A. M. Srb. 1964. *Adaptation*. 2nd Ed. Prentice-Hall. Englewood Cliffs, New Jersey.

Watanabe, T. K. 1969. Frequency of deleterious chromosomes and allelism between lethal genes in Japanese natural populations of *Drosophila melanogaster*. Japan J. Genetics 44:171-187.

Watt, W. B. 1977. Adaptation at specific loci. I. Phosphoglucose isomerase of *Colias* butterflies: Biochemical and population aspects. Genetics 87:177-194.

Weatherley, A. H. 1963. Notions of niche and competition among animals, with special reference to fresh water fish. Nature 197:14-17.

Weed, A. C. 1922. New frogs from Minnesota. Proc. Biol. Soc. Washington 34:107-110.

Wegener, A. 1924. *The Origin of the Continents and Oceans*. 3rd Ed. J. G. A. Skerl. Tr. Dutton. New York.

Wehrhahn, C., and R. W. Allard. 1965. The detection and measurement of the effects of individual genes involved in the inheritance of a quantitative character in wheat. Genetics 51:109-119.

Weinberg, W. 1908. Über den Nachweis der Vererbung beim Menschen. Jahreshafte Verein f. vaterl. Naturkunde in Württemberg 64:368-382.

Wellington, W. G. 1957. Individual differences as a factor in population dynamics: The development of a problem. Canad. J. Zool. 35:293-323.

Wellington, W. G. 1960. Qualitative changes in natural populations during changes in abundance. Canad. J. Zool. 38:289-314.

Wellington, W. G. 1964. Qualitative changes in populations in unstable environments. Canad. Ent. 96:346-451.

Wheeler, W. M. 1923. *Social Life Among the Insects*. Harcourt, Brace. New York.

Wheeler, W. M. 1928. *The Social Insects: Their Origin and Evolution*. Harcourt, Brace. New York.

White, M. J. D. 1954. *Animal Cytology and Evolution*. 2nd Ed. Cambridge University Press. Cambridge.

White, M. J. D. 1969. Chromosomal rearrangements and speciation in animals. Ann. Rev. Genetics 3:75-98.

White, M. J. D. 1973. *Animal Cytology and Evolution*. 3rd Ed. Cambridge University Press. Cambridge.

White, M. J. D. 1978. *Modes of Speciation*. Freeman. San Francisco.

Wickler, W. 1968. *Mimicry in Plants and Animals*. R. D. Martin. Tr. McGraw-Hill. New York.

Wilhelmi, R. W. 1940. Serological reactions and species specificity of some helminths. Biol. Bull. 79:64-90.

Williams, G. C. 1966. *Adaptation and Natural Selection*. Princeton University Press. Princeton, N.J.

Williamson, M. H. 1957. An elementary theory of interspecific competition. Nature 180:422-425.

Williamson, M. H. 1958. Selection, controlling factors and polymorphism. Amer. Nat. 92:329-335.

Wills, C. 1973. In defense of naïve pan-selectionism. Amer. Nat. 107:23-34.

Wills, C., J. Phelps, and R. Ferguson. 1975. Further evidence for selective differences between isoalleles in *Drosophila*. Genetics 79:127-141.

Wilson, M., and E. Pianka. 1963. Sexual selection, sex ratio, and mating system. Amer. Nat. 97:405-406.

Wilson, A. C. 1975. Evolutionary importance of gene regulation. Stadler Genetics Symp. 7:117-134.

Wilson, A. C., and E. M. Prager. 1974. Antigenic comparison of animal lysozymes. In *Lysozyme*. E. F. Osserman, R. E. Canfield, and S. Beychock, Eds. Academic Press. New York.

Wilson, E. O. 1963. Pheromones. Sci. Amer. 208(5):100-114.

Wilson, E. O. 1971. *The Insect Societies*. Harvard University Press. Cambridge.

Wilson, E. O., and W. H. Bossert. 1971. *A Primer of Population Biology*. Sinauer. Stamford, Connecticut.

Wilson, J. T., Ed. 1972. *Continents Adrift*. Freeman. San Francisco.

Wolfenbarger, D. O. 1946. Dispersion of small organisms. Amer. Midland Nat. 35:1-152.

Wolfenbarger, D. O. 1959. Dispersion of small organisms. Lloydia 22:1-105.

Wordsworth, W. 1798. *The Tables Turned*.

Wright, A. H., and A. A. Wright. 1949. *Handbook of Frogs and Toads of the United States and Canada*. 3rd Ed. Comstock. Ithaca, New York.

Wright, S. 1922. Coefficients of inbreeding and relationships. Amer. Nat. 56:330-338.

Wright, S. 1929a. Fisher's theory of dominance. Amer. Nat. 63:274-279.

Wright, S. 1929b. The evolution of dominance. Comment on Dr. Fisher's reply. Amer. Nat. 63:556-561.

Wright, S. 1931. Evolution in Mendelian populations. Genetics 16:97-159.

Wright, S. 1932. The roles of mutation, inbreeding, cross-breeding and selection in evolution. Proc. 6th Internat. Cong. Genetics 1:356-366.

Wright, S. 1934. Physiological and evolutionary theories of dominance. Amer. Nat. 68:24-53.

Wright, S. 1938. Size of population and breeding structure in relation to evolution. Science 87:430-431.

Wright, S. 1943. An analysis of local variability of flower color in *Linanthus parryae*. Genetics 28:139-156.

Wright, S. 1948. On the roles of directed and random changes in gene frequencies in the genetics of populations. Evol. 2:279-294.

Wright, S. 1949a. Population structure in evolution. Proc. Amer. Phil. Soc. 93:471-478.

Wright, S. 1949b. Adaptation and selection. In *Genetics, Paleontology, and Evolution*. G. L. Jepson, G. G. Simpson, and E. Mayr, Eds. Princeton University Press. Princeton, N. J.

Wright, S. 1951. Fisher and Ford on "The Sewall Wright effect." Amer. Sci. 39:452-458.

Wright, S. 1955. Classification of the factors of evolution. Cold Spring Harbor Symp. Quant. Biol. 20:16-24.

Wright, S. 1959. Physiological genetics, ecology of populations, and natural selection. Persp. Biol. Med. 3:107-151.

Wright, S. 1960. Genetics and twentieth century Darwinism: A review and discussion. Amer. J. Human Genet. 12:365-372.

Wright, S. 1964. Pleiotropy in the evolution of structural reduction and of dominance. Amer. Nat. 98:65-69.

Wright, S. 1968a. Dispersion of *Drosophila pseudoobscura*. Amer. Nat. 102:81-84.

Wright, S. 1968b. *Evolution and the Genetics of Populations. Vol. 1. Genetic and Biometric Foundations*. University of Chicago Press. Chicago.

Wright, S. 1969. *Evolution and the Genetics of Populations. Vol. 2. The Theory of Gene Frequencies.* University of Chicago Press. Chicago.

Wright, S. 1970. Random drift and the shifting balance theory of evolution. In *Mathematical Topics in Population Genetics.* K. Kojima, Ed. Springer-Verlag. New York.

Wright, S. 1977. *Evolution and the Genetics of Populations. Vol. 3. Experimental Results and Evolutionary Deductions.* University of Chicago Press. Chicago.

Wright, S. 1978. *Evolution and the Genetics of Populations. Vol. 4. Variability within and among Natural Populations.* University of Chicago Press. Chicago.

Wright, S., T. Dobzhansky, and W. Hovanitz. 1942. Genetics of natural populations. VII. The allelism of lethals in the third chromosome of *Drosophila pseudoobscura.* Genetics 27:363-394.

Wright, S., and W. E. Kerr. 1954. Experimental studies of the distribution of gene frequencies in very small populations of *Drosophila melanogaster.* II. Bar. Evol. 8: 225-240.

Wu, L., A. D. Bradshaw, and D. A. Thurman. 1975. The potential for evolution of heavy metal tolerance in plants. III. The rapid evolution of copper tolerance in *Agrostis stolonifera.* Heredity 34:165-187.

Wynne-Edwards, V. C. 1962. *Animal Dispersion in Relation to Social Behavior.* Oliver and Boyd. Edinburgh.

Yamazaki, T. 1971. Measurement of fitness at the esterase-5 locus in *Drosophila pseudoobscura.* Genetics 67:579-603.

Yanagisawa, L., L. C. Dunn, and D. Bennett. 1961. On the mechanism of abnormal transmission ratios at the T locus in the house mouse. Genetics 46:1635-1644.

Yanofsky, C., H. Berger, and W. J. Brammar. 1969. *In vivo* studies on the genetic code. Proc. XII Internat. Cong. Genetics 3:155-165.

Yardley, D. G., W. W. Anderson, and H. E. Schaffer. 1977. Gene frequency changes at the α-amylase locus in experimental populations of *Drosophila pseudoobscura.* Genetics 87:357-369.

Yerington, A. P., and R. M. Warner. 1961. Flight distances of *Drosophila* determined with radioactive phosphorus. J. Econ. Entomol. 54:425-428.

Yoshikawa, I., and T. Mukai. 1970. Heterozygous effects on viability of spontaneous lethal genes in *Drosophila melanogaster.* Japan J. Genet. 45:443-455.

Zeuner, F. 1963. *A History of Domesticated Animals.* Hutchinson. London.

Zimmering, S., L. Sandler, and B. Nicoletti. 1970. Mechanisms of meiotic drive. Ann. Rev. Genetics 4:409-436.

Zouros, E. 1976. The distribution of enzyme and inversion polymorphism over the genome of *Drosophila*: Evidence against balancing selection. Genetics 83:169-179.

Zouros, E., C. B. Krimbas, S. Tsakas, and M. Loukas. 1974. Genic versus chromosomal variation in natural populations of *Drosophila subobscura.* Genetics 78:1223-1244.

Index

Index